Orthonormal Series Estimators

ORTHONORMAL SERIES ESTIMATORS

Odile Pons

French National Institute for Agricultural Research (INRA), France

W JERSEY • LONDON • SINGAPORE • BEIJING • SHANGHAI • HONG KONG • TAIPEI • CHENNAI • TOKYO

Published by

World Scientific Publishing Co. Pte. Ltd.
5 Toh Tuck Link, Singapore 596224
USA office: 27 Warren Street, Suite 401-402, Hackensack, NJ 07601
UK office: 57 Shelton Street, Covent Garden, London WC2H 9HE

Library of Congress Cataloging-in-Publication Data
Names: Pons, Odile, author.
Title: Orthonormal series estimators / Odile Pons.
Description: Hackensack, NJ : World Scientific, [2020] |
Includes bibliographical references and index.
Identifiers: LCCN 2019052172 | ISBN 9789811210686 (hardcover)
Subjects: LCSH: Series, Orthogonal. | Approximation theory. | Nonparametric statistics. |
AMS: Statistics -- Nonparametric inference -- Estimation. |
Statistics -- Nonparametric inference -- Density estimation.
Classification: LCC QA404.5 .P64 2020 | DDC 515/.243--dc23
LC record available at https://lccn.loc.gov/2019052172

British Library Cataloguing-in-Publication Data
A catalogue record for this book is available from the British Library.

For any available supplementary material, please visit
https://www.worldscientific.com/worldscibooks/10.1142/11563#t=suppl

Printed in Singapore

Preface

Nonparametric functions of a Hilbert space have expansions as series of smooth functions by projection on an orthonormal basis, their estimation is performed like parametric estimation with a large number of parameters. They are easily computed and provide direct estimators of the function. The method applies to a wide class of functions, to deconvolution and to deterministic or stochastic inverse problems. Other applications include the estimation of the mixing density of continuous mixtures of exponential densities, using their Laplace transform, and the mixing density of random Poisson probabilities. The optimal choice of the size of the basis used in the estimation is determined by minimization of an error, by least squares cross-validation or in a penalization approach. The same methods are extended to wavelets.

The series estimators are adapted with estimation of the basis to nonparametric models for regressions on random variables, for multinomial and Poisson variables, hazard functions, drift and variance functions of auto-regressive diffusion processes. Additive nonparametric models and semi-parametric models of varying regression coefficients and projection pursuit are considered with multivariate covariates. Finally, the series estimators are used in nonparametric homogeneity tests and goodness-of-fit tests. In each model we establish the asymptotic properties of the series estimators, they are powerful as compared to other estimation methods, in particular their optimal convergence rate is not reduced as the dimension of the regressor increases.

Odile M.-T. Pons
December 2018

Contents

Chapter 1

Introduction

1.1 Series estimators

The approximation and the estimation of nonparametric functions in a Hilbert space $H = L^2(\mathbb{X}, \mu)$ by projections on an orthonormal basis of functions $(\varphi_k)_{k\geq 0}$ in H are useful in data analysis. Let f be a functions of H and let $a_k = \langle f, \varphi_k \rangle$, then

$$\int \Big\{ f - \sum_{k\geq 0} a_k \varphi_k \Big\}^2 d\mu = \|f\|_2^2 - \sum_{k\geq 0} a_k^2,$$

this proves Bessel's inequality

$$\sum_{k\geq 0} a_k^2 \leq \|f\|_2^2,$$

with the equality if and only if f has a convergent expansion

$$f = \sum_{k\geq 0} a_k \varphi_k.$$

The space V generated by the basis $(\varphi_k)_{k\geq 0}$ has an orthonormal complement $\bar{V}$ in H and the projection $\pi_V f = \sum_{k\geq 0} a_k \varphi_k$ of f on V defines the function $\bar{f} = f - \pi_V f$ of $\bar{V}$. By iterations, the space $\bar{V}$ is the sum of orthonormal spaces generated by orthonormal bases of H and every function f of H has expansions in the sub-spaces of $\bar{V}$. In wavelet expansions, the bases of the sub-spaces of a Hibert space H are simply defined by rescaling and translating the same function. Periodic functions have convergent Fourier expansions which are estimated using estimators of their Fourier coefficients.

The orthonormality of the functions φ_k implies they are linearly independent: for all integers j, $k_1, \ldots, k_j$, and for all constants $c_1, \ldots, c_j$, the equality

$$\sum_{k\in\{k_1,\ldots,k_j\}} c_k\varphi_k = 0$$

entails that its scalar product with $\varphi_{k'}$, for every k' in $\{k_1,\ldots,k_j\}$, is

$$\sum_{k\in\{k_1,\ldots,k_j\}} c_k\langle\varphi_k,\varphi_{k'}\rangle = c_{k'} = 0.$$

Using a polynomial basis such that the kth polynomial φ_k has the degree k is not always relevant for the estimation of functions of H which may be bounded or tend to zero at the end-point of their support. The Hilbert space and the choice of the basis, with a suitable parametrization, enables to estimate a large variety of functions such as densities, regression functions, the drift and the variance of nonparametric diffusion processes. For the regression function of a variable Y on a random variable X, the measure μ on the support $\mathcal{I}_X$ of X determines its distribution function F_X and space H is a subset of $L^2(\mathcal{I}_X, F_X)$, the orthonormal basis of H is estimated using the empirical distribution function of the regressors.

The methods of projections rely on the estimation of a large number of real coefficients defined as the scalar product of the elements of the basis and the function f to be estimated, empirical estimators are generally easily calculated but maximum likelihood or least squares estimators are necessary in more complex models. The quality of the estimation of a regression function is usually measured by the mean integrated squared quadratic error (MISE), it splits as the sum of an error of approximation of the infinite sum $f = \sum_{k\geq 0} a_k\varphi_k$ by the sum of k_n terms

$$f_n = \sum_{k\leq k_n} a_k\varphi_k,$$

where k_n tends to infinity with the number n of observations, and an error of estimation of f_n by an estimator

$$\widehat{f}_n = \sum_{k\leq k_n} \widehat{a}_{nk}\varphi_k$$

or $\widehat{f}_n = \sum_{k\leq k_n} \widehat{a}_{nk}\widehat{\varphi}_{nk}$ with an estimated basis. The approximation error depends on k_n, on the space H and on the properties of the basis, the estimation error is an expression depending on the variance of the estimated coefficients and on k_n.

Other criteria may be considered, the log-likelihood of the observations in regression models with a Gaussian error is related to the MISE error.

More generally, penalties depending on the dimension of the model and on the variance of the variable of interest have been added to quadratic errors.

Their exist numerous methods for the estimation of regular functions, the most used are kernel smoothing empirical estimators, splines estimators, or Bayesian estimators. In the estimation of nonparametric regression functions by splines, the functions are approximated by a linear combination of polynomials with the same degree with knots at arbitrary points. This is a flexible method which reduces the problem to the parametric estimation of a large number of real parameters, they are estimated by minimization of mean squared error with a penalization by the L^1 norm of the vector parameter, which provides the B-splines of third degree. The approximation of a function by n polynomials of degree r has $nr + n - 1$ parameters and its L^p error depends on the order of differentiability of the function and of the dimension of the space (DeVore, 1989). In models with a large number of regressors, the penalization provides an automatic choice of the most important regressors and reduces the dimension of the model. However the approximation by a basis of small degree is restrictive and the behavior at the splines at the infinity is not necessarily suitable. The estimation with splines requires the use of a maximization algorithm and the choice of the parameter of penalization is computationally expensive whereas the kernel and projection methods provide direct expressions of the estimators, furthermore their L^p norms have close approximations.

The properties of kernel estimators rely on differentiability conditions of the estimated function, they operate in the assessment of their bias and therefore in the L^p norms of their estimation error. These conditions are not necessary with the nonparametric methods based on projections on a regular orthonormal basis, they may be used to prove that the series estimators reach the same optimal rate of convergence as the kernel estimators in specific families of functions.

The convergence rate of the kernel estimators depends on the orders of differentiability and integrability of the function, it decreases as the dimension of the functional space increases. This rate is not satisfactory in regression models of large dimension, alternatives are additive nonparametric models and semi-parametric models with a one dimensional nonparametric function for a parametric model of the regressors such as the projection pursuit models (Huber 1985, Hall 1989, Donoho and Johnson 1989).

Series estimators obtained by projections on an orthonormal basis of functions of a Hilbert space have the advantage to perform with the same rate of convergence in multidimensional model as in univariate models and one can use them in most models. Estimators of densities defined by orthogonal functions are not necessarily positive due to the change of signs of the functions of the basis, Crain (1974) proposed to estimate the logarithm of continuous densities by projection on an orthonormal series $(\phi_k)_{k\geq 0}$ as $f(x) = \exp\{\sum_{k=0}^{\infty} c_k\phi_k(x)\}$, on the support of the density. The exponential adjustment of a density is convenient in exponential densities but it may be restrictive, alternatives are modifications of the bases of functions or estimation under constraints.

Exponential and linear models with random parameters are continuous mixture models where the estimation of the unknown mixing distribution of the parameters is a deconvolution problem that we may also solved by projections on an orthonormal basis of functions.

1.2 Mixtures of distributions

Let $\{f_\theta, \theta \in \Theta\}$ be a family of densities with respect to a measure μ on a measurable space $(\mathbb{X}, \mathcal{X})$, considering θ as a random parameter with a continuous distribution function G on the space Θ, the density

$$f_G(x) = \int_{\mathbb{X}} f_\theta(x)\, dG(\theta)$$

is a mixture density on $(\mathbb{X}, \mathcal{X})$ and the distribution function G of θ is the mixing distribution of the density. The mixing distribution is identifiable under the condition

$$f_{G_1} = f_{G_2}\ \mu\text{-a.e. implies } G_1 = G_2\ \mu\text{-a.e.}$$

In the continuous mixture of exponential distributions, the estimation of the mixing distribution G is related to the inverse of its Laplace transform $L_G(t) = \int_0^\infty e^{-\theta t}\, dG(\theta)$ by the deconvolution formula (Feller 1971)

$$G(\theta) = \lim_{t\to\infty} \sum_{k\leq\theta t} \frac{(-1)^k}{k!} t^k L_G^{(k)}(t), \tag{1.1}$$

at every point of continuity of G.

Example 1.1. Let X be an exponential variable with a random parameter θ on $\mathbb{R}_+$ and let g be the probability density of the parameter. A continuous mixture of exponential variables has the distribution function F such that

$$F(x) = 1 - \int_0^\infty e^{-\theta x} g(\theta)\, d\theta, \tag{1.2}$$

so the survival function $\bar{F}(x) = 1 - F(x)$ is the Laplace transform of the density g at x and it has infinitely many derivatives at every $x > 0$. By inversion (1.1) of the Laplace transform, the exponential parameter has the probability distribution depending on the derivatives of the mixture of exponential densities $f(x) = \int f_\theta(x)g(\theta)\,d\theta$

$$G(\theta) = \lim_{x\to\infty} \sum_{k\leq\theta x} \frac{(-1)^k}{k!} x^k \bar{F}^{(k)}(x)$$

where $\bar{F}(x)$ tends to zero as x tends to infinity, hence

$$G(\theta) = \lim_{x\to\infty} \sum_{k\leq\theta x} \frac{1}{k!} x^k f^{(k)}(x). \tag{1.3}$$

Example 1.2. Let X be an exponential variable with a random parameter θ on $\mathbb{R}_+$, with a density function conditionally on θ

$$f_\theta(x) = e^{-\theta T(x) - S(x) - b(\theta)} \tag{1.4}$$

with respect to the Lebesgue measure, depending on a strictly positive statistic $y = T(y)$ such that T is an increasing function on $\mathbb{R}_+$. Let

$$S(X) = \log T'(X)$$

and $b(\theta) = -\log\theta$, the conditional distribution function of X is

$$F_\theta(x) = 1 - e^{-\theta T(x)}.$$

The distribution function of Y given θ is

$$F_{Y,\theta}(y) = F_\theta \circ T^{-1}(y)$$

and its density with respect to the Lebesgue measure is

$$f_{Y,\theta}(y) = \frac{f_\theta \circ T^{-1}(y)}{T' \circ T^{-1}(y)} = e^{-\theta y - S_T(y) - b(\theta)}$$

where $S_T(y) = -\log T' \circ T^{-1}(y)$. Let G be the probability distribution of the parameter θ, the density of the variable Y is

$$f_Y(y) = \frac{1}{T' \circ T^{-1}(y)} \int_0^\infty e^{-\theta y - S_T(y) - b(\theta)}\, dG(\theta).$$

Let $c = \int_0^\infty e^{-b(\theta)}\, dG(\theta)$ and let H be the distribution function having the density with respect to G

$$\frac{dH(\theta)}{dG(\theta)} = c^{-1} e^{-b(\theta)} = (E_G\theta)^{-1}\theta,$$

then

$$f_Y(y) = \frac{c}{T' \circ T^{-1}(y)} \int_0^\infty e^{-\theta y}\, dH(\theta) = \frac{c}{T' \circ T^{-1}(y)} L_H(y),$$

it is proportional to the Laplace transform of H

$$cL_H(y) = e^{-S_T(y)} f_Y(y). \tag{1.5}$$

Assuming that T is an infinitely differentiable statistic, the distribution function H is deduced from the inversion formula (1.1) of the Laplace transform, up to the constant c

$$\int_0^\theta e^{-b(t)}\, dG(t) = cH(\theta) = \lim_{x \to \infty} \sum_{k \leq \theta x} \frac{(-1)^k}{k!} x^k c L_H^{(k)}(x). \tag{1.6}$$

In Example 1.1, the distribution function H satisfies

$$dH(\theta) = \mu^{-1} \theta\, dG(\theta)$$

where $\mu = E_G \theta$ and (1.5) reduces to $f_Y(x) = \mu L_H(x)$.

Example 1.3. Consider the random exponential model with density

$$f_\theta(x) = e^{-\theta x - b(\theta)} \tag{1.7}$$

with respect to a known probability measure F_0, where the constant is $b(\theta) = \log \int e^{-\theta x}\, dF_0(x) = \log L_{F_0}(\theta)$ and θ is a random parameter. By the reparametrization of Example 1.2, the density of the continuous mixture is $f(x) = L_H(x)$ and the expression of the distribution function H is deduced from the inversion formula (1.1) of the Laplace transform.

The estimation of the mixing density of a random parameter is performed with various methods such as the estimation of its moments, parametric or empirical Bayesian estimations with parametric mixing distributions, the minimization of the Kullback distance of mixture probabilities with respect to the mixing distribution, the projection of the mixing density on a basis of orthonormal polynomials.

In discrete mixtures of densities belonging to a parametric family $\mathbb{F}_\Theta = \{f_\theta \in C^2(\Theta), \theta \in \Theta\}$, the mixing distribution has a finite or infinite countable support $\mathbb{K}$ with cardinal K and the density of the mixture distribution is

$$f(x) = \sum_{k \in \mathbb{K}} p_k f_k(x)$$

with the mixing probabilities $p_1, \ldots, p_K$ such that $\sum_{k=1}^K p_k = 1$ and with distinct sub-densities f_k. The parameter vectors $p = (p_k)_{k=1,\ldots,K}$ in $]0,1[^K$ and $\eta = (\theta_k)_{k=1,\ldots,K}$ in Θ^k of the mixture density are identifiable under the condition that the existence of parameters p and p' in $]0,1[^K$, η and η' in Θ^k, with distinct components and such that the equality

$$\sum_{k\in\mathbb{K}} p_k f_{\theta_k} = \sum_{k\in\mathbb{K}} p'_k f_{\theta'_k}$$

implies the existence of a permutation σ of $\{1,\ldots,K\}$ such that $p_\sigma = p'$ and $\eta_\sigma = \eta'$.

The estimators of the parameters under the constraint $\sum_{k=1}^K p_k = 1$ maximize the likelihood of a sample $(X_i)_{i=1,\ldots,n}$ with density

$$f(x) = \sum_{k=1}^{K-1} p_k(f_k - f_K) + f_K.$$

Let $(X_i)_{i=1,\ldots,n}$ be a sample of a X be variable having the density mixture of two exponential densities

$$g(x;p,\theta) = pf(x;\theta_1) + (1-p)f(x;\theta_2),$$

with the survival functions $\bar{F}(x;\theta_j) = e^{-\theta_j x}$ for $j = 1,2$. The likelihood of the sample $L_n = \prod_{i=1}^n g(X_i;p,\theta)$ is maximal at the estimator values $\widehat{\theta}_n$ and $\widehat{p}_n$ of the parameters, they are solutions of the score equations

$$n^{-1}\sum_{i=1}^n g^{-1}(X_i;p,\theta)\{f(X_i;\theta_1) - f(X_i;\theta_2)\} = 0,$$

$$n^{-1}\sum_{i=1}^n g^{-1}(X_i;p,\theta)\dot{f}_{\theta_j}(X_i;\theta_j) = 0,\ j = 1,2,$$

where $\dot{f}_{\theta_j}(x;\theta_j)$ is the first derivative of $f(x;\theta_j)$ with respect to the parameter θ_j. Let $\widehat{G}_n$ be the empirical distribution function of the sample and let $d\widehat{H}_{n,p,\theta} = g_{p,\theta}^{-1} 1_{\{g_{p,\theta}>0\}}\, d\widehat{G}_n$, the second equation is equivalent to

$$\int_{\mathbb{R}_+} e^{-\theta_j x}\, d\widehat{H}_{n,p,\theta}(x) = \theta_j \int_{\mathbb{R}_+} xe^{-\theta_j x}\, d\widehat{H}_{n,p,\theta}(x),$$

for $j = 1,2$. These equations can be solved numerically by an iterative EM algorithm (Estimation-Maximization) for distinct parameters θ_1 and θ_2. The problem extends to a mixture of K densities and to mixtures of parametric densities of $C^2(\mathbb{R})$ under integrability conditions.

1.3 Deconvolution problems

In models of observation with error in a variable X, let

$$Y = X + \varepsilon,$$

where X is an unobserved variable independent of the error variable ε, having a known symmetric density function f_ε. The unknown density of the variable X is the convolution

$$f_X(x) = f_Y \star f_\varepsilon(x) = \int f_\varepsilon(x-y)\, dF_Y(y)$$

of the densities of the variables ε and Y. Let $Y_1, \ldots, Y_n$ be a n-sample of independent and identically distributed observations of the variable Y, with distribution function F_Y, the density f_X has the consistent estimator

$$\widehat{f}_{nX}(x) = n^{-1} \sum_{i=1}^{n} f_\varepsilon(x - Y_i)$$

and its convergence rate is $n^{-\frac{1}{2}}$. More generally, for an error variable ε with a known density f_ε, the density f_X is consistently estimated by

$$\widehat{f}_{nX}(x) = n^{-1} \sum_{i=1}^{n} f_\varepsilon(Y_i - x) \tag{1.8}$$

and the process $n^{\frac{1}{2}}(\widehat{f}_{nX} - f_X)$ converges weakly to a centered Gaussian process $\int f_\varepsilon(y-x)\, dB \circ F_Y(y)$, where B is the standard Brownian bridge on $[0,1]$.

In a model of observation of X with an error having an unknown variance, a sample is available for the variable

$$Y = X + \sigma\varepsilon.$$

The density on $\mathbb{R}$ of the variable $X = Y - \sigma\varepsilon$, $\sigma > 0$, is

$$f_X(x;\sigma) = \int f_\varepsilon\Big(\frac{y-x}{\sigma}\Big)\, dF_Y(y), \tag{1.9}$$

its estimator

$$\widehat{f}_{nX}(x;\sigma) = n^{-1} \sum_{i=1}^{n} f_\varepsilon\Big(\frac{Y_i - x}{\sigma}\Big)$$

depends on the parameter σ and the process $n^{\frac{1}{2}}\{\widehat{f}_{nX}(\cdot;\sigma) - f_X(\cdot;\sigma)\}$ converges weakly to a centered Gaussian variable with variance

$$v(x;\sigma) = n^{-1} \int f_\varepsilon^2\Big(\frac{y-x}{\sigma}\Big)\, dF_Y(y) - n^{-1} f_X^2(x;\sigma),$$

it is supposed finite for every finite variance $\sigma^2 > 0$.

The variance $v(x;\sigma)$ has the consistent empirical estimator

$$\widehat{v}_n(x;\sigma) = n^{-1} \int f_\varepsilon^2\Big(\frac{y-x}{\sigma}\Big)\, d\widehat{F}_{nY}(y) - n^{-1}\widehat{f}_{nX}^2(x;\sigma)$$

and the unknown parameter σ may be defined by minimization of the mean variance

$$K_n(\sigma) = \int_{\mathbb{R}} \widehat{v}_n(x;\sigma)\widehat{f}_{nX}(x;\sigma)\, dx = n^{-1}\sum_{i=1}^n \int_{\mathbb{R}} \widehat{v}_n(x;\sigma) f_\varepsilon\Big(\frac{Y_i - x}{\sigma}\Big)\, dx.$$

Let σ_0^2 be the true variance of the error for the sample, the process $K_n(\sigma) - K_n(\sigma_0)$ converges a.s. to

$$K(\sigma) - K(\sigma_0) = \int_{\mathbb{R}} \Big\{ v(x;\sigma)\frac{f_X(x;\sigma)}{f_X(x;\sigma_0)} - v(x;\sigma_0)\Big\} f_X(x;\sigma_0)\, dx$$

which is zero at σ_0 where it reaches its minimum. The estimator $\widehat{\sigma}_n$ of σ_0 which minimizes the process K_n is therefore a.s. consistent.

The process $n^{\frac{1}{2}}(K_n - K)$ converges weakly to a centered Gaussian process on every compact support which does not include zero. Assuming that f_ε is twice continuously derivable and $K''(\sigma)$ is strictly positive definite in a neighborhood of at σ_0, the derivative with respect to σ

$$U_n(\sigma) = n^{\frac{1}{2}} \frac{\partial (K_n - K)(\sigma)}{\partial \sigma}$$

is such that $U_n(\sigma_0)$ converges weakly to a centered Gaussian variable with finite variance. It follows that $n^{\frac{1}{2}}(\widehat{\sigma}_n - \sigma_0)$ converges weakly to a centered Gaussian variable.

The distribution function of Y is the convolution $F_Y(y,\sigma) = F_X * f_{\sigma\varepsilon}(y)$

$$\begin{aligned} F_Y(y,\sigma) &= \sigma^{-2} \int\int F_Y(y-s) f_\varepsilon\Big(\frac{s-t}{\sigma}\Big) f_\varepsilon\Big(\frac{t}{\sigma}\Big)\, ds\, dt \\ &= \int F_Y(y - \sigma s) f_\varepsilon * f_\varepsilon(s)\, ds \end{aligned}$$

With an error having the normal distribution, the convolution $f_\varepsilon * f_\varepsilon$ is proportional to the normal density and

$$\begin{aligned} F_Y(y,\sigma) &= \int F_Y(y - \sigma s) f_\varepsilon\Big(\frac{s}{\sqrt{2}}\Big)\, ds \\ &= \sqrt{2} \int F_Y(y-t)\, dF_\varepsilon\Big(\frac{t}{\sigma\sqrt{2}}\Big). \end{aligned}$$

In a multidimentional regression model, let $Y_i = A^T X_i + \sigma\varepsilon_i$ be independent observations of a regression model for a real variable Y, for $i = 1, \ldots, n$, where A is vector parameter of $\mathbb{R}^d$, X is an unobserved variable of $\mathbb{R}^d$, independent of the error variable ε, and ε has a known symmetric density function f_ε. The expectation of the variable Y is $\mu = \sum_{j=1}^d a_j \mu_j$ where $\mu_j = EX_{ij}$ and its variance is $\sigma_Y^2 = E\{(X - \mu_X)^T AA^T (X - \mu_X)\} + \sigma^2$, where $\mu_X = EX$. The distribution function of $A^T(X - \mu_X) = Y - \mu - \sigma\varepsilon$ is still determined by (1.9), with an unknown variance σ^2.

A regression with noisy observations of the regressor is a deconvolution problem with $Y = m(X) + \varepsilon$ where $m(X) = E(Y \mid X)$ and X is an unobserved variable independent of the error variable ε with expectation zero. The noisy observed variable is $W = X + \eta$ where the error variable η is independent of X. The density of the variable X is obtained as previously by deconvolution from the known density f_η of the error variable η and from the density of the variable W.

1.4 Regression models

In the linear regression

$$Y_i = \alpha + X_i\beta + \varepsilon_i, \; i = 1, \ldots, n,$$

with vectors of p linearly independent covariates X_i and parameters α in $\mathbb{R}$ and β in $\mathbb{R}^p$, the best linear estimator of the regression coefficient is

$$\widehat{\beta}_n = \Sigma_n^{-1}\{\mathbb{X} - \widehat{E}_n\mathbb{X}\}^T\{\mathbb{Y} - \widehat{E}_n\mathbb{Y}\},$$

where $\mathbb{X}$, and respectively $\mathbb{Y}$, are the $n \times p$ dimensional matrix with lines X_i, and respectively the n dimensional vector with components Y_i, for $i = 1, \ldots, n$, and Σ_n is the empirical variance matrix $\{\mathbb{X} - \widehat{E}_n\mathbb{X}\}^T\{\mathbb{X} - \widehat{E}_n\mathbb{X}\}$ of X, with the empirical mean $\widehat{E}_n\mathbb{X}$. It minimizes the mean squared error $\sum_{i=1}^n (Y_i - \alpha - X_i\beta)^2$. The $p \times p$ dimensional symmetric matrix Σ_n is nonsingular and its inversion may be performed by orthogonalization.

In a misspecified linear model

$$Y = \alpha + AX + BU + \sigma\varepsilon,$$

with $E(\varepsilon \mid X, U) = 0$, a regressor U is unobserved and its distribution is unknown. Let μ be the expectation of U, the conditional expectation of Y given X reduces to the linear model $E(Y \mid X) = \alpha + AX + B\mu$ and the parameter μ is confounded with constant effect of the model with regression variable X. The estimators of α and A are unbiased if the variable U is centered. In a nonparametric regression model

$$Y = \alpha + m(X) + r(U) + \sigma\varepsilon,$$

with an unobserved regressor U and a centered error ε, the conditional expectation of Y given X is

$$E(Y \mid X) = \alpha + m(X) + E\{r(U) \mid X\}.$$

The dependence of X and U implies that $m(X) + E\{r(U) \mid X\}$ is confounded with $m(X)$ and the estimator of m is biased, the same problem appears in parametric regression models.

With a regressor of high dimension p_n increasing with n or larger than n, a diagonalization of the matrix $\mathbb{X}'\mathbb{X}$ enables to restrict the dimension of the problem to the k_n eigenvalues larger than some strictly positive value, i.e. to k_n more significant linear combinations of the components of X. Another formalization of this procedure is to estimate the coefficients of the significant regressors by minimization of a penalized mean squared error

$$Q_n(\beta, \lambda) = \sum_{i=1}^{n}(Y_i - \alpha - \beta' X_i)^2 + \lambda \sum_{j=1}^{p_n} |\beta_j|, \tag{1.10}$$

the estimators of the coefficients are consistent. The method applies to a nonparametric regression model

$$Y = m(X) + \varepsilon,$$

where the regression function m is approximated by a polynomial of large degree and the penalization aims to minimize the number of polynomials. The question is similar for the estimation of a nonparametric regression function using splines of small degree with a large number of observations or with a regressor X of high dimension p_n (Wahba 1990, Meier, Van de Geer and Bühlmann 2009, Van de Geer and Bülhmann 2009). The penalization parameter is chosen by cross-validation, the same method applies to the choice of the number of the function for series estimators.

The regression models with an additive error extend to exponentials families of densities in Generalized Linear Models (GLM) for Bernoulli, multinomial or Poisson variables, with maximum likelihood estimators (Nelder and Wedderburn 1972, McCullagh and Nelder 1983, Hastie and Tibshirani 1990). Series estimators will be defined in several nonparametric generalized linear models, in particular additive models with separable effects of the covariates and additive models with interactions effects of the covariates for regressions and GLM.

Regression models for observations of processes extend the nonparametric regression models, they apply to generalized linear models in exponential families of densities, in proportional hazards and proportional odds ratio models, in diffusion models. In the generalized additive models, the expectation $m(X)$ of a variable Y conditionally on vector of regressors X of $\mathbb{R}^d$ has the form

$$m(X) = g\Big(\alpha + \sum_{j=1}^{p} \mu_j(X_j)\Big),$$

where the functions μ_j are unknown and the link function g^{-1} is known. The link function is usually the identity for a conditional Gaussian variable Y, it is the logarithm for a conditional Poisson variable Y, logit for a conditional binomial and multinomial variable Y.

With a large sample, the log-likelihood includes a penalization proportional to the logarithm of the normalization constant of the density. With splines or polynomial approximations, the estimation is performed by adding an appropriate penalization to the likelihood in order to reduce the dimension of the parameter space.

1.5 Inverse problems

Let H_1 and H_2 be Hilbert functional space and let $A : H_1 \mapsto H_2$ be a linear or nonlinear integral operator, we consider the resolution in H_1 of the inverse integral problem

$$f(x) = Au(x) \tag{1.11}$$

from data on f in H_2, the operator A is known and the unknown function u belongs to H_1. The model extends the regression models and independent observations of f on a grid $(x_i)_{i=1,\dots,n}$ yield the model

$$f(x_i) = Au(x_i) + \varepsilon_i = \langle k(x_i), u\rangle + \varepsilon_i,\ i = 1, \dots, n,$$

with the scalar product of H and where $(\varepsilon_i)_{i=1,\dots,n}$ is a vector of independent and identically distributed variables with expectation zero and a finite variance.

A linear example is the homogeneous Fredholm equation on $H = L^2(I)$, for a real interval $I = [a, b]$ and a known kernel K on I^2

$$Au(x) = \int_I K(x, y)u(y)\, dy. \tag{1.12}$$

A discretization of the interval I into n subintervals $I_i =]x_i, x_{i+1}]$, with length $\delta_n = n^{-1}|I|$ converging to zero as n tends to infinity, provides an approximation by a system of linear equations

$$\begin{aligned} f(x) &= \sum_{j=1}^{n} \int_{I_j} K(x,y)u(y)\,dy \\ &= \delta_n \sum_{j=1}^{n} \{K(x,x_j)u(x_j) + o(1)\} \end{aligned}$$

and the function u is approximated using the vector $U_n = (u(x_i))_{i=1,\dots,n}$ solution of the equations $y_i = \delta_n \sum_{j=1}^{n} K(x_i,x_j)u(x_j)$. Using the estimator of the linear regression model, each component of the parameter U_n has an estimator of order δ_n^{-1} tending to infinity with n, this is an ill posed problem where the inverse of the operator is not bounded.

The operator A is continuous and strongly monotone if the kernel K is strictly positive: there exists a strictly positive constant C depending only on K such that for all u and v of H

$$\langle A(u-v), u-v \rangle \geq C \|u-v\|^2,$$

it follows that A is one-to-one and A^{-1} is a continuous operator, solving the equation (1.11) is well posed problem.

Let $A : H \mapsto K$ be a nonlinear operator, and let and u in H and y satisfying (1.10). The operator A is weakly Fréchet-differentiable on H if for all u and u_0 of H there exists an operator $G(u_0)$ such that for every v of K

$$\|u-u_0\|^{-1} \langle A(u) - A(u_0) - G(u_0).(u-u_0), v \rangle = o(\|u-u_0\|),$$

then, as $\|u-u_0\|$ tends to zero

$$\begin{aligned} \langle A(u) - A(u_0), u-u_0 \rangle = \|u-u_0\| \langle G(u_0).(u-u_0), u-u_0 \rangle \\ +o(\|u-u_0\|^2). \end{aligned}$$

If there exist strictly positive constants C_1 and C_2 depending only on $G(u_0)$ such that $C_1 \leq G(u_0) \leq C_2$, then

$$C_1 \|u-u_0\|^2 \leq \int_I \{A(u) - A(u_0)\}(u-u_0) \leq C_2 \|u-u_0\|^2$$

for every u in a neighborhood of u_0. It follows that A is one-to-one and A^{-1} is continuous and bounded in a neighborhood of u_0. Several methods for the resolution of the equation (1.10) will be considered. With differential operators A, the equation may be often solved by polynomial expansions and each equation defines a function or a family of functions.

1.6 Content of the book

The next chapter presents methods for the estimation of functions of Hilbert spaces by orthonormal bases of functions and test based on these series estimators. For densities and time dependent functions, a basis may be chosen among the collections studied since many years, under the constraint that their support and their behavior on the boundaries are coherent with those of the functions to estimate. The other chapters define new estimators.

The first chapters consider samples of independent and identically distributed variables and the size of the basis used in the estimation depends on the sample size. For processes, the estimators are defined by discretization of their sample path or using the observations of their sample path on an increasing time interval, the size of the basis for the estimation depends then on the observation interval. In the first case, the convergence rates of the estimators depend on the sample size, in the second case they depend on the sampling interval.

Two classes of orthonormal bases are briefly presented in Chapter 2, they are used to estimate the density of real and multivariate samples, and their conditional distribution functions. The asymptotic behavior of the empirical series estimators and weighted estimators is studied, according to the convergence rate of their mean integrated squared errors and the number of functions of the basis used in the estimation is chosen by cross-validation. Likelihood ratio tests for the comparison of the densities of sub-samples and for the validation of parametric or semi-parametric models are based on the series estimators. The estimation of the mixing distribution Poisson variables with a random parameter relies on the same principle though it is not direct and the size of the basis is more restricted. Tests of a Poisson variable against the alternative of a mixture Poisson distribution are considered. Though series estimators of densities were already considered, their weak convergence and tests had not been studied.

Chapter 3 generalizes the series estimators to several models of nonparametric regression functions m of a real variable Y on a vector of explanatory variables X with distribution function F_X, with an independent error. An orthonormal basis of functions of $L^2(F_X)$ is defined as a reparametrization of the Laguerre basis on $\mathbb{R}_+$, the functions of the basis are consistently estimated from a sample of the regressor and they enable to estimate a regression function m of $L^2(F_X)$ by projections. The estimation of the basis modifies the convergence rate of the estimators, they are optimal under specified conditions and we prove their weak convergence to Gaussian processes. The same method applies to nonparametric linear models and nonparametric linear models with interactions, to semi-parametric models like the projection pursuit model where parameters are estimated by mean squares. Tests of homogeneity of sub-samples and goodness of for tests for nonparametric or parametric sub-models are considered.

Chapter 4 deals with nonparametric generalized linear models for discrete variables with Bernoulli, multinomial and Poisson distributions. We define series estimators for the nonparametric regression functions of logistic models of Bernoulli and multinomial variables, and for the intensity function of Poisson variables varying in time or according to regressors. We consider several models and define their estimators by projections on estimated functions of an orthonormal basis of $L^2(F_X)$, they are asymptotically Gaussian like in regression models. We study the asymptotic behavior of the estimators and tests.

Chapter 5 concerns deconvolution problems introduced in Section 1.3 for models where variables are observed with errors, and for mixtures of the exponential densities of Section 1.2, and the resolution of inverse problems presented in Section 1.5 with application to the detection of blurred signals on the line or in the plane. Linear inverse problems are explicitly solved by approximations of the functions of the models on orthonormal bases and by minimization of penalized quadratic risks with constraints on the norm of the solution. Their estimators are similar to estimators of large dimensional linear regression models based on a generalized inverse. The method is extended to nonlinear models by local linearization.

Chapter 6 presents estimators for nonparametric models of hazard functions. We define series estimators for the baseline hazard function of proportional hazards and for nonparametric models of their regression function on covariates, in Cox models with time varying coefficients and for coefficients depending on observed covariates. Frailty models with unobserved covariates shared in sub-populations are mixture models studied under parametric frailty distributions, we consider the estimators of the Gamma frailty model and their asymptotic distributions, they are generalized to frailty models with nonparametric regression on covariates. We determine series estimators for the hazard functions of bivariate times models and for the conditional hazard function of consecutive dependent time variables under right-censoring and tests of sub-models. The estimation is generalized to right-censored and left-truncated nonparametric regression functions.

For counting processes and their intensity, the asymptotic study relies on the behaviour of local martingales as the observation interval increases or with an increasing number of independent samples of observations on the same interval. In semi-parametric models, the estimators of the parameters are defined by maximum likelihood with a partially estimated density.

Time continuous and discrete diffusion processes X_t are studied in Chapter 7, they are Markov processes and their transition probabilities converge to an invariant measure. In models with time dependent or path dependent nonparametric drift and variance functions, we define consistent series estimators. Their convergence rates differ according to their domain. For path dependent functions, the empirical estimator of the invariant distribution F_X defines estimators of an orthonormal basis of $L^2(F_X)$. We consider several models and determine series estimators of their drift and variance functions. In models with stochastic volatility, the estimation of the unobserved variance V_t and of its distribution determine the estimation of an orthonormal basis of $L^2(F_V)$. The estimators are extended the models under random sampling and to ergodic processes. We determine the asymptotic behavior for their series estimators and for tests statistics of sub-models.

The wavelet bases are a complete orthonormal system of spaces $L^2(\mathcal{I})$, for a sub-interval $\mathcal{I}$ of $\mathbb{R}$, and the method of estimation by projection on wavelet bases has become popular, the wavelets based on the Haar function on $[0,1]$ are much studied. Chapter 8 focus on regular wavelets and the estimators of densities, intensities of point processes, regression functions, and diffusions, with generalizations of the wavelets to spaces $L^2(\mathcal{I}_X, F_X)$, for variables or processes X. We study their properties in classes of functions related to the regularity of the wavelets, and their weak convergence.

In the last chapter we study the asymptotic behavior of tests of sub-models for nonparametric and discrete mixtures of densities with a fixed or increasing number of components.

Chapter 2

Series estimators of probability densities

2.1 Introduction

A density f of $L^2(\mathbb{R})$ is expanded as a series by projection on an orthonormal basis of functions and the coefficients of its expansion converge to zero. We consider the empirical estimators of the coefficients and study the integrated mean squared error of the estimated density. In continuous mixtures, the density of a variable X has the form

$$f(x) = \int_\Theta f_\theta(x)\,dG(\theta)$$

where f_θ is the density of the variable conditionally on a random parameter θ belonging to a space Θ and G is the distribution function of θ.

A model of translation $f_\theta(x) = h(x-\theta)$ of a known real density h, with a random translation of θ, implies that the density f of the variable X is the convolution of h and g, the density of G. Expanding g on an orthonormal basis of functions $(\varphi_k)_{k\geq 0}$ tending to zero at the infinity, $g(x) = \sum_{k\geq 0} c_k\varphi_k(x)$, the density of X is

$$f(x) = \sum_{k\geq 0} c_k h * \varphi_k(x).$$

In a model with a random change of scale σ having a density in an interval $S = [a,b]$, with $a > 0$ and b finite, the variable X has the density

$$f(x) = \int_a^b h_\sigma(x)\,dG(\sigma) = -\frac{\sigma}{x}\int_{b^{-1}x}^{a^{-1}x} h(y)g(y^{-1}x)\,dy$$

with a conditional density $f_\sigma(x) = \sigma^{-1}h(\sigma^{-1}x)$. An expansion of the density g of G on an orthonormal basis of functions $(\varphi_k)_{k\geq 0}$ yields

$$f(x) = -\sigma x^{-1}\sum_{k\geq 0} c_k \int_{b^{-1}x}^{a^{-1}x} h(y)\varphi_k(y^{-1}x)\,dy.$$

We present properties of the Laguerre and Hermite bases of functions and their transforms and apply them to the estimation of the mixing density in several classes of distributions. Legendre orthonormal polynomials are defined on the interval $[-1,1]$, they are more restrictive as the sequences of their zeros converge to fixed points as their degree tends to infinity. Poisson variables with a random intensity have often been used, their integrated probabilities with respect to a mixing distribution function define the probabilities of a continuous mixture of Poisson variables. Let N be a Poisson process with a random intensity λ on $\mathbb{R}_+$ and let g be the density probability of the intensity λ. The compound Poisson variable Y has the probabilities

$$\pi(k) = \frac{1}{k!}\int_0^\infty e^{-x}x^k g(x)\,dx,\ k \geq 0. \tag{2.1}$$

The maximum likelihood estimation of the mixing distribution from a sample of independent and identically distributed compound Poisson variables has been studied by Simar (1976). An expansion $g = \sum_{j\geq 0} a_j(g)\varphi_j$ of the mixing density g by projection on an orthonormal basis $(\varphi_j)_{j\geq 0}$ has the coefficients

$$a_j(g) = \int_0^\infty g(x)\varphi_j(x)\,dx$$

satisfying the constraint $\int_0^\infty g(x)\,dx = \sum_{j\geq 0}(-1)^n a_j(g) = 1$. Laguerre's polynomials provide a recursive estimator of g, it is consistent and converges weakly with the optimal rate for densities.

Finally, we study nonparametric tests of comparison of densities, goodness of fit tests based on the series estimators of the densities, and test of a Poisson distribution against the alternative of a mixture Poisson distribution.

2.2 Laguerre's polynomials

Laguerre's orthonormal polynomials are defined by $L_0(x) = 1$ and the derivatives

$$L_n(x) = \frac{e^x}{n!}\frac{d^n}{dx^n}(e^{-x}x^n),\ n \geq 1, \tag{2.2}$$

they are explicitly calculated as

$$L_n(x) = \sum_{j=0}^n \binom{n}{j}\frac{(-1)^j x^j}{j!}. \tag{2.3}$$

Let $L_2(\mathbb{R}_+, \mu_\mathcal{E})$ be the space of square integrable functions with respect to the measure $\mu_\mathcal{E}$ having the exponential density with parameter 1 with respect to the Lebesgue measure in $\mathbb{R}_+$. In $L_2(\mathbb{R}_+, \mu_\mathcal{E})$, the scalar product of functions f and g is

$$\langle f, g\rangle = \int_0^\infty f(x)g(x)e^{-x}\,dx.$$

Laguerre's polynomials satisfy the recurrence formula

$$(n+1)L_{n+1}(x) - (x-2n-1)L_n(x) + nL_{n-1}(x) = 0, \; n \in \mathbb{N}, \tag{2.4}$$

with the initial values $L_0(x) = 1$ and $L_1(x) = 1 - x$, and a differential recurrence equation

$$\begin{aligned} &xL_n'(x) - nL_n(x) + nL_{n-1}(x) = 0, \\ &L_n(0) = 1, \; n \in \mathbb{N}. \end{aligned} \tag{2.5}$$

Equation (2.4) entails that for every $n \geq 1$

$$\begin{aligned} &\int_0^\infty xL_n(x)L_{n-1}(x)(x)e^{-x}\,dx = n, \\ &\int_0^\infty xL_n(x)L_{n+1}(x)(x)e^{-x}\,dx = n+1 \end{aligned}$$

and Equation (2.5) entails

$$\begin{aligned} &\int_0^\infty xL_n'(x)L_{n-1}(x)(x)e^{-x}\,dx = -n, \\ &\int_0^\infty xL_n(x)L_n(x)(x)e^{-x}\,dx = n. \end{aligned}$$

The orthonormality and the initial value imply that for every integer $n \geq 1$

$$\int_0^\infty L_n(x)e^{-x}\,dx = \langle L_0, L_n\rangle = 0.$$

By definition (2.2), for every integer n, L_n is a polynomial of degree n with the value 1 at zero. Laguerre's polynomials are equivalent to differentiation operators. Let f be a function of $C^n(\mathbb{R}_+)$ such that $\lim_{x\to\infty} f(x)x^k e^{-x} = 0$ for $k = 1, \ldots, n$, then its scalar products with Laguerre's polynomials are calculated with integrations by parts

$$\begin{aligned} \langle f, L_n\rangle &= \frac{1}{n!}\int_{\mathbb{R}_+} f(x)(x^n e^{-x})^{(n)}\,dx \\ &= \frac{(-1)^n}{n!}\int_{\mathbb{R}_+} f^{(n)}(x)x^n e^{-x}\,dx = \frac{(-1)^n}{n!}\langle f^{(n)}(x), x^n\rangle. \end{aligned} \tag{2.6}$$

From (2.6), the scalar product of real polynomials with the functions L_n are expressed in terms of the gamma functions $\Gamma_{n+1} = \int_{\mathbb{R}_+} x^n e^{-x}\,dx = n!$.

Proposition 2.1. *For every $x \geq 0$ and for all integers k and n*

$$\langle x^n, L_n(x)\rangle = (-1)^n n! = (-1)^n \Gamma_{n+1},$$
$$\langle x^{k+n}, L_n(x)\rangle = (-1)^n \frac{(k+n)!^2}{k!n!},\; k > 0,$$
$$\langle x^k, L_{k+n}(x)\rangle = 0,\; n > 0.$$

Parseval's equality for an orthonormal basis of functions applies to every function $f = \sum_{n\geq 0} a_n L_n$ of $L^2_{\mathcal{E}}(\mathbb{R}_+)$

$$\|f\|_{2,\mathcal{E}} = \int_0^\infty f^2(x)e^{-x}\,dx = \Big\{\sum_{n\geq 0} a_n^2\Big\}^{\frac{1}{2}}. \tag{2.7}$$

Laguerre's transform $Lf(x) = \sum_{k\geq 0} a_k L_k(x)$ converges if its norm $\|f\|_2$ is a convergent series. Let $\|f\|_2$ be the L^2-norm of a function f with respect to Lebesgue's measure. Under the condition $\sum_{n\geq 0} a_n^2(f)$ finite, the partial sum $S_n(f) = \sum_{k\leq n} a_k(f)L_k(x)$ converges to f if and only if the error of approximation $R_n(f) = f - S_n(f) = \sum_{k>n} a_k(f)L_k(x)$ converges to zero.

As x tends to infinity, $L_n(x)$ tends to infinity for n even and $-L_n(x)$ tends to infinity for n odd. Laguerre's polynomials have been modified as functions converging to zero at the infinity, let

$$\varphi_n(x) = \sqrt{2}e^{-x}L_n(2x),\; x \geq 0. \tag{2.8}$$

The set of functions $(\varphi_n)_{n\geq 0}$ is an orthonormal basis with respect to the Lebesgue measure on $\mathbb{R}_+$, with $\varphi_0(x) = \sqrt{2}e^{-x}1_{\{x\geq 0\}}$ and

$$\int_0^\infty \varphi_n^2(x)\,dx = 2\int_0^\infty e^{-2x}L_n^2(2x)\,dx = 1.$$

They converge to zero at the infinity and they are integrable on $\mathbb{R}_+$, with $\int_0^\infty \varphi_n(x) = 1$ if n is even and $\int_0^\infty \varphi_n(x) = -1$ if n is odd.

For every k, the maximum of $x^k e^{-x}$ is achieved at $x = k$ with the value $k^k e^{-k}$ and the maximum of $(k!)^{-1}x^k e^{-x}$ is bounded, it follows that the functions φ_n are uniformly bounded on $\mathbb{R}_+$ and they converge to zero as x tends to infinity.

By the recurrence equation (2.4), the first derivative of the functions $(\varphi_n)_n$ satisfy $\varphi_n(0) = 1$ and for every $n \geq 1$

$$x\varphi_n'(x) = (2n - x)\varphi_n(x) - 2n\varphi_{n-1}(x). \tag{2.9}$$

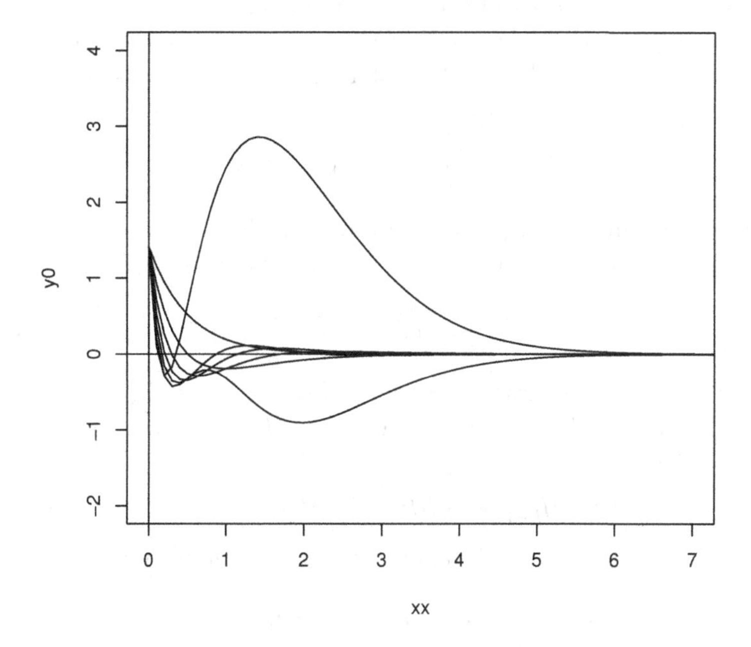

Fig. 2.1 Laguerre's polynomials φ_1 to φ_6.

Laguerre's polynomials are generalized by a change of their support and scaled, with real parameters α and $\lambda > 0$, for $x \geq \alpha$

$$L_n\Big(\frac{x-\alpha}{\lambda}\Big) = \sqrt{\lambda}\frac{e^{\frac{x-\alpha}{\lambda}}}{n!}\frac{d^n}{dx^n}\Big\{e^{-\frac{x-\alpha}{\lambda}}\Big(\frac{x-\alpha}{\lambda}\Big)^n\Big\},\ n \geq 1, \tag{2.10}$$

with the initial value $L_n(\alpha) = 1$, and $L_0 \equiv 1$. This transform yields expansions of densities on a translated support $[\alpha, +\infty[$.

A real function f defined on the interval $I = [\alpha, +\infty[$, converging to zero at the infinity and belonging to $L^2(\mathbb{R}_+)$ has a projection on the transformed orthonormal basis with real parameters α and λ

$$\varphi_n(x; \alpha, \lambda) = \sqrt{2\lambda}e^{-\frac{x-\alpha}{\lambda}} L_n\Big(2\frac{x-\alpha}{\lambda}\Big),\ x \geq \alpha.$$

The parameter scale λ is used for modelling the density of a variable defined on the interval I and with an arbitrary expectation.

2.3 Hermite's polynomials

Let $L^2(\mathbb{R}, \mu_{\mathcal{N}})$ be the space of square integrable functions with respect to the measure $\mu_{\mathcal{N}}$ having the normal density $f_{\mathcal{N}}$ with respect to Lebesgue's measure in $\mathbb{R}$. The scalar product of functions f and g belonging to $(L_2(\mathbb{R}), \mu_{\mathcal{N}})$ is

$$\langle f, g\rangle_{\mu_{\mathcal{N}}} = \int_{\mathbb{R}} f(x)g(x)f_{\mathcal{N}}(x)\,dx.$$

Hermite's polynomials $(H_n)_{n\in\mathbb{N}}$ is an orthonormal basis of $(L_2(\mathbb{R}), \mu_{\mathcal{N}})$ defined by the nth derivatives of the normal density function $f_{\mathcal{N}}$

$$H_k(t) = (-1)^k \frac{d^k e^{-\frac{t^2}{2}}}{dt^k} \frac{e^{\frac{t^2}{2}}}{\sqrt{k!}} \tag{2.11}$$

the functions H_{2k} are symmetric at zero and the functions H_{2k+1} are odd, they are polynomial of degree k for every integer k. They satisfy the recurrence equation

$$H_{k+1}(x) = xH_k(x) - H_k'(x), \tag{2.12}$$

for every $k \geq 1$, with $H_0 = 1$. The recurrence (2.12) for every $n \geq 1$ entails

$$\int_0^\infty H_k'(x)H_k(x)\,f_{\mathcal{N}}(x)\,dx = \int_0^\infty xH_k^2(x)\,f_{\mathcal{N}}(x)\,dx,$$

$$\int_0^\infty H_k'(x)H_{k+1}(x)\,f_{\mathcal{N}}(x)\,dx = -1.$$

A normal density has expansion an on the basis $(H_k)_k$, with the coefficients

$$a_k = \frac{(-1)^k}{\sqrt{k!}} \int_{\mathbb{R}} \{f_{\mathcal{N}}(x)\}^{(k)}\,dx = \langle H_k, H_0\rangle_{f_{\mathcal{N}}} = 0,\ k \geq 1.$$

The polynomials $|H_k|$ tend to infinity as n tends to infinity, they are multiplied by the normal density and normalized for the estimation of functions converging to zero at the infinity. For $k \geq 0$, the functions

$$\mathcal{H}_k(x) = \frac{e^{-\frac{x^2}{2}} H_k(\sqrt{2}x)}{(4\pi)^{\frac{1}{4}}},\ k \geq 1, \tag{2.13}$$

are orthonormal with respect to Lebesgue's measure and they converge to zero at the infinity.

Let f be a real density function on $\mathbb{R}$, having a converging expansion $f(x) = \sum_{k\geq 0} a_k \mathcal{H}_k(x)$, the coefficients of its expansion are

$$a_k = \int_{\mathbb{R}} f(x)\mathcal{H}_k(x)\,dx = \pi^{\frac{1}{4}} \langle f(x), H_k(\sqrt{2}x)\rangle_{f_{\mathcal{N}}}.$$

For the normal density, it follows that for every $k \geq 1$

$$a_k = \frac{1}{(4\pi)^{\frac{1}{4}}} \int_{\mathbb{R}} f_{\mathcal{N}}(\sqrt{2}x)\mathcal{H}_k(\sqrt{2}x)\,dx = 0.$$

For every k, the maximum of $x^k e^{-\frac{x^2}{2}}$ is achieved at $x = k^{\frac{1}{2}}$ with the value $(e^{-1}k)^{\frac{k}{2}}$ and the maximum of $(k!)^{-\frac{1}{2}} x^k e^{-\frac{x^2}{2}}$ is bounded, it follows that the functions $\mathcal{H}_n$ are bounded on $\mathbb{R}_+$. More precisely, there exist constants C_∞ and $C_M \leq 1 + \frac{1}{2}M^{\frac{5}{2}}$ (Szëgo, 1959) such that

$$\begin{aligned} |\mathcal{H}_k(x)| &\leq C_\infty (k+1)^{-\frac{1}{12}},\, x \in \mathbb{R}, \\ |\mathcal{H}_k(x)| &\leq C_M (k+1)^{-\frac{1}{4}},\, x \in [-M, M]. \end{aligned} \tag{2.14}$$

By the recurrence (2.12), the first derivative of the function $\mathcal{H}_k$ satisfies the recurrence equation

$$\mathcal{H}'_k(x) = \sqrt{2}\mathcal{H}_{k+1}(x) - x(\sqrt{2}+1)\mathcal{H}_k(x), \tag{2.15}$$

and the pth derivative of a function $\mathcal{H}_k$ is a linear combination of Hermite functions. For every integer p, we have

$$\mathcal{H}_k^{(p)} = \sum_{s=m}^{p} \alpha_k^{s,p} \mathcal{H}_{k+s},$$

with the bounds (Walter, 1977)

$$\begin{aligned} |\alpha_k^{s,p}| &\leq K_p (k+p)^{\frac{p}{2}}, \\ s &\leq m = -(p \wedge k). \end{aligned} \tag{2.16}$$

2.4 Series estimator of densities on $\mathbb{R}$

Let $L^2(\mathbb{R}, \mu)$ be the space of square integrable functions with respect to the measure μ. The scalar product of functions f and g in $L^2(\mathbb{R}, \mu)$ is

$$\langle f, g \rangle_\mu = \int_{\mathbb{R}} f(x)g(x)\,d\mu(x)$$

and the norm of a function f is $\|f\|_\mu = \{\int_{\mathbb{R}} f^2(x)\,d\mu(x)\}^{\frac{1}{2}}$. Let $(\psi_k)_{k\geq 0}$ be an orthonormal basis of functions in $L^2(\mathbb{R}, \mu)$ endowed with the norm $\|\cdot\|_\mu$. Every function f of $L^2(\mathbb{R}), \mu)$ has an expansion as the infinite series

$$S_f(x) = \sum_{k \geq 0} a_k \psi_k(x)$$

with coefficients defined by the scalar product of f and ψ_k

$$a_k = \langle f, \psi_k \rangle_\mu.$$

The norm of S_f is

$$\|S_f\|_\mu = \Big(\sum_{k\geq 0} a_k^2\Big)^{\frac{1}{2}},$$

the series S_f converges in $L^2(\mathbb{R},\mu)$ to f if $(\sum_{k\leq 0} a_k^2)^{\frac{1}{2}}$ is finite and the inequalities of the Hilbert spaces apply to the basis $(\psi_k)_{k\geq 0}$. The squared approximation error of a function f of $(L^2(\mathbb{R},\mu)$ by a finite sum

$$f_n(x) = \sum_{k=0}^{K_n} a_k \psi_k(x)$$

is $\|f - f_n\|_\mu = \sum_{k\geq K_n+1} a_k^2$ and it converges to zero as K_n tends to infinity for $\|f\|_\mu$ finite.

Let $X_1, \ldots, X_n$ be a n-sample of a variable X with density f with respect to μ, f belonging to $L^2(\mathbb{R},\mu)$ and let F be the distribution function of a variable X with density f with respect to μ. The coefficients

$$a_k = \int_{\mathbb{R}} \psi_k(x)\, dF(x)$$

are estimated from the empirical distribution function $\widehat{F}_n$ of the sample by

$$\widehat{a}_{kn} = \int_{\mathbb{R}} \psi_k(x)\, d\widehat{F}_n(x). \tag{2.17}$$

The density f is approximated by the empirical estimator

$$\widehat{f}_n(x) = \sum_{k=0}^{K_n} \widehat{a}_{kn} \psi_k(x) \tag{2.18}$$

of the function f_n. Let $\nu_n = n^{\frac{1}{2}}(\widehat{F}_n - F)$ be the empirical process of the sample.

Lemma 2.1. *For every integer k, the estimator $\widehat{a}_{kn}$ is a.s. consistent and the variable*

$$n^{\frac{1}{2}}(\widehat{a}_{kn} - a_k) = \int_{\mathbb{R}} \psi_k(x)\, d\nu_n(x)$$

converges weakly to the centered Gaussian variable $A_k = \int_{\mathbb{R}} \psi_k(x)\, dW_1 \circ F(x)$ defined by the standard Brownian bridge W_1 and with the asymptotic variance

$$v_k = \int_{\mathbb{R}^2} \psi_k(x)\psi_k(y)\, d\{F(x \wedge y) - F(x)F(y)\}.$$

The distribution of the variable A_k is also written as $\int_0^1 \psi_k \circ F^{-1}\, dW_1$ with the quantile function F^{-1}.

The mean integrated squared error of the density estimator is the sum

$$\text{MISE}_n(f) = E\|\widehat{f}_n - f_n\|_2^2 + \|f - f_n\|_2^2$$

of the squared mean integrated estimation error $E\|\widehat{f}_n - f_n\|_2^2$ of the function f_n and the square error of approximation of the density by a function f_n

$$\|f_n - f\|_2^2 = \sum_{k=K_n+1}^{\infty} a_k^2$$

which converges to zero as K_n tends to infinity. The squared estimation error

$$\begin{aligned} E\|\widehat{f}_n - f_n\|_2^2 &= \sum_{k=0}^{K_n} E(\widehat{a}_{kn} - a_k)^2 \\ &= n^{-1} \sum_{k=0}^{K_n} \int_{\mathbb{R}} \psi_k^2(x)\, d[F(x)\{1 - F(x)\}] \end{aligned}$$

is the integral of the variance of $\widehat{f}_n$, its convergence rate is $n^{-1}K_n$. As n and K_n tend to infinity, the size K_n defining the approximating function f_n is optimum for the L^2 norm if both errors have the same order so K_n is chosen in order that

$$K_n = o(n), \quad \|f_n - f\|_2 = O(n^{-\frac{1}{2}} K_n^{\frac{1}{2}}). \tag{2.19}$$

Under the condition (2.19), the estimator $\widehat{f}_n$ is L^2-consistent.

Example 2.1. An exponential density f_θ, with parameter $\theta > 0$, has an expansion on Laguerre's basis (2.8) and the coefficients of its expansion are

$$a_{k\theta} = 2\sqrt{2}\frac{\theta}{\theta + 1}\Big(\frac{\theta - 1}{\theta + 1}\Big)^k.$$

The squared approximation error of f_θ is

$$\|f_n - f_\theta\|_2^2 = 2\theta\Big(\frac{\theta - 1}{\theta + 1}\Big)^{2(K_n+1)}$$

and the condition (2.19) is fulfilled by choosing K_n such that

$$\alpha_\theta^{K_n} = o(K_n n^{-1}),$$

with $\alpha_\theta = (\theta - 1)(\theta + 1)^{-2}$ strictly smaller than one. Gamma and Weibull densities have similar properties.

Example 2.2. Let f be a density of the Sobolev space

$$W_{2s} = \Big\{ f \in L^2(\mathbb{R}, \mu) : \sum_{k \geq 0} k^{2s} a_k^2(f) < \infty \Big\}, \tag{2.20}$$

for $s > 1$, then there exists a constant C such that $\|f - f_n\|_2^2 \leq CK_n^{-2s}$ and the optimal K_n which minimizes the MISE has the order

$$K_{opt} = 0(n^{\frac{1}{2s+1}}).$$

The condition (2.19) is fulfilled with $n^{-1}K_{opt} = 0(n^{-\frac{2s}{2s+1}}) = o(1)$ and $K_{opt}^{-2s} = O(n^{-1}K_{opt})$.

Proposition 2.2. *Under the condition (2.19), the error* $E\|\widehat{f}_n - f\|_2$ *for an estimator of a density* f *of* $L^2(\mathbb{R}, \mu)$ *defined by (2.18) and (2.17) equals the optimal convergence rate of a kernel estimator for* f *in* C^s, $s \geq 1$ *if* $K_n = O(n^{\frac{1}{2s+1}})$.

Proof. Let $K_n = n^\alpha$, with $0 < \alpha < 1$, then $n^{-\frac{1}{2}} K_n^{\frac{1}{2}} = n^{\frac{\alpha-1}{2}}$. The optimal convergence rate of a kernel estimator of a density f in C^s has the order $n^{-\frac{s}{2s+1}}$, this is the order of the convergence rate $K_n^{\frac{1}{2}} n^{-\frac{1}{2}}$ of $E\|\widehat{f}_n - f_n\|_2$ if $\alpha = (2s+1)^{-1}$. Under the condition (2.19), the approximation error has the same rate. □

In the Sobolev space (2.20), the projection based estimator has a same convergence rate as the optimal rate for kernel estimators.

The error $E\|\widehat{f}_n - f\|_2$ of an estimator defined with $K_n = n^\alpha$ and $0 < \alpha < (2s+1)^{-1}$ would have a faster convergence rate than the kernel estimator with the optimal bandwidth however reducing K_n increases the bias of the estimator.

The bias of $\widehat{f}_n$ is $f - f_n$ and its mean integrated squared error is

$$\mathrm{MSE}_n(f)(x) = E\{\widehat{f}_n(x) - f_n(x)\}^2 + \{f(x) - f_n(x)\}^2$$

where the first term is the variance of $\widehat{f}_n(x)$

$$\begin{aligned} V_n(x) &= E\{\widehat{f}_n(x) - f_n(x)\}^2 \\ &= n^{-1} \sum_{k=0}^{K_n} \sigma_{n,k}^2 \psi_k^2(x) + n^{-1} \sum_{k' \neq k = 0}^{K_n} \sigma_{n,kk'}^2 \psi_k(x)\psi_{k'}(x) \end{aligned}$$

where $\sigma_{n,k}^2 = n \operatorname{Var}(\widehat{a}_{kn})$ and $\sigma_{kk'n}^2 = n \operatorname{Cov}(\widehat{a}_{kn}, \widehat{a}_{k'n})$. They converge to $\sigma_k^2 = \sigma_{kk}^2$ and

$$\sigma_{kk'}^2 = \int_{\mathbb{R}^2} \psi_k(x)\psi_{k'}(y)\, d\{F(x \wedge y) - F(x)F(y)\}.$$

Proposition 2.3. *Under the condition (2.19), the estimator $\widehat{f}_n$ defined by projection on an uniformly bounded basis of functions $(\psi_k)_k$ of $L^2(\mathbb{R},\mu)$ is such that $(K_n^{-1}n)^{\frac{1}{2}}(\widehat{f}_n - f_n)$ converges weakly to a centered Gaussian process with finite variance*

$$V(x) = \sum_{k,k'\geq 0} \sigma^2_{kk'}\psi_k(x)\psi_{k'}(x).$$

Proof. For every sequence $(a_n)_{n\geq 1}$ tending to infinity, the expectation of $a_n(f-\widehat{f}_n)$ converges to the limit of $g_n = a_n(f-f_n)$, it satisfies $\int |g_n| \leq (\int g_n^2)^{\frac{1}{2}}$ which converges to zero if the order of a_n is larger than the order of $(K_n^{-1}n)^{\frac{1}{2}}$ by the convergence rate of the MISE and under the condition (2.19), in that case $|g_n|$ converges to zero almost everywhere. By the same arguments, the variance of $a_n(\widehat{f}_n - f)$ converges to zero almost everywhere if the order of a_n is larger than the order of $(K_n^{-1}n)^{\frac{1}{2}}$. The weak convergence of the process $(K_n^{-1}n)^{\frac{1}{2}}(\widehat{f}_n - f_n)$ is established by the convergence of its variance function $K_n^{-1}nV_n$ to V and by the weak convergence of Lemma 2.1 for the variables $n^{\frac{1}{2}}(\widehat{a}_{kn} - a_k)$, $k \geq 0$. □

Under the conditions (2.19), the process $(K_n^{-1}n)^{\frac{1}{2}}(\widehat{f}_n - f)$ converges weakly to an uncentered Gaussian process with expectation the limit of $(K_n^{-1}n)^{\frac{1}{2}}(f_n - f)$ depending on the basis and with variance V. If the bias is a $o((K_n^{-1}n)^{\frac{1}{2}})$ like in Example 2.1, the limiting Gaussian process is centered.

The knowledge of bounds for the orthonormal basis improves the evaluation of the estimation error. With Hermite polynomials (2.13) and by (2.14), the estimated coefficients of the estimator $\widehat{f}_n$ of a density f of $L^2(\mathbb{R})$ have variances

$$E(\widehat{a}_{kn} - a_k)^2 \leq 2n^{-1}C_\infty(k+1)^{-\frac{1}{6}},$$

for all $r > 0$ and $k \leq m_n = O(n^{\frac{1}{r}})$.

For a density f of $C^r(\mathbb{R})$ such that $(x-D)^r f$ belongs to $L^2(\mathbb{R})$, with the differential operator D, and for every integer $p < r$, the pth derivative of f is estimated by projection on $K_n = O(n^{\frac{1}{r}})$ derivatives $\mathcal{H}_k^{(p)}$ of Hermite functions and the MISE error for the estimator of $f^{(p)}$ depends on the regularity of the density. Walter (1977) proved that

$$\text{MISE}(\widehat{f}_n^{(p)}) = O(n^{\frac{p}{r}+\frac{5}{6r}-1}).$$

Let F be a distribution function with jumps at points x_j and a continuous part with density f_0 with respect to a dominant measure μ, it is written as

$$F(x) = \sum_{j=1}^{J_n} p_j \delta_{x_j}(x) + p_0 F_0(x)$$

with $p_0 = 1 - \sum_{j=1}^{J_n} p_j$ and $p_j = \Delta F(x_j)$. The empirical distribution function $\widehat{F}_n$ has multiple jumps at x_j with the mass

$$\Delta \widehat{F}_n(x_j) = \widehat{p}_{nj}.$$

and an estimator $\widehat{p}_{0n}$ of p_0 is deduced.

The scalar product $\langle f, \psi_k \rangle_\mu = \int \psi_k \, dF_0$ of the continuous part of F is the weighted integral

$$a_k = p_0 \int_0^\infty \psi_k(x) f_0(x) \, d\mu(x).$$

The coefficient a_k has the estimator $\widehat{a}_{nk}$ previously defined and the density f_0 is estimated by $\widehat{f}_{0n}$ such that

$$\widehat{p}_{0n} \widehat{f}_{0n}(x) = \sum_{k=1}^{K_n} \widehat{a}_{nk} \psi_k(x),$$

where $K_n = o(n)$ and tends to infinity as n tends to infinity.

A real density f defined on the interval $I = [\alpha, +\infty[$ and belonging to $L^2(\mathbb{R}_+)$ has a projection on the transformed orthonormal Laguerre's basis with real parameters α and λ

$$\varphi_k(x; \alpha, \lambda) = \sqrt{2\lambda} e^{-\frac{x-\alpha}{\lambda}} L_k\Big(2\frac{x-\alpha}{\lambda}\Big), \ x \geq \alpha.$$

With unknown parameters α and λ, the coefficients a_k of the expansion $f(x) = \sum_{k \geq 0} a_k \varphi_k(x; \alpha, \lambda)$ have estimators $\widehat{a}_{kn}(\alpha, \lambda)$ and the parameters are estimated by minimization of a cumulated difference of the estimators defined for the empirical distribution functions

$$D_n(\alpha, \lambda) = \int \Big\{ \widehat{F}_n(x) - \sum_{k \leq m_n} \widehat{a}_{kn}(\alpha, \lambda) \int_{-\infty}^x \varphi_k(y; \alpha, \lambda) \, dy \Big\}^2 dx$$

or

$$\rho_n(\alpha, \lambda) = \int \Big\{ \widehat{F}_n(x) - \sum_{k \leq m_n} \widehat{a}_{kn}(\alpha, \lambda) \int_{-\infty}^x \varphi_k(y; \alpha, \lambda) \, dy \Big\} d\widehat{F}_n(x),$$

which converge a.s. to

$$D(\alpha,\lambda) = \int \Big\{F(x) - F(x;\alpha,\lambda)\Big\}^2\,dx$$

and respectively

$$\rho(\alpha,\lambda) = \int \Big\{F(x) - F(x;\alpha,\lambda)\Big\}\,dF(x).$$

The estimators $\widehat{\alpha}_n$ and $\widehat{\lambda}_n$ are therefore a.s. consistent and their convergence rates to a centered Gaussian distribution is $n^{-\frac{1}{2}}$.

Let $(\mathbb{F}, \|\cdot\|_2)$ be a space of densities on a subset I of $\mathbb{R}$, provided with the norm $L^2(I)$ and let $T(f)$ be an integral functional of f in $C^2(\mathbb{F})$, of the form

$$T(f) = \int_I \phi(f, f', , f^{(k)})$$

such as $T(f) = \int_I f^{(k)2}(x)\,dx$, for $k \geq 2$, and the Fisher information $I(f) = \int_I f^{-1}(x) f^{'2}(x)\,dx$ which are used in asymptotic statistics. Estimators $\widehat{T}_n(f) = T(\widehat{f}_n)$ have been defined with the derivatives of a kernel estimator of f (Singh 1987), they may be defined with an estimator $\widehat{f}_n$ based on the projection of the empirical distribution function of a sample with density f on an orthonormal basis of functions. Let $h_k = (f, f', , f^{(k)})^T$ and let $\widehat{h}_{n,k} = (\widehat{f'_n}, \ldots, \widehat{f_n^{(k)}})^T$ be the estimator of h_k defined on an orthonormal basis. By the recurrence equation of the Laguerre and Hermite functions, the expansions of the derivatives of f are linear combination of functions of these bases, and the convergence rate of their estimators is the same as for $\widehat{f}_n$. From the differentiability of T, we have

$$\widehat{T}_n(f) - T(f) = \Big\{\int_I (\widehat{h}_{n,k} - h_k)\phi'(h_k)\Big\}\{1 + o_{a.s.}(1)\},$$

and by the weak convergence of the coefficients of the series expansion of $\widehat{f}_n$ and its derivatives, the process $n^{\frac{1}{2}} K_n^{-\frac{1}{2}}\{\widehat{T}_n(f) - T(f)\}$ converges weakly to a Gaussian process with finite mean and variance functions.

The MISE of the estimator $\widehat{T}_n(f)$ is the sum of the approximation error

$$\|T(f_n) - T(f)\|_2^2 = \Big[\int_I \{\phi(h_{n,k}) - \phi(h_k)\}\Big]^2 \leq \|\phi(h_{m,k}) - \phi(h_k)\|_2^2,$$

and the error of estimation has the bound

$$\begin{aligned} E\|\widehat{T}_n - T(f_n)\|_2^2 &= E\Big[\int_I \{\phi(\widehat{h}_{n,k}) - \phi(h_{m,k})\}\Big]^2 \\ &\leq E\int_I \{(\widehat{h}_{n,k} - h_{m,k})^T \phi'(h_{m,k})\}^2 + o(1) \\ &\leq (k+1)\sum_{j=0}^{k} \int_I \{E(\widehat{f}_n^{(j)} - f_m^{(j)})^2 \phi_j^{'2}(h_{m,k})\} + o(1), \end{aligned}$$

where ϕ'_j is the first derivative of ϕ with respect to its jth component. Since the bounds split the norms as sums of the norms for the components of the scalar product, the errors have the same orders as the errors of $\widehat{f}_n$.

2.5 Cross-validation for a density

The number K_n of functions of an orthonormal basis used in the approximation of a density f of L^2 by its projections was estimated by minimization of $\mu_n(K_n) = \text{MISE}(\widehat{f}_n)$ which tends to infinity if $K_n = n^\alpha$ with $\alpha \geq 1$. By the orthonormality of the basis, the integrated squared error of $\widehat{f}_n$ is the sum

$$\begin{aligned} I_n(K_n) &= \int \{\widehat{f}_n(x) - f(x)\}^2\, dx \\ &= \sum_{k=0}^{K_n} (\widehat{a}_{kn} - a_k)^2 + \sum_{k \geq K_n+1} a_k^2. \end{aligned}$$

For every $k \leq K_n$, $\widehat{a}_{kn} - a_k = 0_p(n^{-\frac{1}{2}})$ and the first sum is a $0_p(n^{-1}K_n) = o_p(1)$, the second sum is a $o(1)$ and it may be evaluated precisely according to the choice of the basis. The minimization of $I_n(K_n)$ is asymptotically equivalent to the minimization of

$$\begin{aligned} I_n(K_n) - \int f^2 &= \int \widehat{f}_n^2(x)\, dx - 2\int \widehat{f}_n(x)\, dF(x) \\ &= \sum_{k=0}^{K_n} \widehat{a}_{kn}^2 - 2\sum_{k=0}^{K_n} a_k \widehat{a}_{kn}. \end{aligned}$$

Let $\widehat{f}_{n,j}$ be the projection based estimator of the density defined with the sub-sample without the observation X_j, Hall (1987) estimated the integral $I_n(K_n) - \int f^2$ as

$$CV_n(K_n) = n^{-1} \sum_{j=1}^{n} \int \widehat{f}_{n,j}^2(x)\, dx - 2n^{-1} \sum_{j=1}^{n} \widehat{f}_{n,j}(X_j),$$

CV_n is asymptotically equivalent to

$$J_n(m) = \sum_{k=0}^{K_n} \widehat{a}_{kn}^2 - \frac{2}{n(n-1)} \sum_{k=0}^{K_n} \sum_{j \neq i=1}^{n} \psi_k(X_i)\psi_k(X_j). \tag{2.21}$$

The expectation of $J_n(m)$ is

$$\begin{aligned} EJ_n(K_n) &= \sum_{k=0}^{K_n} E\widehat{a}_{kn}^2 - 2\sum_{k=0}^{K_n} a_k^2 \\ &= \sum_{k=0}^{K_n} \mathrm{Var}(\widehat{a}_{kn}) - \sum_{k=0}^{K_n} a_k^2 \\ &= EI_n(K_n) - \int f^2(x)\,dx. \end{aligned}$$

An estimator $\widehat{K}_n$ of K_n is defined by minimization of J_n. The difference of $I_n - J_n$ is written as

$$\begin{aligned} I_n(K_n) - J_n(K_n) = 2\sum_{k=0}^{K_n}\Big\{\frac{1}{n}\sum_{i=1}^{n}\psi_k(X_i)\Big\}\Big\{\frac{1}{n-1}\sum_{j\neq i=1}^{n}\psi_k(X_j) - a_k\Big\} \\ + \int f^2(x)\,dx \end{aligned}$$

and it converges in probability to $\int f^2(x)\,dx$ as n tends to infinity. It follows that the minima of I_n and J_n are asymptotically equivalent in probability and $\widehat{K}_n$ is asymptotically optimal i.e.

$$\frac{I_n(\widehat{K}_n)}{\inf_{K_n} I_n(K_n)} \to 1$$

in probability as n tends to infinity.

The asymptotic optimality of the cross-validation estimator for kernel estimators of densities has been proved by Hall (1983), Stone (1984) and by Hall (1987) for weighted orthogonal series estimators $\widehat{f}_{n,b} = \sum_{k=0}^{\infty} b_k\widehat{a}_{kn}\phi_k$, where b_k tends to zero as k tends to infinity and a summability condition is fulfilled, such as $\sum_{k=0}^{\infty}(k+1)^{-a}|b_k|$ finite for some a in $]0,1[$. For the Hermite polynomials $(\mathcal{H}_k)_k$, the exponent is $a = \frac{1}{2}$. These properties apply to the estimator of Section 2.6.

2.6 Weighted series density estimators

Wahba (1981) considered be an orthonormal basis of functions $(\psi_k)_{k\geq 0}$ in $L^2(\mathbb{R})$ provided with Lebesgue's measure, such that there exist constants $m_0 > 1$ and M_0 for which

$$\sum_{k\geq 0} k^{-2m}\psi_k^2(x) \leq M_0 < \infty, \tag{2.22}$$

for every integer $m \geq m_0$ and for every x. Let $f = \sum_{k\geq 0} a_k \psi_k$ be a density of the space

$$W_{2m} = \Big\{ f \in L^2(\mathbb{R}, \mu) : \sum_{k\geq 0} k^{2m} a_k^2(f) < \infty \Big\},$$

for $m \geq m_0$, it satisfies $\sum_{k\geq 0} \lambda_k^{-1} a_k^2 \leq C_1$, for some constant C_1 and with $\lambda_k = k^{-2m}$, then

$$\|f\|_2^2 \leq M_0 C_1$$

by the Cauchy–Schwarz inequality. The density is estimated by

$$\widehat{f}_{nm\nu}(x) = \sum_{k=1}^{n} \frac{\lambda_k}{\nu + \lambda_k} \widehat{a}_{nk} \psi_k(x) \tag{2.23}$$

with $\widehat{a}_{nk} = n^{-1} \sum_{i=1}^{n} \psi_k(X_i)$ and weights proportional to λ_k, with $\nu > 0$ constant with respect to k. The choice of an optimal parameter ν enables to reach the optimal convergence rate of the estimator as K_n is replaced by n.

Proposition 2.4. *Under the conditions (2.22) and $\nu_n = O(n^{-\frac{2m}{2m+1}})$, the estimator of a density f of W_{2m} is such that* $\mathrm{MISE}(\widehat{f}_{nm\nu_n}) = O(n^{-\frac{2m}{2m+1}})$.

Proof. The integrated squared error of the estimator is

$$\begin{aligned}
\mathrm{ISE}_{nm}(\nu_n) &= \int \{\widehat{f}_{nm\nu_n}(x) - f(x)\}^2 \, dx \\
&= \sum_{k=1}^{n} \Big\{ \frac{\lambda_k}{\nu_n + \lambda_k} (\widehat{a}_{nk} - a_k) - \frac{\nu_n}{\nu_n + \lambda_k} a_k \Big\}^2 + \sum_{k=n+1}^{\infty} a_k^2,
\end{aligned}$$

it has the expectation

$$\mathrm{MISE}_{nm}(\nu_n) = \sum_{k=1}^{n} \frac{\lambda_k^2 v_k}{n(\nu_n + \lambda_k)^2} + \sum_{k=1}^{n} \frac{\nu_n^2 a_k^2}{(\nu_n + \lambda_k)^2} + \sum_{k=n+1}^{\infty} a_k^2$$

which is denoted $\mathrm{MISE}_{nm}(\nu_n) = S_{1n} + S_{2n} + S_{3n}$. The variance of $\widehat{a}_{nk}$ under the conditions (2.22) is $n^{-1} v_k$ such that

$$v_k = \int \psi_k^2 \, d\{F(1 - F)\} < \sup |f| \leq (M_0 C_1)^{\frac{1}{2}}.$$

The condition (2.22) entails

$$\begin{aligned}
S_{1n} &= \sum_{k=1}^{n} \frac{\lambda_k^2 v_k}{n(\nu_n + \lambda_k)^2} < C n^{-1} \sum_{k=1}^{n} \frac{1}{(k^{2m} \nu_n + 1)^2} \\
&< C n^{-1} \nu_n^{-\frac{1}{2m}} \int_0^{\infty} \frac{dx}{(1 + x^{2m})^2},
\end{aligned}$$

with $x = \nu_n^{\frac{1}{2m}} k$ in the integral, and

$$S_{2n} = \sum_{k=1}^{n} \frac{\nu_n^2 a_k^2}{(\nu_n + \lambda_k)^2} < \nu_n \sum_{k=1}^{n} \lambda_k^{-1} a_k^2 < \nu_n C_1,$$

$$S_{3n} = \sum_{k=n+1}^{\infty} a_k^2 \leq C_2 n^{-2m},$$

for a constant $C_2 > 0$. Let $\nu_n = n^{-\alpha}$, $\alpha > \frac{1}{2}$, then S_{1n} and S_{2n} have the same order if and only if $\alpha = 2m(2m+1)^{-1}$ and the main terms of $\text{MISE}_{nm}(\nu_n)$ have the order $n^{-\frac{2m}{2m+1}}$, as n tends to infinity. □

The advantage of the estimator $\widehat{f}_{nm\nu}$ is that the residual $f - f_n$ is smaller with a projection on n rather than $o(n)$ functions ψ_k but the estimated coefficients are biased and

$$f(x) - E\widehat{f}_{nm\nu_n}(x) = \sum_{k>n} a_k \psi_k(x) + \sum_{k=1}^{n} \frac{1}{1 + \nu_n^{-1} k^{-m}} a_k \psi_k(x),$$

it converges to zero, as n tends to infinity, with $\nu_n = o(1)$.

The estimator (2.23) of the density is a projection on n functions of the basis, replacing n by an arbitrary $N_n > n$ does not change the order of S_{1n} and S_{2n}, and the order of $S_{3n} = O(N_n^{-2m})$ is reduced.

2.7 Tests for densities

We consider the comparison of two densities f_1 and f_2 from the observations of two samples $(X_{1i})_{i=1,\dots,n_1}$ under f_1 and $(X_{2i})_{i=1,\dots,n_2}$ under f_2. Let $n = n_1 + n_2$ be the total sample size such that $p_{1n} = p_n = n^{-1} n_1$ converges to $p_1 = p$ in $]0,1[$ and $p_{2n} = 1 - p_n = n^{-1} n_2$ converges to $p_2 = 1 - p$, as n tends to infinity, the whole sample has the mixture density $f = pf_1 + (1-p)f_2$ with a known mixing probability. Let $\widehat{f}_n = p_{1n}\widehat{f}_{1n} + p_{2n}\widehat{f}_{2n}$ with the series estimators $\widehat{f}_{jn}$, $j = 1, 2$, defined by (2.18).

A test statistic is defined by the logarithm of the estimated log-likelihood ratio of the densities under the alternative of two distinct densities and the hypothesis $H_0 : f_1 = f_2$ that the sub-samples have the same density f_0, let

$$\widehat{T}_n = K_n^{-1} \left\{ \sum_{i=1}^{n_1} \log \frac{\widehat{f}_{1n}(X_i)}{\widehat{f}_n(X_i)} + \sum_{i=n_1+1}^{n} \log \frac{\widehat{f}_{2n}(X_i)}{\widehat{f}_n(X_i)} \right\}. \tag{2.24}$$

By Proposition 2.3 the estimators $\widehat{f}_{jn}$ satisfy

$$\widehat{f}_{jn} = f_j + a_{n_j}^{\frac{1}{2}} B_{jn},$$

where $a_{n_j} = n_j^{-1} K_{n_j}$, $K_{n_j} = n_j^{\alpha}$ with the optimal $0 < \alpha < 1$, and the process B_{jn} converges weakly to a Gaussian process B_j, for $j = 0, 1, 2$, where $n_0 = n$. Due to the convergence rates of the estimators, the log-likelihood ratio statistic is normalized by $K_n^{-\frac{1}{2}} = n^{-\frac{\alpha}{2}}$.

Let $(\eta_n)_{n\geq 1}$ and $(\gamma_n)_{n\geq 1}$ be sequences of uniformly bounded functions of $L^2(\mathbb{X}, f_0)$, converging uniformly to non zero functions η and γ, as n tends to infinity, and such that $\int_{\mathbb{X}} \eta f_0 \, dx = 0$, $\int_{\mathbb{X}} \gamma f_0 \, dx = 0$ and $\int_{\mathbb{X}} \eta\gamma f_0 \, dx = 0$. We consider a test of the hypothesis H_0 against a sequence of local alternatives A_n defined by the densities of the sub-samples

$$A_n : \begin{cases} f_{1n}(t) = f_{2n}(t)\{1 + a_{n_1}^{\frac{1}{2}} \eta_n(t)\}, \\ f_{2n}(t) = f_0(t)\{1 + a_{n_2}^{\frac{1}{2}} \gamma_n(t)\}, \ t \in \mathbb{R}, \end{cases}$$

they converge to f_0 with the same rate as the estimators.

Proposition 2.5. *Under H_0 the log-likelihood ratio statistic has an expansion $\widehat{T}_n = -\frac{1}{2} Z_{0n} + o_p(1)$ where Z_{0n} converges weakly to*

$$Z_0 = \sum_{j=1}^{2} \int_{\mathcal{I}_X} \frac{(B_0 - B_j)^2}{2 f_0^2} \, dF_{X_j}.$$

Under the local alternatives A_n, $\widehat{T}_n = -\frac{1}{2} Z_n + o_p(1)$ where Z_n converges weakly to a limit Z strictly different from Z_0. Under fixed alternatives, $\widehat{T}_n$ diverges.

Proof. Let $\widehat{f}_{jn} = f_{jn} + U_{jn}$ where $U_{jn} = a_{n_j}^{\frac{1}{2}} B_{jn} = \widehat{f}_{jn} - f_{jn} = O_p(a_{n_j}^{\frac{1}{2}})$ and $f_{jn} - f_0 = O(a_{n_j}^{\frac{1}{2}})$, for $j = 0, 1, 2$. Under H_0, the logarithm of the estimated likelihood ratio of the sub-samples is approximated by a second order Taylor expansion, for $j = 1, 2$

$$\log \frac{\widehat{f}_{jn}}{\widehat{f}_n} = \widehat{f}_n^{-1}\{(f_{jn} - f_0) + U_{jn} - U_{0n}\}$$
$$- \frac{\widehat{f}_n^{-2}}{2}\{U_{jn} U_{0n} + f_{jn} - f_0\}^2 + o(a_n)$$

which entails an asymptotic expansion of the log-likelihood ratio statistic.

Under H_0, the densities are $f_{jn} = f_0$ and

$$U_{0n} - U_{jn} = \widehat{f}_n - \widehat{f}_{jn} = \sum_{k=K_{n_j}+1}^{K_n} \widehat{a}_{kn} \psi_k,$$

with $K_n - K_{n_1} = n^\alpha(1-p^\alpha)+o(n^\alpha)$, its convergence rate is $a_n^{\frac{1}{2}}$. We obtain

$$\sum_{i=1}^{n_1} \log \frac{\widehat{f}_{1n}}{\widehat{f}_n}(X_i) = -\sum_{i=1}^{n_1} \frac{U_{0n}(X_i) - U_{1n}(X_i)}{f_0(X_i)}\{1+o_p(1)\}$$
$$-\sum_{i=1}^{n_1} \frac{\{U_{0n}(X_i) - U_{1n}^2(X_i)\}^2}{2f_0^2(X_i)}\{1+o_p(1)\} + o(K_n)$$

$$\sum_{i=n_1+1}^{n} \log \frac{\widehat{f}_{2n}}{\widehat{f}_n}(X_i) = -\sum_{i=n_1+1}^{n} \frac{U_{0n}(X_i) - U_{2n}(X_i)}{f_0(X_i)}\{1+o_p(1)\}$$
$$-\sum_{i=n_1+1}^{n} \frac{\{U_{0n}(X_i) - U_{2n}(X_i)\}^2}{2f_0^2(X_i)}\{1+o_p(1)\} + o(K_n).$$

The expectations of the variables in the first sums are asymptotically equivalent to

$$E\int \{\widehat{f}_{jn}(x) - \widehat{f}_n(x)\}\,dx = o(1),$$

where $E\widehat{f}_{jn}(x) = \sum_{k=0}^{K_{n_j}} a_k\psi_k(x)$ converges to f_0 for $j=0,1,2$. The asymptotic distribution of $\widehat{T}_n$ follows from the weak convergence of the variables

$$Y_{jn} = (n_j a_n)^{-\frac{1}{2}} \sum_{i=1}^{n_j} \frac{\widehat{f}_{jn}(X_{ji}) - \widehat{f}_n(X_{ji})}{f_0(X_{ji})}$$
$$= \int \frac{p_j^{\frac{\alpha-1}{2}} B_{jn} - B_{0n}}{f_0}\,d\nu_{jn} + o(1)$$

where ν_{jn} is the empirical process of the jth sub-sample. The variable $Y_{0n} = Y_{1n} + Y_{2n}$ converges weakly under H_0 to a centered Gaussian variable $Y_0 = \sum_{j=1,2}\int f_0^{-1}(p_j^{\frac{\alpha-1}{2}} B_j - B_0)\,d\nu_j$ with finite variance, and the first order term of the expansion of $\widehat{T}_n$ converges to zero due to the normalization by K_n^{-1}. The second order term of the expansions is

$$Z_{0n} = K_n^{-1}\sum_{j=1}^{2}\sum_{i=1}^{n_j} \frac{\{U_{0n}(X_{ji}) - U_{jn}(X_{ji})\}^2}{2f_0^2(X_{ji})}$$
$$= \sum_{j=1}^{2}\int_{\mathcal{I}_X} \frac{(B_{0n} - B_{jn})^2}{2f_0^2}\,d\widehat{F}_{nX_j} \tag{2.25}$$

and it converges weakly to Z_0. Under fixed alternatives, $\widehat{T}_n$ is asymptotically equivalent to

$$K_n^{-1}\left\{\sum_{i=1}^{n_1}\log\frac{f_1}{f_0}(X_i) + \sum_{i=n_1+1}^{n}\log\frac{f_2}{f_0}(X_i)\right\}$$

and it diverges. Under the local alternatives A_n, we have

$$\begin{aligned}\frac{f_{1n}-f_0}{f_0} &= a_{n_1}^{\frac{1}{2}}\eta_n + a_{n_2}^{\frac{1}{2}}\gamma_n + o(a_n^{\frac{1}{2}})\\ &= a_n^{\frac{1}{2}}\{p^{\frac{\alpha-1}{2}}\eta_n + (1-p)^{\frac{\alpha-1}{2}}\gamma_n + o(1)\},\\ \frac{f_{2n}-f_0}{f_0} &= a_{n_2}^{\frac{1}{2}}\gamma_n = a_n^{\frac{1}{2}}\{(1-p)^{\frac{\alpha-1}{2}}\gamma_n + o(1)\},\end{aligned}$$

they are denoted $a_n^{\frac{1}{2}}\{\gamma_{jn}+o(1)\}$ and the variables Y_{jn} are replaced by

$$\begin{aligned}\widetilde{Y}_{jn} &= (n_j a_n)^{-\frac{1}{2}}\sum_{i=1}^{n_j}\int \frac{p_j^{\frac{\alpha-1}{2}}(B_{jn}+\gamma_{jn}) - B_{0n}}{f_0}\,d\nu_{jn} + o(1)\\ &= \int \{p_j^{\frac{\alpha-1}{2}}(B_{jn}+\gamma_{jn}) - B_{0n}\}\,d\nu_{jn} + o(1).\end{aligned}$$

The variable $Y_n = \widetilde{Y}_{1n}+\widetilde{Y}_{2n}$ converges weakly under the local alternatives to an uncentered Gaussian variable $Y = \sum_{j=1,2}\int f_0^{-1}\{p_j^{\frac{\alpha-1}{2}} B_j+\gamma_j)-B_0\}\,d\nu_j$. The second order term of the expansions is

$$\begin{aligned}Z_n &= K_n^{-1}\sum_{j=1}^{2}\sum_{i=1}^{n_j}\frac{\{(B_{jn}+\gamma_{jn})(X_{ji}) - B_{0n}(X_{ji})\}^2}{2f_0^2(X_{ji})}\\ &= \sum_{j=1}^{2}\int_{\mathcal{I}_X}\frac{(B_{jn}+\gamma_{jn}-B_{0n})^2}{2f_0^2}\,d\widehat{F}_{nX_j}\end{aligned}$$

and it converges in probability to a limit Z different from Z_0. □

The estimation of the variance V_0 of the variable Y_0 of Proposition 2.5 is equivalent to the estimation of the limit of Z_{0n} defined by (2.25). Under H_0, The process $U_{jn}+U_{0n} = \widehat{f}_{jn}+\widehat{f}_n - 2f_0$ is estimated by $\widehat{f}_{jn}-\widehat{f}_n$ and the limit of Z_0 is estimated by

$$\widehat{V}_{0n} = K_n^{-1}\sum_{j=1}^{2}\sum_{i=1}^{n_j}\frac{\{U_{jn}(X_{ji}) - U_{0n}(X_{ji})\}^2}{2\widehat{f}_n^2(X_{ji})},$$

then a test based on the statistic $\widehat{T}_n$ may be defined as a test based on $\widehat{V}_{0n}$.

Proposition 2.5 extends to a test of comparison of d samples, for every $d \geq 2$, with the test statistic

$$\widehat{T}_n = K_n^{-1}\sum_{j=1}^{d}\sum_{i=1}^{n_j}\log\frac{\widehat{f}_{jn}(X_{ji})}{\widehat{f}_n(X_{ji})}.$$

It applies to several other cases. Let f_θ be a symmetric density with a center of symmetry at an unknown parameter θ which is estimated by the empirical mean $\widehat{\theta}_n$ of a sample $(X_i)_{i=1,\dots,n}$ with density f_θ. Considering the sub-sample $(X_{1i})_{i=1,\dots,n_1}$ of the observations smaller than $\widehat{\theta}_n$, with density $f_1(x) = f(\theta + x)$ and the sub-sample $(X_{2i})_{i=1,\dots,n_2}$ of the observations smaller than $\widehat{\theta}_n$, with density $f_2(x) = f(\theta - x)$. A test for symmetry for a density is performed like a test of homogeneity, with the hypothesis $H_0 : f_1 = f_2$.

The same approach is used for a goodness of fit test for the density of a sample $(X_i)_{i=1,\dots,n}$, with an hypothesis $H_0 : f = f_0$ against an alternative $f \neq f_0$ or local alternatives, with the statistic

$$T_n = K_n^{-1} \sum_{i=1}^n \log \frac{\widehat{f}_n(X_i)}{f_0(X_i)}, \tag{2.26}$$

its asymptotic behavior is similar to the behavior of $\widehat{T}_n$ in Proposition 2.5 with a single sample, where the terms U_{0n} and B_{0n} are omitted.

The validity of a parametric family of densities $\mathbb{F}_\Theta = \{f_\theta \in C^2(\Theta), \theta \in \Theta\}$, defined by an open bounded convex subset Θ of $\mathbb{R}^d$, is assertened by a likelihood ratio test of the hypothesis $H_0 : f$ belongs to $\mathbb{F}_\Theta$ against alternatives that f does not belong to $\mathbb{F}_\Theta$, with the statistic

$$T_{1n} = K_n^{-1} \sum_{i=1}^n \log \frac{\widehat{f}_n(X_i)}{f_{\widehat{\theta}_n}(X_i)}, \tag{2.27}$$

where $\widehat{\theta}_n$ is the maximum likelihood estimator of the unknown parameter value of the density $f_0 = f_{\theta_0}$ of the sample under H_0. As the convergence rate of $\widehat{\theta}_n$ is $n^{-\frac{1}{2}}$, $f_{\widehat{\theta}_n}$ converges to f_0 with the same rate and

$$\begin{aligned} T_{1n} &= K_n^{-1} \sum_{i=1}^n \left\{ \log \frac{\widehat{f}_n(X_i)}{f_0(X_i)} - \log \frac{f_{\widehat{\theta}_n}(X_i)}{f_0(X_i)} \right\} \\ &= K_n^{-1} \sum_{i=1}^n \log \frac{\widehat{f}_n(X_i)}{f_0(X_i)} + o(1). \end{aligned}$$

The statistic T_{1n} has an expansion similar to T_n under H_0 and alternatives, where the process U_{0n} is omitted.

Let $\mathbb{F}$ be a semi-parametric family $\mathbb{F}_{\Theta,\mathcal{G}} = \{f_{\theta,g}, \theta \in \Theta, g \in \mathcal{G}\}$ defined on an open bounded convex parameter set Θ and a family $\mathcal{G}$ of functions of $C^2(\mathcal{I})$, such that the distribution function of a sample $(X_i)_{i=1,\ldots,n}$ is written as

$$F_\theta = G \circ H_\theta^{-1}$$

with a known monotone function $H_\theta(x)$ in $C^2(\Theta)$ uniformly with respect to x in $\mathcal{I}$ and in $C^2(\mathcal{I})$ for every θ in Θ. For every θ, the variables X_i are transformed as variables $Y_{\theta,i} = H_\theta^{-1}(X_i)$ with the distribution function $G = F \circ H_\theta$ and the unknown density $g = h_\theta\, f_\theta \circ H_\theta^{-1}$. Let $\widehat{g}_{n\theta}$ be a series estimator of the sample $(Y_{\theta,i})_i$, the log-likelihood $l_n(\theta) = \sum_{i=1}^n \log f_\theta(X_i)$ of the sample is estimated by

$$\widehat{l}_n(\theta) = \sum_{i=1}^n \{\log \widehat{g}_{n\theta}(Y_{\theta,i}) - \log h_\theta(Y_{\theta,i})\}$$

which is maximum at the estimator $\widehat{\theta}_n$ of the parameter value θ_0 and the density g is estimated by $\widehat{g}_n = \widehat{g}_{n\widehat{\theta}_n}$. The estimator $\widehat{\theta}_n$ is $n^{-\frac{1}{2}}$-consistent and $\widehat{g}_n$ has the convergence rate $n^{-\frac{1}{2}} K_n^{\frac{1}{2}}$.

A goodness of fit test of validity for a semi-parametric family $\mathbb{F}_{\Theta,\mathcal{G}}$ is performed with a log-likelihood ratio statistic comparing the nonparametric density estimator $\widehat{f}_n$ of the sample and its estimator $f_{n\widehat{\theta}_n}$ in $\mathbb{F}_{\Theta,\mathcal{G}}$, it is equivalent to a test based on the statistic $\widehat{T}_n$.

2.8 Estimation of a density on $\mathbb{R}^d$

Let X be a random variable defined on a subset $\mathcal{I}_X$ of $\mathbb{R}^d$, with density f_X of $L^2(\mathcal{I}_X, \mu)$ and let $X_1, \ldots, X_n$ be a n-sample of a variable X. We assume that there exists an orthonormal basis of functions $(\psi_k)_{k\geq 0}$ in $L^2(\mathcal{I}_X, \mu)$ such that f has the expansion

$$f(x) = \sum_{k_1,\ldots,k_d \geq 0} a_{k_1\ldots k_d} \psi_{k_1}(x_1) \cdots \psi_{k_d}(x_d).$$

It is approximated by $f_n(x) = \sum_{k_1=0}^{K_{1n}} \cdots \sum_{k_d=0}^{K_{dn}} a_{k_1,\ldots,k_d} \psi_{k_1}(x_1) \cdots \psi_{k_d}(x_d)$ with the coefficients $a_{k_1\ldots k_d} = \int_{\mathcal{I}_X} \psi_{k_1}(x_1) \cdots \psi_{k_d}(x_d)\, dF_X(x)$. The coefficients are estimated like (2.17) with the empirical distribution function $\widehat{F}_n$ of the sample

$$\widehat{a}_{nk_1\ldots k_d} = \int_{\mathcal{I}_X} \psi_{k_1}(x_1) \cdots \psi_{k_d}(x_d)\, d\widehat{F}_n(x),$$

with $k = (k_1, \ldots, k_d)$. They satisfy the converge of Lemma 2.1 with the rate $n^{-\frac{1}{2}}$. The density estimator

$$\widehat{f}_n = \sum_{k_1=0}^{K_{1n}} \cdots \sum_{k_d=0}^{K_{dn}} \widehat{a}_{n,k_1,\ldots,k_d} \psi_{k_1} \cdots \psi_{k_d}$$

has the mean integrated squared error

$$\text{MISE}_n(f) = E\|\widehat{f}_n - f_n\|_2^2 + \|f - f_n\|_2^2$$

where the squared error of approximation of the density is

$$\|f - f_n\|_2^2 = \sum_{k_1 > K_{1n}} \cdots \sum_{k_d > K_{dn}} a_k^2$$

and it converges to zero as n tends to infinity. The mean squared estimation error is

$$\begin{aligned}
E\|\widehat{f}_n - f_n\|_2^2 &= \sum_{k_1=0}^{K_{1n}} \cdots \sum_{k_d=0}^{K_{dn}} E(\widehat{a}_{kn} - a_k)^2 \\
&= n^{-1} \sum_{k_1=0}^{K_{1n}} \cdots \sum_{k_d=0}^{K_{dn}} \int_{\mathcal{I}_X \times \mathcal{I}_X} \psi_{k_1}(x_1)\psi_{k_1}(y_1) \cdots \psi_{k_d}(x_d)\psi_{k_d}(y_d) \\
&\quad \times E\{W \circ F(dx_1, \ldots, dx_d)\, W \circ F(dy_1, \ldots, dy_d)\} \\
&= 0(n^{-1} K_{1n} \cdots K_{dn})
\end{aligned}$$

where W is the Brownian bridge on $[0,1]$, with covariance function $E(W_s W_t) = s \wedge t - st$. The MISE is minimum as the errors have the same order and we assume that the following condition is satisfied

$$K_{1n} \cdots K_{dn} = o(n), \quad \|f_n - f\|_2 = O(n^{-\frac{1}{2}}(K_{1n} \cdots K_{dn})^{\frac{1}{2}}). \tag{2.28}$$

Example 2.1 extends to an exponential density on $\mathbb{R}_+^d$ and (2.28) is fulfilled. In Example 2.2. the space W_{2s}, $s > 1$, is modified as

$$W_{2s,d} = \Big\{ f \in L^2(\mathbb{R}^d, \mu) : \sum_{k_1, \ldots, k_d \geq 0} a_{k_1, \ldots, k_d}^2 k_1^{2s} \cdots k_d^{2s} < \infty \Big\}, \tag{2.29}$$

then there exists a constant C such that $\|f - f_n\|_2^2 \leq C(K_{1n} \cdots K_{dn})^{-2s}$. The optimal sizes $K_{1n} \cdots K_{dn}$ which minimize $O((K_n)^{-2s}) + O(n^{-1}K_n)$, with $K_n = K_{1n} \cdots K_{dn}$, have orders satisfying

$$K_{n,opt} = 0(n^{\frac{1}{2s+1}})$$

and we can chose them so that $K_{jn} = 0(n^{\frac{1}{d(2s+1)}})$ for $j = 1, \ldots, d$. The condition (2.28) is fulfilled with the optimal order for the squared approximation error $0(n^{-\frac{2s}{2s+1}}) = o(1)$ in $W_{2s,d}$ and $K_{n,opt}^{-2s} = O(n^{-1}K_{n,opt})$.

Proposition 2.6. *Under the condition (2.28), the MISE error* $E\|\widehat{f}_n - f\|_2$ *for an estimator of a density* f *of* $L^2(\mathbb{R}^d, \mu)$ *and* $W_{2s,d}$, $s \geq 1$, *has the optimal convergence rate* $0(n^{\frac{s}{2s+1}})$ *if* $K_{1n} \cdots K_{dn} = O(n^{\frac{1}{2s+1}})$.

Proposition 2.7. *Under the conditions (2.28), the process* $(K_{1n}\cdots K_{dn})^{-\frac{1}{2}}n^{\frac{1}{2}}(\widehat{f}_n - f)$ *converges weakly to a Gaussian process with expectation the limit of* $(K_{1n}\cdots K_{dn})^{-\frac{1}{2}}n^{\frac{1}{2}}(f_n - f)$ *depending on the basis and with variance the limit of the mean squared estimation error.*

Let (X,Y) be a random variable on a subset $\mathcal{I}_X \times \mathcal{I}_Y$ of $\mathbb{R}^{d_1} \times \mathbb{R}^{d_2}$, with joint density f_{XY} and marginal densities f_X and f_Y. The estimation of the conditional density $f_{Y|X}(;x) = f_{XY}(x,y)f_X^{-1}(x)$ of a variable Y conditionally on X is the ratio of the estimators of the bivariate density and of the marginal density of X, with the convention $\frac{0}{0} = 0$

$$\widehat{f}_{nY|X}(y;x) = \widehat{f}_{nXY}(x,y)\widehat{f}_{nX}^{-1}(x)$$

its convergence rate has the order $\alpha_n = n^{\frac{1}{2}}(K_{1n}K_{2n})^{-\frac{1}{2}} + n^{\frac{1}{2}}K_n^{-\frac{1}{2}}$. Under the conditions (2.19) and (2.28), its optimal convergence rate has the order $0(n^{\frac{s}{2s+1}})$ like $\widehat{f}_{nXY}$ and $\widehat{f}_{nX}^{-1}$.

Proposition 2.8. *Under the conditions (2.28), the process* $\alpha_n(\widehat{f}_n - f)$ *converges weakly to a Gaussian process with expectation the limit of* $\alpha_n(f_n - f)$ *depending on the basis and with variance the limit of the mean squared estimation error.*

2.9 Estimation of a compound Poisson distribution

Let Y be a Poisson variable with a random intensity λ conditionally on the variable λ with the mixing density g, with respect to Lebesgue's measure. The compound Poisson variable Y has the mixture probabilities

$$\pi(k) = \frac{1}{k!}\int_0^\infty e^{-x}x^k g(x)\,dx,\ k \geq 0, \tag{2.30}$$

Its expectation is $\mu = E\lambda$ and its variance is

$$\operatorname{Var} Y = \operatorname{Var}\{E(Y \mid \lambda)\} + E\{\operatorname{Var}(Y \mid \lambda)\} = \operatorname{Var}\lambda + E\lambda,$$

empirical estimators of $E\lambda$ and $\operatorname{Var}\lambda$ are deduced from a sample of the variable Y.

Considering the intensity λ as a random variable on $\mathbb{R}_+$, its density g is estimated by projection on Laguerre's orthonormal basis of functions $(\varphi_j)_{j\geq 0}$ of $L^2(\mathbb{R}_+)$ defined by (2.8), it has the expansion $g = \sum_{j\geq 0} a_j(g)\varphi_j$ with the coefficients $a_j(g) = \int_0^\infty g(x)\varphi_j(x)\,dx$ such that $\sum_{j\geq 0} a_j^2(g) = \|g\|_2^2$.

The expansion of g entails

$$\begin{aligned}\pi(k) &= \frac{\sqrt{2}}{k!}\sum_{j\geq 0} a_j(g)\int_0^\infty e^{-2x}x^k L_j(2x)\,dx\\ &= \frac{\sqrt{2}}{2^{k+1}k!}\sum_{j\geq 0} a_j(g)\int_0^\infty e^{-x}x^k L_j(x)\,dx.\end{aligned}$$

For $k = 0$, the integral $\int_0^\infty e^{-x}L_j(x)\,dx = \int_0^\infty e^{-x}L_0(x)L_j(x)\,dx$ is zero. Applying the property

$$\int_0^\infty e^{-x}x^k L_j(x) = 0,$$

for every $j > k$, the expression of the probabilities of Y reduces to finite sums

$$\pi(k) = \frac{\sqrt{2}}{2^{k+1}k!}\sum_{j=1}^{k} a_j(g)\int_0^\infty e^{-x}x^k L_j(x)\,dx$$

with the integrals

$$\begin{aligned}\frac{1}{k!}\int_0^\infty e^{-x}x^k L_j(x)\,dx &= \frac{(-1)^j}{k!j!}\int_0^\infty (x^k)^{(j)}e^{-x}x^j\,dx\\ &= \frac{(-1)^j}{(k-j)!j!}\int_0^\infty e^{-x}x^k\,dx\\ &= (-1)^j\binom{k}{j},\end{aligned}$$

which implies

$$\pi(k) = \frac{\sqrt{2}}{2^{k+1}}\sum_{j=1}^{k}(-1)^j\binom{k}{j}\,a_j(g). \tag{2.31}$$

The coefficients a_k, $k \geq 1$, are deduced recursively from the difference $2^k\pi(k) - 2^{k-1}\pi(k-1)$ starting from $a_0 = 0$

$$\begin{aligned}2^k\pi(k) - 2^{k-1}\pi(k-1) &= \frac{1}{\sqrt{2}}\Big\{-a_1 + (k-1)a_2 - \frac{(k-1)(k-2)}{2}a_3 + \cdots\\ &\quad +(-1)^{k-1}(k-1)a_{k-1} + (-1)^k a_k\Big\}\\ &= \frac{1}{\sqrt{2}}\sum_{j=0}^{k-1}(-1)^{j+1}\binom{k-1}{j}\,a_{j+1}.\end{aligned} \tag{2.32}$$

Let $C_k = -2^{k-1}(1,-2)$, $P_k = (\pi(k-1), \pi(k))^T$ and $A_k = (a_1, \ldots, a_k)^T$, (2.32) has the form $D_k A_k = C_k P_k$ with a strictly positive definite triangular matrix D_k, and it is solved as

$$A_k = D_k^{-1} C_k P_k, \tag{2.33}$$

in particular a_k is a sum of k terms

$$\begin{aligned}
a_1 &= -\sqrt{2}\{2\pi(1) - \pi(0)\}, \\
a_2 &= 2\sqrt{2}\{2\pi(2) - 2\pi(1) + \pi(0)\}, \\
a_3 &= -\sqrt{2}\{8\pi(3) - 12\pi(2) + 6\pi(1) - \pi(0)\}, \\
a_4 &= 2^3\sqrt{2}\{2\pi(4) - 4\pi(3) + 3\pi(2) - \pi(1) - \pi(0)\}, \text{ etc.}
\end{aligned}$$

Let N_n be the largest value observed in a sample of n independent observations $Y_1, \ldots, Y_n$ of the variable Y, it tends to infinity as n tends to infinity and it has the probability $\pi(N_n)$ which converges to zero as n tends to infinity. The sample has the empirical frequencies

$$\widehat{\pi}_n(k) = n^{-1} \sum_{i=1}^n 1_{\{Y_i = k\}}$$

for $k = 0, \ldots, N_n$, and the coefficients a_k, $k = 0, \ldots, N_n$, are estimated by plugging the empirical frequencies into the relationship (2.32), starting from $\widehat{a}_{n0} = \widehat{\pi}_n(0)$. The density function g is approximated by

$$g_n = \sum_{j=0}^{N_n} a_j(g)\varphi_j$$

and it is consistently estimated by

$$\widehat{g}_n = \sum_{j=0}^{N_n} \widehat{a}_{nj}\varphi_j,$$

its bias $b_n = g - E(\widehat{g}_n \mid N_n)$, conditionally on N_j, is

$$b_n = \sum_{j=N_n+1}^{\infty} a_j\varphi_j.$$

For $k = 0, \ldots, N_n$, the variables $n^{\frac{1}{2}}\{\widehat{\pi}_n(k) - \pi(k)\}$ converge to centered Gaussian variables with variances $\pi(k)\{1-\pi(k)\}$ and covariances $-\pi(k)\pi(k')$. Then the estimators $\widehat{a}_{nk}$ are consistent and the variables $n^{\frac{1}{2}}\{\widehat{a}_{nk} - a_k\}$ converge weakly to centered Gaussian variables with finite variances σ_k^2 and covariances $\sigma_{kk'}$, for k and $k' = 1, \ldots, N_n$.

For a density g of $L^2(\mathbb{R}_+)$, the series $\sum_{k\geq 0} a_k^2$ converges to $\int_0^\infty g^2$ and $\sum_{k\geq N_n} a_k^2(g)$ converges to zero as n tends to infinity. The variance of $n^{\frac{1}{2}}\{\widehat{g}_n(x) - g_n(x)\}$ conditionally on N_n is

$$v_n(x) = \sum_{k=0}^{N_n} \sigma_k^2 \varphi_k^2(x) + \sum_{k'\neq k=0}^{N_n} \sigma_{kk'}\varphi_k(x)\varphi_{k'}(x),$$

with the variance σ_k^2 of $n^{\frac{1}{2}}\{\widehat{a}_{nk} - a_k\}$ deduced from the variance of $n^{\frac{1}{2}}\{\widehat{\pi}_n(k) - \pi(k)\}$ by the inversion of (2.32), and the covariance between $n^{\frac{1}{2}}\{\widehat{g}_n(x) - g_n(x)\}$ and $n^{\frac{1}{2}}\{\widehat{g}_n(y) - g_n(y)\}$ has the same form replacing $\varphi_k^2(x)$ by $\varphi_k(x)\varphi_k(y)$ and the square of the sum by the product of the sums at x and y.

The integral

$$V_n = \int_0^\infty v_n(x)\, dx = \sum_{k=0}^{N_n} \sigma_k^2$$

is defined by the asymptotic variances of the variables $n^{\frac{1}{2}}\{\widehat{a}_{nk} - a_k\}$. As n tends to infinity, $N_n^{-1}V_n$ converges to a finite and non zero limit V. The mean integrated squared error of the estimator $\widehat{g}_n(x)$ conditionally on N_n is

$$\begin{aligned} E\|\widehat{g}_n - g\|_2^2 &= E\|\widehat{g}_n - g_n\|_2^2 + \|g - g_n\|_2^2 \\ &= n^{-1}V_n + \sum_{k=N_n+1}^{\infty} a_k^2, \end{aligned}$$

it converges to zero as n tends to infinity and the convergence rate of the norm $E\|\widehat{g}_n - g_n\|_2$ of the estimator is a $O(n^{-\frac{1}{2}}N_n^{\frac{1}{2}}) = o(1)$ as n tends to infinity.

Proposition 2.9. *The process $N_n^{-\frac{1}{2}}n^{\frac{1}{2}}(\widehat{g}_n - g_n)$ converges weakly, conditionally on N_n, to a centered Gaussian process with a finite covariance function.*

Proof. The order of the squared estimation error implies that the integrated variance $N_n^{-1}\int V_n(x)\,dx$ is bounded, the limit of $N_n^{-1}V_n(x)$ is therefore bounded and it is strictly positive. The weak convergence of the finite dimensional distributions of the process $N_n^{-\frac{1}{2}}n^{\frac{1}{2}}(\widehat{g}_n - g_n)$ is established by the convergence of the squared estimation error $E\|\widehat{g}_n - g_n\|_2^2$ and by the weak convergence of the empirical estimators $\widehat{\pi}_{nk}$. The tightness of the process $N_n^{-\frac{1}{2}}n^{\frac{1}{2}}(\widehat{g}_n - g_n)$ is a property of the functions φ_k in $C_b^2(\mathbb{R}_+)$. $\square$

For a density g in the Sobolev space W_s defined by (2.20), the approximation error $\|g - g_n\|_2^2 = O(N_n^{-2s})$ converges to zero as n tends to infinity. It follows that the process

$$W_n = N_n^{-\frac{1}{2}} n^{\frac{1}{2}} (\widehat{g}_n - g_n) \tag{2.34}$$

converges weakly, conditionally on N_n, to a Gaussian process with a finite covariance function. Furthermore

$$\begin{aligned} P(N_n > n^{\frac{1}{2m+1}}) &= = 1 - P^n(Y \le n^{\frac{1}{2m+1}}) \\ &= 1 - (1 - n^{-\frac{1}{2m+1}} \mu)^n \end{aligned}$$

where $EY = \mu$, and this probability converges to zero as n tends to infinity. Conditionally on $N_n < m_{opt}$, the sequence $N_n^{-\frac{1}{2}} n^{\frac{1}{2}} (g_n - g)$ diverges.

Like in Section 2.6, a weighted series estimator of a density g in a space W_{2m} (2.23) may be considered, let

$$\widehat{g}_{nm\nu}(x) = \sum_{k=1}^{N_n} \frac{\lambda_k}{\nu_n + \lambda_{km}} \widehat{a}_{nk} \varphi_k(x) \tag{2.35}$$

with the estimators (2.33) of the coefficients and with the parameters ν_n and λ_{km} depending on m. Proposition 2.4 is modified according to the sum of N_n terms in the series.

Proposition 2.10. *Under the conditions (2.22) and $\nu_n = O(n^{-\frac{2m}{2m+1}})$, the estimator of a density g of W_{2m} is such that* $\mathrm{MISE}(\widehat{f}_{nm\nu_n}) = O(N_n^{-m})$.

Proof. The proof follows the same arguments as for Proposition 2.4, the mean integrated squared error of the estimator is

$$\mathrm{MISE}_{nm}(\nu_n) = \sum_{k=1}^{N_n} \frac{\lambda_k^2 v_k}{n(\nu_n + \lambda_k)^2} + \sum_{k=1}^{N_n} \frac{\nu_n^2 a_k^2}{(\nu_n + \lambda_k)^2} + \sum_{k=N_n+1}^{\infty} a_k^2$$

still denoted $\mathrm{MISE}_{nm}(\nu_n) = S_{1n} + S_{2n} + S_{3n}$. The variance of $\widehat{a}_{nk}$ is $n^{-1} v_k = O(n^{-1})$ and

$$\begin{aligned} S_{1n} &= \sum_{k=1}^{N_n} \frac{\lambda_k^2 v_k}{n(\nu_n + \lambda_k)^2}, \\ S_{2n} &= \sum_{k=1}^{N_n} \frac{\nu_n^2 a_k^2}{(\nu_n + \lambda_k)^2} < \nu_n \sum_{k=1}^{N_n} \lambda_k^{-1} a_k^2 = O(\nu_n), \\ S_{3n} &= \sum_{k=N_n+1}^{\infty} a_k^2 = O(N_n^{-m}), \end{aligned}$$

in W_{2m}. The sums S_{1n} and S_{2n} have the same bound as in Proposition 2.4 and they are $O(\nu_n)$ with $\nu_n = O(n^{-\frac{2m}{2m+1}})$. For the last term, the probability $P(N_n^{-m} < \nu_n) = P(N_n > n^{\frac{2}{2m+1}})$ converges to zero as n tends to infinity. □

2.10 Tests for random Poisson processes

Let Y be a mixture Poisson variable with mixing a distribution G for a random parameter λ. The expectation and the centered moments of Y conditionally on λ are λ whereas its conditional moments are

$$\begin{aligned}
\mu_1 &= EY = E_G\lambda = \int_{\mathbb{R}_+} \lambda\, dG(\lambda),\\
\mu_2 &= \operatorname{Var} Y = \mu_1 + \operatorname{Var}_G(\lambda),\\
\mu_3 &= E\{(Y-\mu_1)^3\} = \mu_1 + 3\operatorname{Var}_G(\lambda) + E_G\{(\lambda-\mu_1)^3\},
\end{aligned}$$

and so on. A comparison of the distributions of X and Y may be sequentially performed by a comparison of the empirical estimators of their centered moments which is equivalent to tests of the hypothesis $H_0 : \mu_k = \mu_1$ for $k = 2, 3$, equivalently $H_0 : E_G\{(\lambda - \mu_1)^k\} = 0$, for $k = 2, 3$. The moments μ_k of the variable Y are estimated by $\widehat{\mu}_{n1} = \bar{Y}_n$ and $\widehat{\mu}_{nk} = n^1 \sum_{i=1}^n (Y_i - \widehat{\mu}_{n1})^k$, they are a.s. consistent with the convergence rate $n^{-\frac{1}{2}}$. Under the hypothesis H_0, the variables $M_{nk} = n^{\frac{1}{2}}(\widehat{\mu}_{nk} - \widehat{\mu}_{n1})$ converge to centered Gaussian variables and they diverge under fixed alternatives. Under local alternatives with $\mu_{nk} = \mu_1 + n^{-\frac{1}{2}} m_{nk}$, the variables M_{nk} converge to uncentered Gaussian variables.

The estimation of the mixing distribution G of the random Poisson probabilities provides a goodness of fit test of the hypothesis H_0 of model (2.30). Let $\widetilde{\pi}_n(k)$ be the estimator defined by plugging the estimated mixing density $\widehat{g}_n$ in the mixture probabilities $\pi(k)$. Let π_0 be the true probability distribution of the process Y, the empirical process $\widehat{W}_n = n^{\frac{1}{2}}(\widehat{\pi}_n - \pi_0)$, $k = 0, \ldots, N_n$, converges weakly to a centered Gaussian process $\widehat{W}_0$. A test statistic of the hypothesis H_0 is defined as

$$T_n = N_n^{-1} \sum_{i=1}^n \{\widetilde{\pi}_n(Y_i) - \widehat{\pi}_n(Y_i)\}^2,$$

it is a mean squared difference of processes $N_n^{-\frac{1}{2}} \widehat{W}_n$ and $\widetilde{W}_n = N_n^{-\frac{1}{2}} n^{\frac{1}{2}}(\widetilde{\pi}_n - \pi_0)$ which converges weakly to a process $\widetilde{W}_0$ by Proposition 2.9.

Proposition 2.11. *Under the hypothesis H_0, the statistic T_n converges weakly to $T_0 = \int \widetilde{W}_0^2 \, dF_Y$. Under fixed alternatives, it diverges.*

Proof. The statistic T_n has the approximation

$$T_n = n^{-1} \sum_{i=1}^{n} \widetilde{W}_n^2(Y_i) + o_p(1) = \int_{\mathbb{R}_+} \widetilde{W}_n^2 \, d\widehat{F}_{nY} + o_p(1)$$

where $\widehat{F}_{nY}$ is the empirical distribution of the sample of Poisson variables and its weak convergence follows from Proposition 2.9. □

We consider a sequence of alternatives $(A_n)_n$ such that the mixing density and the probability distribution of Y under A_n are

$$g_n(x) = g_0(x)\{1 + N_n^{\frac{1}{2}} n^{-\frac{1}{2}} u_n(x)\},$$
$$\pi_n(k) = \pi_0(k) + n^{-\frac{1}{2}} v_n(k), k \geq 0,$$

where (u_n, v_n) converges uniformly to a non zero limit (u, v) as n tends to infinity. Let $\pi_{u_n}(k) = \frac{1}{k!} \int_0^\infty e^{-x} x^k u_n(x) \, dx$, then we obtain the approximation

$$N_n^{-\frac{1}{2}} n^{\frac{1}{2}} \{\widetilde{\pi}_n(k) - \widehat{\pi}_n(k)\} = W_n + \pi_{u_n}(k) + o_p(1)$$

and it converges weakly to the process W_0 shifted according to the limit $\pi_u(k)$ of the integral $\pi_{u_n}(k)$.

Proposition 2.12. *Under the alternative A_n the statistic T_n converges weakly to $T_A = \int (W_0 + \pi_u)^2 \, dP_Y$.*

Le Cam and Traxler (1978) studied the log-likelihood ratio test of the hypothesis $H_0 : P(\lambda), \lambda > 0$, against a Poisson mixture, we restrict the test to local alternatives. Let P_n be the product probability measure of n independent and identically distributed variables $Y_1, \ldots, Y_n$ with a Poisson distribution and let Q_n be their probability measure under a mixture Poisson distribution. The logarithm of the likelihood ratio is

$$l_n(\lambda, G) = \log \frac{dQ_n}{dP_n} = \sum_{i=1}^{n} \log \phi_n(Y_i)$$

where

$$\phi_n(k) = E_G \Big\{ e^{-(T_n - \lambda)} \frac{T_n^k}{\lambda^k} \Big\}$$

where the variable T_n has the mixing distribution function G_n under local alternatives A_n of the hypothesis H_0. The maximum likelihood estimator

$\widehat{\lambda}_n = \bar{Y}_n$ of the parameter value λ_0 is $n^{-\frac{1}{2}}$-consistent under H_0, and we assume that under local alternatives T_n belongs to a $n^{-\frac{1}{2}}$-neighborhood of λ_0. Under the local alternatives A_n, let $\mu_n = E_{G_n} T_n$ such that $n^{\frac{1}{2}}(\mu_n - \lambda_0)$ converges to a strictly positive limit μ and the variable $n^{\frac{1}{2}}(T_n - \mu_n)$ converges weakly to a centered variable L_0 under A_n. As $E_{G_n}\bar{Y}_n = \mu_n$, the variable $n^{\frac{1}{2}}(\widehat{\lambda}_n - \mu_n)$ converges weakly under G_n to a centered Gaussian variable $L_{\bar{Y}}$, this implies the weak convergence under G_n to the convolution of distributions

$$n^{\frac{1}{2}}(T_n - \widehat{\lambda}_n) \Rightarrow L_0 \star L_{\bar{Y}}.$$

Under the hypothesis H_0, T_n reduces to λ_0 and the variable $n^{\frac{1}{2}}(\widehat{\lambda}_n - \mu_n)$ converges weakly to $L_{\bar{Y}} - \mu$.

Proposition 2.13. *The test statistic*

$$T_n = 2\sum_{i=1}^n \log E_{G_n} \exp[-(T_n - \lambda_n) + Y_i \log\{1 + \lambda_n^{-1}(T_n - \lambda_n)\}]$$

converges weakly under the hypothesis H_0 to $T_0 = \lambda_0^{-1}(L_{\bar{Y}} - \mu)^2$. Under the local alternatives A_n, T_n converges weakly to $T_G = \lambda_0^{-1} L_{\bar{Y}}^2$.

Proof. For λ_n in a neighborhood of λ_0, the log-likelihood ratio is expanded as

$$\begin{aligned} l_n(\lambda_n, G_n) &= \sum_{i=1}^n \log E_{G_n} \exp[-(T_n - \lambda_n) + Y_i \log\{1 + \lambda_n^{-1}(T_n - \lambda_n)\}] \\ &= \sum_{i=1}^n \log E_{G_n} \exp\Big\{(Y_i - \lambda_n)\frac{T_n - \lambda_n}{\lambda_n} - Y_i \frac{(T_n - \lambda_n)^2}{2\lambda_n^2} \\ &\quad + o_p((T_n - \lambda_n)^2)\Big\} \end{aligned}$$

where the expectation is for T_n under G_n. Let $\widetilde{T}_n = n^{\frac{1}{2}}(T_n - \widehat{\lambda}_n)$ and let

$L_n = n^{\frac{1}{2}}(\widehat{\lambda}_n - \mu_n)$, the log-likelihood at $\widehat{\lambda}_n$ has the approximation

$$\begin{aligned} l_n(\widehat{\lambda}_n, G_n) &= \sum_{i=1}^n \log E_{G_n} \exp\Big\{(Y_i - \widehat{\lambda}_n)\frac{n^{-\frac{1}{2}}\widetilde{T}_n}{\widehat{\lambda}_n} - Y_i\frac{n^{-1}\widetilde{T}_n^2}{2\widehat{\lambda}_n^2} + o_p(n^{-1})\Big\} \\ &= \sum_{i=1}^n \log E_{G_n}\Big\{1 + (Y_i - \widehat{\lambda}_n)\frac{n^{-\frac{1}{2}}\widetilde{T}_n}{\widehat{\lambda}_n} - Y_i\frac{n^{-1}\widetilde{T}_n^2}{2\widehat{\lambda}_n^2} \\ &\qquad +(Y_i - \widehat{\lambda}_n)^2\frac{n^{-1}\widetilde{T}_n^2}{2\widehat{\lambda}_n^2} + o_p(n^{-1})\Big\} \\ &= \sum_{i=1}^n \log\Big\{1 - (Y_i - \widehat{\lambda}_n)\frac{n^{-\frac{1}{2}}L_n}{\widehat{\lambda}_n} - Y_i\frac{n^{-1}E_{G_n}(\widetilde{T}_n^2)}{2\widehat{\lambda}_n^2} \\ &\qquad +(Y_i - \widehat{\lambda}_n)^2\frac{n^{-1}E_{G_n}(\widetilde{T}_n^2)}{2\widehat{\lambda}_n^2} + o_p(n^{-1})\Big\}, \end{aligned}$$

where the variable $V_n = E_{G_n}(\widetilde{T}_n^2)$ is a $O_p(1)$. Expanding the logarithm provides the approximation

$$\begin{aligned} -l_n(\widehat{\lambda}_n, G_n) &= n^{-\frac{1}{2}}\sum_{i=1}^n (Y_i - \widehat{\lambda}_n)\frac{L_n}{\widehat{\lambda}_n} + n^{-1}\sum_{i=1}^n Y_i\frac{E_{G_n}(\widetilde{T}_n^2)}{2\widehat{\lambda}_n^2} \\ &\qquad -n^{-1}\sum_{i=1}^n (Y_i - \widehat{\lambda}_n)^2\frac{E_{G_n}(\widetilde{T}_n^2) - L_n^2}{2\widehat{\lambda}_n^2} + o_p(n^{-1}) \\ &= \frac{E_{G_n}(\widetilde{T}_n^2)}{2\widehat{\lambda}_n} - n^{-1}\sum_{i=1}^n (Y_i - \widehat{\lambda}_n)^2\frac{E_{G_n}(\widetilde{T}_n^2) - L_n^2}{2\widehat{\lambda}_n^2} + o_p(n^{-1}) \end{aligned}$$

under the hypothesis H_0, the variable $l_n(\widehat{\lambda}_n, G_n)$ is asymptotically equivalent to

$$l_n = \frac{L_n^2}{2\lambda_0}$$

where L_n converges to $L_0 = L_{\bar{Y}} - \mu$ and $T_n = 2l_n$ converge weakly to $T_0 = L_0^2\lambda_0^{-1}$.

Under the local alternatives A_n, the variables $\widehat{\lambda}_n$ and $n^{-1}\sum_{i=1}^n (Y_i - \widehat{\lambda}_n)^2$ are a.s. asymptotically equivalent to μ_n, and $l_n(\widehat{\lambda}_n, G_n)$ is asymptotically equivalent to

$$l_{G_n} = \frac{L_n^2}{2\mu_n}$$

where L_n converges weakly to $L_{\bar{Y}}$ and $2l_{G_n}$ converge weakly to T_G. □

Proposition 2.13 proves that the statistic T_n is powerful for a test of H_0 under local alternatives.

2.11 Mixtures of Poisson processes

Let $(Y_t)_{t\geq 0}$ be a conditional Poisson process with a random intensity λ having the mixing density g with respect to the Lebesgue measure on $\mathbb{R}_+$. The distribution of N is described by the probabilities

$$\pi_t(k) := P(Y_t = k) = \frac{t^k}{k!}\int_0^\infty e^{-\lambda t}\lambda^k g(\lambda)\,d\lambda$$

for every integer k.

An expansion of g on Laguerre's orthonormal basis of functions (2.8) entails

$$\pi_t(k) = \frac{\sqrt{2}t^k}{k!}\sum_{j\geq 0} a_j(g)\int_0^\infty e^{-\lambda(t+1)}\lambda^k L_j(2\lambda)\,d\lambda$$

and by the expression (2.3) we obtain

$$\begin{aligned}
\pi_t(k) &= \frac{\sqrt{2}t^k}{k!}\sum_{j\geq 0} a_j(g)\sum_{l=0}^{j}\binom{j}{l}\frac{(-2)^l}{l!}\int_0^\infty e^{-\lambda(t+1)}\lambda^{k+l}\,d\lambda\\
&= \frac{\sqrt{2}t^k}{k!}\sum_{j\geq 0} a_j(g)\sum_{l=0}^{j}\binom{j}{l}\frac{(-2)^l}{(t+1)^{k+l+1}l!}\int_0^\infty e^{-x}x^{k+l}\,dx\\
&= \frac{\sqrt{2}t^k}{k!}\sum_{j\geq 0} c_{kj}(t)a_j(g),
\end{aligned}$$

$$c_{kj}(t) = \sum_{l=0}^{j}\binom{j}{l}\binom{k+l}{k}\frac{(-2)^l}{(t+1)^{k+l+1}},$$

the functions $c_{kj}(t)$ have the lower bound $(t+1)^{-k}$. Solving these equations provides expressions of the coefficients a_j as linear combinations of the functions $\pi_t(k)$ with coefficients depending on the functions $c_{kj}(t)$, for all t and k.

Let $(Y_i)_{i=1,\dots,n}$ be a sample of n independent observations of the process Y observed on a time interval $[0,T]$. The probabilities $\pi_T(k)$ have the empirical estimators

$$\widehat{\pi}_{nT}(k) = n^{-1}\sum_{i=1}^{n} 1_{\{Y_{Ti}=k\}},$$

they are a.s. consistent and the variables $n^{\frac{1}{2}}\{\widehat{\pi}_{nT}(k) - \pi_T(k)\}$ converge weakly to centered Gaussian variables. Estimators $\widehat{a}_{nj}$ of the coefficients

a_j are deduced and the variables $n^{\frac{1}{2}}(\widehat{a}_{nj} - a_j)$ converge weakly to centered Gaussian variables, for every integer j.

Let N_n be the maximum number of jumps of the variables Y_{Ti}, for $i = 1, \ldots, n$, it tends to infinity as the observation interval increases and the order of the approximation can be chosen equal to N_n. The mixing density function g is approximated by $g_n = \sum_{j=0}^{N_n} a_j(g)\varphi_j$, its approximation error $\|g - g_n\|_2^2 = \sum_{j>N_n} a_j^2$ converges to zero.

The function g_n is consistently estimated by

$$\widehat{g}_n = \sum_{j=0}^{N_n} \widehat{a}_{nj}\varphi_j,$$

the asymptotic behavior of $N_n^{-\frac{1}{2}} n^{\frac{1}{2}}(\widehat{g}_n - g_n)$ is given by Proposition 2.9. Let g be in W_{2m} and let $([0, T_n])_n$ be an increasing sequence of observation intervals such that $N_n > n^{\frac{1}{2m+1}}$ with a probability converging to one, then the process $N_n^{-\frac{1}{2}} n^{\frac{1}{2}}(\widehat{g}_n - g)$ converges weakly to a Gaussian process.

Under the conditions of Proposition 2.10, the MISE of the weighted series estimator (2.35) for the density g in W_{2m} has the same asymptotic behavior as the MISE of (2.23) in Proposition 2.4.

Chapter 3

Estimation of nonparametric regression functions

3.1 Introduction

Let $(Y_i, X_i)_{i=1,\ldots,n}$ be a sample of n independent and identically distributed variables with a distribution function $F_{X,Y}$ on a subset $\mathcal{I}_{XY}$ of $\mathbb{R}$, and with marginal distribution functions F_X on $\mathcal{I}_X$ and F_Y on $\mathcal{I}_Y$, and let $(\varepsilon_i)_{i=1,\ldots,n}$ be a sample of errors with conditional mean $E(\varepsilon_i \mid X_i) = 0$ and variance σ_ε^2. The condition of a finite variance for the variable Y_i implies $E(Y_i^2) = E\{m^2(X_i)\} + \sigma_\varepsilon^2$ and the regression function belongs to $L^2(\mathcal{I}_X, F_X)$. In a regression model, the expectation of a response variable Y is related to a vector of explanatory variables X through parametric or nonparametric models with unknown scalar or functionals parameters.

The nonparametric regression model

$$Y_i = m(X_i) + \varepsilon_i,\ i = 1, \ldots, n,$$

is defined by a real function m such that

$$E(Y_i \mid X_i) = m(X_i),$$

$\mathrm{Var}(Y_i \mid X_i) = \sigma^2$ in the homoscedastic case and

$$\mathrm{Var}(Y_i \mid X_i) = \sigma^2(X_i)$$

in the heteroscedastic case.

Under the assumption of uniformly distributed random variables X_i on $[a, b]$, an estimator of a continuous nonparametric regression function m may be defined by projections on an orthonormal basis of functions of $L^2([a, b], \mu)$, with the Lebesgue measure μ. The Fourier sine and cosine series determine orthonormal trigonometric polynomials on $[-1, 1]$

$$S_k(x) = \sin(k\theta),\ -\frac{\pi}{2} \leq \theta \leq \frac{\pi}{2},$$
$$T_k(x) = \cos(k\theta),\ 0 \leq \theta \leq \pi,$$

they provide expansions of the function m as $m(r, \cos\theta) = r\sum_{k\geq 0} a_k S_k(x)$ or $m(r, \cos\theta) = r\sum_{k\geq 0} b_k T_k(x)$, with the Fourier cosine coefficients of the function m. Legendre series of polynomials are oscillatory functions on $[-1, 1]$ with resonance effects on the edges, they are solutions of second order differential equations. For a function of $L^2([a, b], \mu)$ having a convergent series expansion, the coefficients are estimated like to those of a density with respect to the Lebesgue measure and their properties are similar. More generally, kernel nonparametric estimators

$$\widehat{m}_{nh}(x) = \frac{\sum_{i=1}^n Y_i K_h(X_i - x)}{\sum_{i=1}^n K_h(X_i - x)}$$

of a continuous regression function m have been widely studied. In the heteroscedastic case, the variance function $\sigma^2(x)$ has the kernel estimator

$$\widehat{\sigma}^2_{nh}(x) = \frac{\sum_{i=1}^n \{Y_i - \widehat{m}_{nh}(X_i)\}^2 K_h(X_i - x)}{\sum_{i=1}^n K_h(X_i - x)},$$

they are consistent under the usual conditions for the kernel K, the bandwidth h and the regression function.

For a continuous regression function m of $L^2(\mathcal{I}_X, F_X)$, we consider here an estimator based on the projection of the function m on an orthonormal basis of functions $(\psi_k)_{k\geq 0}$ of $L^2(\mathcal{I}_X, F_X)$. The advantage of this method over the kernel estimation relies on the fact that the coefficients are expressed as integrals of the product of the functions of the basis and regression functions which does not requires the existence and the estimation of a density as the kernel regression estimator, furthermore the convergence of the kernel estimator requires differentiability conditions. The nonparametric regression models are linearized by projection on a basis and the regression coefficients have been estimated by inversion of the empirical variance matrix of known functions of the regressors (Gao et al. 2002).

In the estimation of a regression function by projection on an orthonormal basis of functions of $L^2(F_X)$, the coefficients of the expansions have direct and simple empirical estimators, the method applies to multivariate models and semi-parametric models. The main problem is the choice of the basis according to the distribution of the regressors.

Several kind of orthonormal sets of functions may be considered. Let k be the degree of the polynomial ψ_k

$$\varphi_k(x) = \sum_{j=0}^k a_{k,j} x^j, \ k \geq 0,$$

the coefficients are determined by the orthogonality properties $\int \varphi_k \varphi_n \, dF_X = 0$ for every $n \leq k$, and $\int \varphi_k^2 \, dF_X = 0$, starting from $\varphi_0 \equiv 1$. Let $\mu = EY$ and $\sigma_X^2 = \operatorname{Var} X$, the first polynomials are

$$\begin{aligned}
\varphi_1(x) &= \frac{\mu - x}{\sigma_X}, \\
\varphi_2(x) &= \frac{b}{\sigma_X} \Big\{ -\frac{x^2 \sigma_X^2}{E(X^3) - \mu E(X^2)} + x + \frac{E^2(X^2) - \mu E(X^3)}{E(X^3) - \mu E(X^2)} \Big\}, \\
b^{-2} &= \operatorname{Var}(X^2) - \frac{(\sigma_X^2 - \mu^2) E^2(X^2) + E^3(X^3)}{E(X^3) - \mu E(X^2)}.
\end{aligned}$$

The $k+1$ coefficients of φ_k are solutions of k linear equations and a nonlinear equation, for $j = 0, \ldots, k-1$

$$\begin{aligned}
0 &= a_{k,k} E(X^{k+j}) + a_{k,k-1} E(X^{k+j-1}) + \ldots + a_{k,0} E(X^j), \\
1 &= \int_{\mathcal{I}_X} \varphi_k^2 \, dF_X.
\end{aligned}$$

Other orthonormal sets of functions of $L^2(\mathcal{I}_X, F_X)$ are built by transformation of a known orthonormal basis. The transform (2.10) may be used to extend the support of the regressor and to adjust the expectation of the variables with the value at zero of the polynomials. Donoho and Johnstone (1989) defined a basis in polar coordinates on $\mathbb{R}^2$ using the product Laguerre's polynomials depending on the radius and the complex exponential of the angle, and other product transformations on $\mathbb{R}^2$.

In the following we define orthonormal transformations of Laguerre's polynomials in $L^2(\mathcal{I}_X, F_X)$, where I_X is a subset of $\mathbb{R}^d$, for every $d \geq 1$. Let

$$H = -\log(1 - F_X),$$

the density f_X is expressed as

$$f_X(x) = h(x) e^{-H(x)}$$

with the survival function $\bar{F}_X(x) = 1 - F_X(x) = e^{-H(x)}$ and the hazard function $h(x) = f_X(x) \bar{F}_X^{-1}(x)$ related to the density f_X. For every X, the values of the function H belong to $\mathbb{R}_+$, $H(x)$ is zero at zero and it tends to infinity as x tends to infinity.

For $k \geq 1$, Laguerre's polynomials L_k are orthonormal with respect to the exponential distribution function, they are transformed as

$$\psi_k(x) = L_k \circ H(x), \tag{3.1}$$

they satisfy $\psi_0 \equiv 1$ and by a change of variable

$$\begin{aligned}
\langle \psi_k, \psi_k \rangle_{F_X} &= \int_{\mathcal{I}_X} \psi_k^2(x)\, dF_X(x) = \int_{\mathcal{I}_X} L_k^2 \circ H(x) h(x) e^{-H(x)}\, dx \\
&= \int_0^\infty L_k^2(x) e^{-x}\, dx = 1, \\
\langle \psi_k, \psi_{k'} \rangle_{F_X} &= \int_{\mathcal{I}_X} L_k \circ H(x) L_{k'} \circ H(x) h(x) e^{-H(x)}\, dx \\
&= \int_0^\infty L_k(x) L_{k'}(x) e^{-x}\, dx = 0,
\end{aligned}$$

for all integers k and k', the functions ψ_k are therefore orthonormal with respect to F_X. By definition, the distribution function F_X and the functions ψ_k have the same degree of differentiability on $\mathcal{I}_X$ for L_k in $C^\infty(\mathbb{R}_+)$.

3.2 Orthonormal series estimators

Let $\mathcal{I}_X$ be the support of F_X in $\mathbb{R}$ and let τ_X be the end-point of the support of F_X in $\mathcal{I}$, the cumulative hazard function $H(x) = -\log\{1 - F_X(x)\}$ is finite on every sub-interval where F_X is strictly lower than 1 and it tends to infinity at τ_X where $F_X(\tau_X) = 1$. By projection on the orthonormal basis of functions $(\psi_k)_{k \geq 0}$ of $L^2(\mathbb{R}, F_X)$, the regression function m has the expansion

$$m = \sum_{k \geq 0} b_k \psi_k$$

with the coefficients

$$b_k = \int_{\mathbb{R}} m(x) \psi_k(x)\, dF_X(x) = E\{m(X)\psi_k(X)\},$$

for $k \geq 1$, and $b_0 = \int_{\mathbb{R}} m(x)\psi_0(x)\, dF_X(x) = Em(X)$.

As $E\{Y_i \psi_k(X_i) \mid X_i\} = E\{m(X_i)\psi_k(X_i) \mid X_i\}$ for every integer k, we have

$$\begin{aligned}
E\{Y_i \psi_k(X_i)\} &= \int_{\mathcal{I}_X} m(x) \psi_k(x)\, dF_X(x) \\
&= \sum_{l \geq 0} b_l \int_{\mathcal{I}_X} \psi_k(x) \psi_l(x)\, dF_X(x) = b_k.
\end{aligned}$$

The function ψ_k depends on the distribution function F_X through the cumulative hazard function H, it has the empirical estimator

$$\widehat{\psi}_{nk}(x) = L_k \circ \widehat{H}_n(x) \tag{3.2}$$

where the function H is estimated on $\mathbb{R}$ by

$$\widehat{H}_n(x) = \int_0^x 1_{\{\widehat{F}_{nX}(t)<1\}} \frac{d\widehat{F}_{nX}(t)}{1-\widehat{F}_{nX}(t)} \tag{3.3}$$

with the empirical distribution function $\widehat{F}_{nX}$ of X. If F_X is continuous near τ_X, $X_{n:n} < \tau_X$ for every n and the uniform a.s. consistency of $\widehat{F}_{nX}$ on $\mathcal{I}_X$ entails the uniform a.s. consistency of $\widehat{H}_n$ on every sub-interval of $\mathcal{I}_X$ where $\widehat{F}_{nX} < 1$, as n tends to infinity. Breslow and Crowley (1974) established a bound for the difference between $\widehat{\Lambda}_n$ and $-\log\{\widehat{\bar{F}}_n\}$ for censored observations. Let $\widehat{\bar{F}}_{nX} = 1 - \widehat{F}_{nX}(t)$.

Proposition 3.1. *If* $t < T_{n:n}$,

$$0 < -\log\{\widehat{\bar{F}}_n(t)\} - \widehat{\Lambda}_n(t) < \frac{\widehat{F}_{nX}(t)}{n\widehat{\bar{F}}_{nX}(t)}$$

and it is a $O(n^{-1})$. *On every interval* $[0,a]$ *such that* $a < T_{n:n}$ *the process* $B_n = n^{1/2}(\widehat{F}_n - F)$ *converges weakly to a centered Gaussian process with finite variance function.*

Proposition 3.2. *On every interval* $[0,a]$ *such that* $a < T_{n:n}$, *the process* $n^{\frac{1}{2}}\{\widehat{H}_n(x) - H(x)\}$ *converges weakly to a centered Gaussian process with covariance function* $c_H(x,y) = \int_0^{x\wedge y} \bar{F}^{-2}\, dF$.

Lemma 3.1. *For every integer* $k \geq 1$, *on every sub-interval of* $\mathcal{I}_X$ *where* $F_X(x) < 1$, *the estimator* $\widehat{\psi}_{nk}$ *is a.s. uniformly consistent and the process* $n^{\frac{1}{2}}(\widehat{\psi}_{nk} - \psi_k)$ *converges weakly to a centered Gaussian process with a bounded covariance function*

$$c_{\psi_k}(x,y) = \{L_k' \circ H(x)\}\,\{L_k' \circ H(y)\}\, c_H(x,y).$$

Proof. On a sub-interval $[0,a]$ of $\mathcal{I}_X$ where $F_X(x) < 1$, the estimators $\widehat{\psi}_{nk} = L_k \circ \widehat{H}_n$ are a.s. uniformly consistent. By differentiability of L_k, the process $n^{\frac{1}{2}}\{\widehat{\psi}_{nk}(x) - \psi_k(x)\}$ has the approximation

$$\begin{aligned} n^{\frac{1}{2}}\{\widehat{\psi}_{nk}(x) - \psi_k(x)\} &= n^{\frac{1}{2}}\{L_k \circ \widehat{H}_n(x) - L_k \circ H(x)\} \\ &= n^{\frac{1}{2}}\{\widehat{H}_n(x) - H(x)\}\, L_k' \circ H(x)\,\{1 + o_p(1)\} \end{aligned}$$

and Proposition 3.2 implies that the process $n^{\frac{1}{2}}\{\widehat{\psi}_{nk}(x) - \psi_k(x)\}$ converges weakly on $[0,a]$ to a centered Gaussian process. The asymptotic covariance function of $n^{\frac{1}{2}}\{\widehat{\psi}_{nk}(x) - \psi_k(x)\}$ is $c_{\psi_k}(x,y)$ and it is bounded on $[0,a]$, for every integer k. □

The bias of $\widehat{\psi}_{nk}$ is obtained from a second order expansion

$$b(\widehat{\psi}_{nk}) = E(\widehat{\psi}_{nk}) - \psi_k = \text{Var}(\widehat{H}_n)L'' \circ H = O(n^{-1}).$$

For $i = 1, \dots, n$, let $\widehat{\psi}_{nki}$ be the estimator of ψ_k defined with the sample $X_1, \dots, X_n$ where the ith order statistic $X_{(n:i)}$ is omitted. The coefficients b_k are estimated by

$$\widehat{b}_{nk} = (n-1)^{-1} \sum_{i=1}^{n-1} Y_{(i)} \widehat{\psi}_{nki}(X_{(n:i)}), \tag{3.4}$$

where the observation $X_{(n:i)}$ is omitted in the estimator $\widehat{\psi}_{nki}$ and the variables $\widehat{\psi}_{nki}(X_{(n:i)})$, $i = 1, \dots, n-1$, are bounded as n tends to infinity, $Y_{(i)}$ is the response variable to the regressors $X_{(n:i)}$. By Lemma 3.1 and the independence of $\widehat{\psi}_{nki}$ and $X_{(n:i)}$, the bias of estimator (3.4) is such that

$$\begin{aligned} b_k - E\widehat{b}_{nk} &= \int_{\mathcal{I}_X} m(x)\{\psi(x) - E\widehat{\psi}_{nk,i})(x)\}\, dF_X(x) \\ &= \int_{\mathcal{I}_X} m(x)\, \text{Var}\{\widehat{H}_{nk,i})(x)\} L_k'' \circ H(x)\, dF_X(x) = 0(n^{-1}), \end{aligned}$$

and $\widehat{b}_{nk}$ is a.s. consistent. Furthermore $E(Y^2 \mid X) = \sigma_\varepsilon^2 + m^2(X)$ and

$$v_k = E\{Y^2\psi_k^2(X)\} - b_k^2 = \sigma_\varepsilon^2 + \text{Var}\{m(X)\psi_k(X)\} \tag{3.5}$$

by the orthonormality of the basis.

Let $\mathcal{I}_{nX}$ be the restriction of the interval $\mathcal{I}_X$ to $\{x \in \mathcal{I}_X, x < X_{(n:n)}\}$ and let $b_{nk} = E\{m(X)\psi_k(X)1_{\{\mathcal{I}_{nX}\}}(X)\}$.

Lemma 3.2. *For every integer $k \geq 1$, the variable $n^{\frac{1}{2}}(\widehat{b}_{nk} - b_{nk})$ converges weakly to a centered Gaussian variable with a strictly positive variance*

$$\begin{aligned} V_k = \sigma_\varepsilon^2 &+ \int_{\mathcal{I}_X} m^2(x)\psi_k^2(x)\, d\,[F(x)\{1 - F(x)\}] \\ &+ \int_{\mathcal{I}_X \times \mathcal{I}_X} m(x)m(x')c_{\psi_k}(x, x')\, dF_X(x)\, dF_X(x'). \end{aligned}$$

Proof. Let F_{XY} be the distribution function of (X, Y), let $\widehat{F}_{nXY}$ be its empirical estimator, and let $\nu_{nXY} = n^{\frac{1}{2}}(\widehat{F}_{nXY} - F_{XY})$, the estimators

$\widehat{\psi}_{nk}(x)$ are defined on $\mathcal{I}_{nX}$. The variable $B_{nk} = n^{\frac{1}{2}}(\widehat{b}_{nk} - b_{nk})$ is written as

$$\begin{aligned} B_{nk} &= n^{\frac{1}{2}}\Big\{\int_{\mathcal{I}_{nX}\times\mathcal{I}_Y} y\widehat{\psi}_{nk}(x)\,\widehat{F}_{nXY}(dx,dy) \\ &\quad -\int_{\mathcal{I}_{nX}\times\mathcal{I}_Y} y\psi_k(x)\,F_{XY}(dx,dy)\Big\} \\ &= \int_{\mathcal{I}_{nX}\times\mathcal{I}_Y} y\widehat{\psi}_{nk}(x)\,\nu_{nXY}(dx,dy) \\ &\quad +n^{\frac{1}{2}}\int_{\mathcal{I}_{nX}\times\mathcal{I}_Y} y\{\widehat{\psi}_{nk}(x)-\psi_k(x)\}\,F_{XY}(dx,dy) \end{aligned}$$

and its weak convergence follows from Lemma 3.1. Its limiting variance is the limit of $V_n = \sigma_\varepsilon^2 + \mathrm{Var}\,[m(X_{(i)})\{\widehat{\psi}_{nki}(X) - \psi_k(X_{(i)})\}]$ and the second term converges to the limit of the sum

$$\begin{aligned} &\int_{\mathcal{I}_X} m^2(x)\psi_k^2(x)\,d\,[F(x)\{1-F(x)\}] \\ &+\int_{\mathcal{I}_X\times\mathcal{I}_X} m(x)m(x')c_{n\psi_k}(x,x')\,dF_X(x)\,dF_X(x') \end{aligned}$$

where $c_{n\psi_k}(x,x')$ is defined on $\mathcal{I}_{nX}\times\mathcal{I}_{nX}$, it is finite, for every k. □

By Lemmas 3.1 and 3.2, the processes $n^{\frac{1}{2}}(\widehat{b}_{nk}\widehat{\psi}_{nk} - b_k\psi_k)$ converge weakly to centered Gaussian processes with finite and non degenerated covariance functions, for every integer k.

Let $K_n = o(n)$ tend to infinity as n tends to infinity, the series of the regression function m is approximated by

$$m_n = \sum_{k=0}^{K_n} b_k\psi_k.$$

The $L^2(\mathbb{R}, F_X)$ norm of m_n converges to $\|m\|_2 = \sum_{k=0}^{\infty} b_k^2$ and the squared error of approximation

$$\|m - m_n\|_2^2 = \sum_{k>K_n} b_k^2$$

tends to zero as n tends to infinity. The function m is then estimated by the process

$$\widehat{m}_n(x) = \sum_{k=0}^{K_n} \widehat{b}_{nk}\widehat{\psi}_{nk}(x), \tag{3.6}$$

on every sub-interval of $\mathcal{I}_X$ where $\widehat{F}_{nX}(x) < 1$. This estimator is defined by the estimators (3.2) of the functions ψ_k and (3.4) for the coefficients, it is a.s. consistent. The bias of $\widehat{m}_n$ is such that

$$m - E\widehat{m}_n = \sum_{k>K_n} b_k\psi_k + \sum_{k=0}^{K_n}\{b_k\psi_k - E(\widehat{b}_{nk}\widehat{\psi}_{nk})\}$$

and by Lemma 3.1, $E\{\widehat{b}_{nk}\widehat{\psi}_{nk}(x)\} - b_k\psi_k(x)$ develops as

$$\mathrm{Cov}(\widehat{b}_{nk}, \widehat{\psi}_{nk}(x)) + \psi_k(x)E\{m(X_{(n:i)})(\widehat{H}_{n,i} - H)^2(X_{(n:i)})L'' \circ H(X_{(n:i)})\}$$
$$+b_kE\{(\widehat{H}_{n,i} - H)^2(X_{(n:i)})L'' \circ H(X_{(n:i)})\} = O(n^{-1}).$$

The mean integrated squared error in $L^2(\mathbb{R}, F_X)$ of the regression function estimator is the sum

$$\mathrm{MISE}_n(m) = E\|\widehat{m}_n - m_n\|_2^2 + \|m - m_n\|_2^2$$

of the squared estimation error $E\|\widehat{m}_n - m_n\|_2^2$ and the squared error of the approximation by K_n functions of the basis which converges to zero.

Lemma 3.3. *Let K_n tend to infinity as n tends to infinity, then*

$$E\|\widehat{m}_n - m_n\|_2 = O(n^{-\frac{1}{2}}K_n),$$

it converges to zero for $K_n = o(n^{\frac{1}{2}})$.

Proof. The squared estimation error of the function m_{K_n} develops as

$$E\|\widehat{m}_n - m_n\|_2^2 = E\int_{\mathcal{I}_X}\Big\{\sum_{k\le K_n}(\widehat{b}_{nk}\widehat{\psi}_{nk} - b_k\psi_k)\Big\}^2 dF_X$$
$$= E\sum_{k\le K_n}\Big\{(\widehat{b}_{nk} - b_k)^2 + \int_{\mathcal{I}_X}\widehat{b}_{nk}^2(\widehat{\psi}_{nk} - \psi_k)^2\, dF_X$$
$$+2(\widehat{b}_{nk} - b_k)\widehat{b}_{nk}\int_{\mathcal{I}_X}(\widehat{\psi}_{nk} - \psi_k)\psi_k\, dF_X\Big\}$$
$$+2E\sum_{k\ne k'\le K_n}(\widehat{b}_{nk} - b_k)\widehat{b}_{nk'}\int_{\mathcal{I}_X}\widehat{\psi}_{nk'}\psi_k\, dF_X$$
$$+E\sum_{k\ne k'\le K_n}\widehat{b}_{nk}\widehat{b}_{nk'}\int_{\mathcal{I}_X}(\widehat{\psi}_{nk} - \psi_k)(\widehat{\psi}_{nk'} - \psi_{k'})\, dF_X.$$

For every $k \le K_n$, $E(\widehat{b}_{nk} - b_k)^2$ is a $O(n^{-1})$ and $\widehat{b}_{nk} - b_k = O_p(n^{-\frac{1}{2}})$, by the weak convergence of Lemma 3.1

$$\int_{\mathcal{I}_X}(\widehat{\psi}_{nk} - \psi_k)(\widehat{\psi}_{nk'} - \psi_{k'})\, dF_X = O_p(n^{-1})$$

and the expectation of its limit is a $O(n^{-1})$ for all integers k and k'. Then $\int_{\mathcal{I}_X}(\widehat{\psi}_{nk}-\psi_k)^2\,dF_X = O_p(n^{-1})$ and, from the orthogonality of the functions ψ_k and $\psi_{k'}$, for $k' \neq k$

$$\begin{aligned}\int_{\mathcal{I}_X} \widehat{\psi}_{nk'}\psi_k\,dF_X &= \int_{\mathcal{I}_X} (\widehat{\psi}_{nk'} - \psi_{k'})\psi_k\,dF_X \\ &= \Big\{\int_{\mathcal{I}_X} (\widehat{H}_n - H)\,L'_k \circ H\,\dot{\psi}_{k'}\,dF_X\Big\}\Big\{1 + o_p(1)\Big\} \\ &= O_p(n^{-\frac{1}{2}}),\end{aligned}$$

it follows that $E\|\widehat{m}_n - m_n\|_2^2 = O(K_n^2 n^{-1})$. □

As n tends to infinity, the number K_n defining the approximating function m_n is optimum as the errors have the same order, therefore K_n is chosen so that

$$K_n = o(n^{\frac{1}{2}}), \quad \|m - m_n\|_2 = O(n^{-\frac{1}{2}}K_n). \tag{3.7}$$

The order of the approximation error depends on the basis through the order of the integrals of the functions $m(x)\psi_k(x)$ which determines the convergence rate of $\sum_{k>K_n}\{\int m(x)\psi_k(x)\,dF_X\}_k^2$.

Example 3.1. Let m be a regression function in the space

$$W_{4s} = \Big\{m \in L^2(\mathcal{I}_X, F_X) : \sum_{k\geq 0} k^{4s}b_k^2(m) < \infty\Big\}, \tag{3.8}$$

for $s > 1$, then there exists a constant C such that $\|m - m_K\|_2^2 \leq CK^{-4s}$ and the optimal K which minimizes the MISE has the order

$$K_{opt} = 0(n^{\frac{1}{2(2s+1)}}).$$

The condition (3.7) is fulfilled with $n^{-1}K_{opt}^2 = 0(n^{-\frac{2s}{2s+1}}) = o(1)$ and $K_{opt}^{-2s} = O(n^{-1}K_{opt}^2)$.

Proposition 3.3. *The convergence rate of the MISE for the estimator $\widehat{m}_n$ of a regression function m defined by (2.18) and (2.17) equals the optimal convergence rate of a kernel estimator for m in C^s, $s \geq 2$, for $K_n = O(n^{\frac{1}{2(2s+1)}})$.*

Proof. Let $K_n = n^\alpha$, with $0 < \alpha < \frac{1}{2}$, then $n^{-\frac{1}{2}}K_n = n^{\frac{2\alpha-1}{2}}$. The optimal convergence rate for a nonparametric estimator of a regression function m in C^s has the order $n^{-\frac{s}{2s+1}}$, this is the order of the convergence rate $K_n n^{-\frac{1}{2}}$ of $E\|\widehat{m}_n - m_n\|_2$ if $\alpha = \{2(2s+1)\}^{-1}$. □

If the convergence rate of K_n is smaller than $n^{\frac{1}{2(2s+1)}}$, then the MISE of $\widehat{m}_n$ is smaller than optimal rate $n^{-\frac{s}{2s+1}}$.

If the distribution function F_X is known, it is not necessary to estimate the functions ψ_k and the convergence rate of the estimator $\widehat{m}_n$ is $K_n^{\frac{1}{2}} n^{-\frac{1}{2}}$. The optimal convergence rate for a kernel estimator is then achieved with $\alpha = (2s+1)^{-1}$, like in Proposition2.2 for a density.

Proposition 3.4. *Under the condition (3.7), the process* $K_n^{-1} n^{\frac{1}{2}} (\widehat{m}_n - m)$ *converges weakly to a Gaussian process with finite expectation and variance functions.*

Proof. The expectation of the process $\widehat{m}_n$ is

$$\begin{aligned} E\widehat{m}_n(x) &= \sum_{k=0}^{K_n} \int_{\mathcal{I}_X} m(s) E\{\widehat{\psi}_{nki}(s)\widehat{\psi}_{nk}(x)\} \, dF_X(s) \\ &= \sum_{k=0}^{K_n} \psi_k(x) \int_{\mathcal{I}_X} m(s)\psi_k(s) \, dF_X(s) + O(K_n n^{-1}) \\ &= m_n(x) + O(K_n n^{-1}). \end{aligned}$$

The weak convergence of the finite dimensional distributions of the process $K_n^{-1} n^{\frac{1}{2}} (\widehat{m}_n - m_n)$ is established by the convergence of the variance of linear combinations of the process at any vector $(x_1, \ldots, x_p)$ and by the central limit theorem. The functions ψ_k belong to $C_b^2(\mathbb{R}_+)$ and this entails the tightness of the process, this implies the weak convergence of $K_n^{-1} n^{\frac{1}{2}} (\widehat{m}_n - m_n)$ to a centered Gaussian process. The approximation error $(\sum_{k=K_n+1}^{\infty} b_k^2)^{\frac{1}{2}}$ has the rate of convergence $n^{-\frac{1}{2}} K_n$ (cf. proof of Lemma 3.3), hence $K_n^{-1} n^{\frac{1}{2}} (m_n - m)$ converges to a finite limit.

The variance of $K_n^{-1} n^{\frac{1}{2}} (\widehat{m}_n - m)$ is asymptotically equivalent to nK_n^{-2} times the sum over k and $k' \leq K_n$ of the covariances of $(\widehat{b}_{nk} - b_k)\,\psi_k + (\widehat{\psi}_{nk} - \psi_k)\, b_k$ and $(\widehat{b}_{nk'} - b_{k'})\,\psi_{k'} + (\widehat{\psi}_{nk'} - \psi_{k'})\, b_{k'}\}$, its limit is finite. □

In the heteroscedastic model, the variance of the error conditionally on X is $\sigma^2(X)$ and the variance of the estimator $\widehat{b}_{nk}$ has the approximation

$$\begin{aligned} \text{Var}\, \widehat{b}_{nk} &= E\{\sigma^2(X_i)\widehat{\psi}_{nk}^2(X_i)\} + \text{Var}\{m(X_i)\widehat{\psi}_{nk}(X_i)\} \\ &= E\{\sigma^2(X_i)\psi_k^2(X_i)\} + \text{Var}\{m(X_i)\psi_k(X_i)\} + 0(n^{-1}). \end{aligned}$$

The variables $n^{\frac{1}{2}} (\widehat{b}_{nk} - b_k)$ converge weakly to centered Gaussian variables with strictly positive variances

$$v_k^2 = E\{\sigma^2(X_i)\psi_k^2(X_i)\} + \text{Var}\{m(X_i)\psi_k(X_i)\}.$$

The rates of Lemma 3.3 and Propositions 3.3 and 3.4 are unchanged. The variance function has the expansion $\sigma^2(x) = \sum_{k\geq 0} c_k\psi_k(x)$, it is approximated by a function

$$\sigma_n^2(x) = \sum_{k=0}^{L_n} c_k\psi_k(x)$$

where $L_n = o(n)$ and tends to infinity as n tends to infinity. The quadratic approximation error is a $o(1)$. The coefficients c_k are estimated by

$$\widehat{c}_{nk} = (n-1)^{-1}\sum_{i=1}^{n-1}\{Y_{(i)} - \widehat{m}_{nh}(X_{(n:i)})\}^2\widehat{\psi}_{nki}(X_{(n:i)}), \tag{3.9}$$

using the notations of (3.4), and the function σ^2 is consistently estimated by

$$\widehat{\sigma}_n^2(x) = \sum_{k=0}^{L_n} \widehat{c}_{nk}\widehat{\psi}_{nk}(x). \tag{3.10}$$

If $E\{Y - m(X)\}^4$ is finite, the variables $n^{\frac{1}{2}}(\widehat{c}_{nk} - c_k)$ converge weakly to centered Gaussian variables with variances $E\{Y - m(X)\}^4 + E[\{Y - m(X)\}\widehat{\psi}_{nk}(X_i)]^4 = 0(1)$. The estimator $\widehat{\sigma}_n^2$ has the same form as the estimator of the regression function and, under the same conditions and by the same arguments

$$E\|\widehat{\sigma}_n^2 - \sigma^2\|_2 = O(n^{-\frac{1}{2}}L_n)$$

and the process $L_n^{-1}n^{\frac{1}{2}}(\widehat{\sigma}_n^2 - \sigma^2)$ converges weakly to a Gaussian process with a finite expectation and variance functions.

In the same setting, let $E(Y \mid X) = m(X)$ be a regression function of Y on a real regressor X observed with an error, in the model

$$Y = m(X) + \varepsilon_1,$$
$$Z = X + \varepsilon_2,$$

with independent centered symmetric error variables ε_1 and ε_2. The variables Y and Z are observed but not X and the distribution function of the error variables is supposed to be known. By (1.8), the survival function of X has the estimator

$$\widehat{\bar{F}}_{nX}(x) = n^{-1}\sum_{i=1}^{n} \bar{F}_2(x - Z_i)$$

where F_2 is the distribution function of ε_2, this estimator and the formula (3.3) determine a consistent estimator $\widehat{H}_n$ of the hazard function of X on every subset of the support $\mathcal{I}_X$ of F_X where $\widehat{F}_{nX} = 1 - \widehat{\bar{F}}_{nX}$ is strictly positive, the estimators $\widehat{\psi}_{nk}$ of the functions ψ_k follow from (3.2). The convergence properties of the estimated basis are the same as Lemma 3.1.

The projection of the regression function m on the basis of functions $(\psi_k)_{k\geq 0}$ is estimated by (3.6)

$$\widehat{m}_n = \sum_{k=0}^{K_n} \widehat{b}_{nk}\widehat{\psi}_{nk}(x),$$
$$\widehat{b}_{nk} = \int y\widehat{\psi}_{nk}(x)\, d\widehat{F}_n(x,y)x,$$

with the joint empirical distribution function of (X, Y)

$$\widehat{F}_n(x,y) = n^{-1}\sum_{i=1}^{n} F_2(x - Z_i)1_{\{Y_i\leq y\}}.$$

The convergence rate of the MISE of the estimator $\widehat{m}_n$ and Propositions 3.3 and 3.4 are not modified.

3.3 Weighted series estimators

Let $(\psi_k)_{k\geq 0}$ be an orthonormal basis of functions of $L^2(\mathbb{R}, F_X)$ such that the condition (2.22) is fulfilled and let m be a regression function of the space $W_{4s} = \{m \in L^2(\mathbb{R}) : \sum_{k\geq 0} k^{4s}b_k^2(m) < \infty\}$, for $s \leq m_0$. Like in Section 2.6, a weighted estimator of m is defined by projection on the estimated basis $(\widehat{\psi}_{nk})_{k\geq 0}$ as

$$\widehat{m}_n(x) = \sum_{k=0}^{n} \frac{\lambda_k}{\nu_n + \lambda_k}\widehat{b}_{nk}\widehat{\psi}_{nk}(x), \tag{3.11}$$

with the estimators (3.4) of the coefficients, with $\lambda_k = k^{-s}$ such that and ν_n converges to zero as n tends to infinity. In W_{4s}, there exists a constant C such that $\|m - m_k\|^2 \leq Ck^{-4s}$.

Proposition 3.5. *Under the conditions (2.22) and $\nu_n = O(n^{-\frac{2s}{2s+1}})$, the estimator of a regression function m of W_s satisfies*

$$\text{MISE}(\widehat{m}_n) = O(n^{-\frac{2s}{2s+1}}).$$

Proof. The MISE of the estimator $\widehat{m}_n$ is expanded like in the proof of Lemma 3.3

$$\begin{aligned}
\text{MISE}(\widehat{m}_n) &= E\int \{\widehat{m}_n(x) - m(x)\}^2 \, dF_X(x) \\
&= E\int \Big[\sum_{k=1}^n \frac{\lambda_k}{\nu_n+\lambda_k}\{\widehat{b}_{nk}\widehat{\psi}_{nk}(x) - b_k\psi_k(x)\Big\} \\
&\qquad - \frac{\nu_n}{\nu_n+\lambda_k} b_k\psi_k(x)\Big]^2 \, dF_X(x) + \sum_{k=n+1}^\infty b_k^2 \\
&= E\sum_{k=1}^n \Big[\frac{\lambda_k^2}{(\nu_n+\lambda_k)^2}\Big\{(\widehat{b}_{nk}-b_k)^2 + \widehat{b}_{nk}^2 \int_{\mathcal{I}_X} (\widehat{\psi}_{nk}-\psi_k)^2 \, dF_X \\
&\qquad + 2(\widehat{b}_{nk}-b_k)\widehat{b}_{nk}\int_{\mathcal{I}_X}(\widehat{\psi}_{nk}-\psi_k)\psi_k \, dF_X\Big\} \\
&\quad -2\frac{\lambda_k\nu_n}{(\nu_n+\lambda_k)^2}\Big\{(\widehat{b}_{nk}-b_k)b_k + b_k\widehat{b}_{nk}\int(\widehat{\psi}_{nk}-\psi_k)\psi_k \, dF_X\Big\} \\
&\quad + \frac{\nu_n^2}{(\nu_n+\lambda_k)^2} b_k^2\Big] + \sum_{k=n+1}^\infty b_k^2 \\
&\quad + E\sum_{k\neq k'=1}^n \Big[\frac{\lambda_k\lambda_{k'}}{(\nu_n+\lambda_k)(\nu_n+\lambda k')} \\
&\qquad \cdot\Big\{\widehat{b}_{nk}\widehat{b}_{nk'}\int_{\mathcal{I}_X}(\widehat{\psi}_{nk}-\psi_k)(\widehat{\psi}_{nk'}-\psi_{k'}) \, dF_X \\
&\qquad + 2(\widehat{b}_{nk}-b_k)\widehat{b}_{nk'}\int_{\mathcal{I}_X}(\widehat{\psi}_{nk'}-\psi_{k'})\psi_k \, dF_X\Big\} \\
&\quad -2\frac{\lambda_k\nu_n}{(\nu_n+\lambda_k)(\nu_n+\lambda k')}\widehat{b}_{nk}b_{k'}\int_{\mathcal{I}_X}\widehat{\psi}_{nk}\psi_{k'} \, dF_X\Big].
\end{aligned}$$

By Lemmas 3.1 and 3.2 and by the condition (2.22), the variances of the estimators $\widehat{b}_{nk}$, and respectively $\widehat{\psi}_{nk}$, have approximations $n^{-1}v_{b,k} + o(n^{-1})$, and respectively $n^{-1}v_{b\psi,k} + o(n^{-1})$, and their covariances have approximations of the same order $n^{-1}v_{b\psi,k} + o(n^{-1})$ where

$$nv_{b\psi,k}(x) = E\Big\{W_n(x)\int \psi_k \, dW_n\Big\} g' \circ \bar{F}(x) L' \circ H(x).$$

The orthogonality of Laguerre's polynomials entails that the estimated functions of the basis are asymptotically orthogonal as well the estimated coefficients, for all distinct integers k and k' we obtain

$$E \int_{\mathcal{I}_X} (\widehat{\psi}_{nk} - \psi_k)(\widehat{\psi}_{nk'} - \psi_{k'})\, dF_X = o(1),$$
$$E\{(\widehat{b}_{nk} - b_k) \int_{\mathcal{I}_X} (\widehat{b}_{nk'} - b_{k'})\, dF_X\} = o(1),$$
$$E\{(\widehat{b}_{nk} - b_k) \int_{\mathcal{I}_X} (\widehat{\psi}_{nk'} - \psi_{k'})\, dF_X\} = o(1).$$

The error MISE($\widehat{m}_n$) is then bounded by

$$\begin{aligned}
&\sum_{k=1}^{n} \Big[\frac{\lambda_k^2}{n(\nu_n + \lambda_k)^2} \Big\{ v_{b,k} + b_k^2 \int_{\mathcal{I}_X} v_{b\psi,k}\, dF_X \\
&\quad + 2\widehat{b}_{nk} \int_{\mathcal{I}_X} v_{b\psi,k}\psi_k\, dF_X \Big\} \{1 + o(1)\} \\
&\quad - 2\frac{\lambda_k \nu_n}{(\nu_n + \lambda_k)^2} \Big\{ b_k(\widehat{b}_{nk} - b_k) \int \widehat{\psi}_{nk}\psi_k\, dF_X \Big\} + \frac{\nu_n^2}{(\nu_n + \lambda_k)^2} b_k^2 \Big] \\
&+ E \sum_{k \neq k'=1}^{n} \Big[\frac{\lambda_k \lambda_{k'}}{(\nu_n + \lambda_k)(\nu_n + \lambda k')} \\
&\quad \cdot \Big\{ \widehat{b}_{nk}\widehat{b}_{nk'} \int_{\mathcal{I}_X} (\widehat{\psi}_{nk} - \psi_k)(\widehat{\psi}_{nk'} - \psi_{k'})\, dF_X \\
&\quad + 2(\widehat{b}_{nk} - b_k)\widehat{b}_{nk'} \int_{\mathcal{I}_X} (\widehat{\psi}_{nk'} - \psi_{k'})\psi_k\, dF_X \Big\} \\
&- 2\frac{\lambda_k \nu_n}{(\nu_n + \lambda_k)(\nu_n + \lambda k')} \widehat{b}_{nk} b_{k'} \int_{\mathcal{I}_X} \widehat{\psi}_{nk}\psi_{k'}\, dF_X \Big] + \sum_{k=n+1}^{\infty} b_k^2.
\end{aligned}$$

The bounds of Proposition (2.4) apply to MISE($\widehat{m}_n$) and each term of the expansion has the bound

$$\begin{aligned}
&\sum_{k=1}^{n} \frac{\lambda_k^2}{n(\nu_n + \lambda_k)^2} \Big\{ v_{b,k} + b_k^2 \int_{\mathcal{I}_X} v_{b\psi,k}\, dF_X + 2\widehat{b}_{nk} \int_{\mathcal{I}_X} v_{b\psi,k}\psi_k\, dF_X \Big\} \\
&= O(n^{-1}\nu_n^{-\frac{1}{2s}}), \\
&\sum_{k=1}^{n} \frac{\nu_n^2}{(\nu_n + \lambda_k)^2} b_k^2 = O(\nu_n),
\end{aligned}$$

$$
\begin{aligned}
&\sum_{k=1}^{n} \frac{\lambda_k \nu_n}{(\nu_n+\lambda_k)^2} E\Big\{b_k(\widehat{b}_{nk}-b_k)\int \widehat{\psi}_{nk}\psi_k \, dF_X\Big\} \\
&\leq \Big\{\sum_{k=1}^{n} \frac{\lambda_k^2 v_{b,k}}{n(\nu_n+\lambda_k)^2}\Big\}^{\frac{1}{2}} \Big[\sum_{k=1}^{n} \frac{\nu_n^2 b_k^2}{(\nu_n+\lambda_k)^2}\{1+o(1)\}\Big]^{\frac{1}{2}} \\
&= O(n^{-1}\nu_n^{-\frac{1}{2s}}), \\
&\sum_{k=n+1}^{\infty} b_k^2 = O(n^{-4s}).
\end{aligned}
$$

By the Cauchy–Schwarz inequality, the sums on $k \neq k'$ are bounded by the products on the square roots of the sums on a single index, so the first two sums are $O(n^{-1}\nu_n^{-\frac{1}{2s}})$ and the last sums is a $O(n^{-2s})$. By the order of ν_n, all terms of the sum have the same order and we obtain $\text{MISE}(\widehat{m}_n) = O(n^{-\frac{2s}{2s+1}})$. □

3.4 Cross-validation for regression functions

The integrated squared error of the estimator $\widehat{m}_n$ is the sum

$$
\begin{aligned}
I_n(K_n) &= \int_{\mathcal{I}_X} \{\widehat{m}_n(x) - m(x)\}^2 \, dF_X(x) \\
&= \sum_{k=0}^{K_n} (\widehat{b}_{kn} - b_k)^2 + \sum_{k>K_n} b_k^2.
\end{aligned}
$$

For every $k \leq K_n$, $\widehat{b}_{kn} - b_k = 0_p(n^{-\frac{1}{2}})$ and the first sum is a $0_p(n^{-1}K_n)$, $I_n(K_n)$ is a $o(1)$. The cross-validation for a regression function generalizes the method developed for the density, the minimization of I_n with respect to K_n is asymptotically equivalent to the minimization of

$$
\begin{aligned}
I_n(K_n) - \int_{\mathcal{I}_X} m^2 \, dF_X &= \int_{\mathcal{I}_X} \widehat{m}_n^2(x) \, dF_X(x) - 2\int_{\mathcal{I}_X} \widehat{m}_n(x) m(x) \, dF_X(x) \\
&= \sum_{j,k=0}^{K_n} \widehat{b}_{j,n}\widehat{b}_{kn} \int_{\mathcal{I}_X} \widehat{\psi}_{nj}(x)\widehat{\psi}_{nk}(x) \, dF_X(x) \\
&\quad -2\sum_{j,k=0}^{K_n} b_j \widehat{b}_{kn} \int_{\mathcal{I}_X} \widehat{\psi}_{nj}(x)\psi_k(x) \, dF_X(x).
\end{aligned}
$$

By the weak convergence of Lemma 3.1, for all integers j and k

$$
\int_{\mathcal{I}_X} (\widehat{\psi}_{nj} - \psi_j)(\widehat{\psi}_{nk} - \psi_k) \, dF_X = O_p(n^{-1})
$$

and the orthogonality of the functions ψ_j and ψ_k, for $j \neq k$, implies $\int_{\mathcal{I}_X} \widehat{\psi}_{nj}\psi_k \, dF_X = O_p(n^{-\frac{1}{2}})$, then $\int_{\mathcal{I}_X} \widehat{\psi}_{nj}(x)\widehat{\psi}_{nk}(x)\, dF(x) = 1_{\{j=k\}} + O_p(n^{-\frac{1}{2}})$ and

$$I_n(K_n) - \int_{\mathcal{I}_X} m^2 \, dF_X = \sum_{k=0}^{K_n} \widehat{b}_{kn}^2 - 2\sum_{k=0}^{K_n} b_k \widehat{b}_{kn} + O_p(n^{-\frac{1}{2}}).$$

Let $\widehat{m}_{n,j}$ be the projection based estimator of the regression function defined with the sub-sample without the observation $(Y_{(j)}, X_{(n:j)})$, the integral $I_n(K_n) - \int_{\mathcal{I}_X} m^2 \, dF_X$ is estimated by

$$CV_n(K_n) = (n-1)^{-1} \sum_{j=1}^{n-1} \widehat{m}_{n,j}^2(X_{(n:j)}) - 2(n-1)^{-1} \sum_{j=1}^{n-1} Y_{(j)} \widehat{m}_{n,j}(X_{(n:j)}),$$

for every $j = 1, \ldots, n-1$, the estimator CV_n is asymptotically equivalent to

$$J_n(K_n) = \sum_{k=0}^{K_n} \widehat{b}_{kn}^2 - \frac{2}{n(n-1)} \sum_{k=0}^{K_n} \sum_{j \neq i=1}^{n-1} Y_{(i)} Y_{(j)} \widehat{\psi}_{kn,ij}(X_{(n:i)}) \widehat{\psi}_{kn,j}(X_{(n:j)}). \tag{3.12}$$

The expectation of J_n is

$$\begin{aligned} EJ_n(K_n) &= \sum_{k=0}^{K_n} E\widehat{b}_{kn}^2 - 2\sum_{k=0}^{K_n} b_k^2 + o(1) \\ &= \sum_{k=0}^{K_n} \mathrm{Var}(\widehat{b}_{kn}) - \sum_{k=0}^{K_n} b_k^2 + o(1) \\ &= EI_n(K_n) - \int_{\mathcal{I}_X} m^2 \, dF_X + o(1). \end{aligned}$$

An estimator $\widehat{m}_n$ of m is defined as a series of $\widehat{K}_n$ terms by minimization of J_n. The difference of $I_n - J_n$ is written as

$$\begin{aligned} I_n(K_n) - J_n(K_n) = 2\sum_{k=0}^{K_n} &\Big\{\frac{1}{n}\sum_{i=1}^{n-1} Y_{(i)} \widehat{\psi}_{kn}(X_{(n:i)})\Big\} \\ &\times \Big\{\frac{1}{n-1} \sum_{j \neq i=1}^{n-1} Y_{(j)} \widehat{\psi}_{kn}(X_{(n:j)}) - b_k\Big\} + \int_{\mathcal{I}_X} m^2 \, dF_X \end{aligned}$$

and it converges in probability to $\int_{\mathcal{I}_X} m^2 \, dF_X$ as n tends to infinity. It follows that the minima of I_n and J_n are asymptotically equivalent in probability and $\widehat{m}_{\widehat{K}_n}$ is asymptotically optimal i.e.

$$\frac{I_n(\widehat{m}_{\widehat{K}_n})}{\inf_K I_n(K)} \to 1$$

in probability as n tends to infinity.

3.5 Projection pursuit regression

Let Y be a real variable and let $X = (X_1, \ldots, X_d)^T$ be a d dimensional vector of regressors with a distribution function F_X on a set $\mathcal{I}_X$. The projection pursuit regression model

$$E(Y \mid X) = m(\theta^T X) = m_\theta(X) \tag{3.13}$$

is defined by an univariate regression function m of $C^2(\mathbb{R})$ and a parameter θ in a subset Θ of $\mathbb{R}^d$, with $\theta^T x = \sum_{j=1,\ldots,d} \theta_j x_j$. Under the probability P_0 of the observations, the parameter value is θ_0 and $E_0(Y \mid X) = m_0(\theta_0^T X)$. The real variable $X_\theta = \theta^T X$ has a distribution function $F_{X,\theta}$ with an empirical estimator $\widehat{F}_{nX,\theta}$ based on a sample $(X_i)_{i=1,\ldots,n}$ of X and depending on the parameter θ by (3.13). At θ_0, the empirical distribution function $\widehat{F}_{n,X}$ of the sample estimates F_X, at θ let $F_{X,\theta}$ be the distribution function of $\theta^T X$ and let $\widehat{F}_{n,X,\theta} = \widehat{F}_{n,\theta^T X}$ be its empirical estimator.

For every θ, let $(\psi_{k,\theta})_{k\geq 0}$ be the orthonormal basis of functions of $L^2(\mathbb{R}, F_{X,\theta})$ defined by Laguerre's polynomials L_k as

$$\psi_{k,\theta} = L_k \circ H_\theta$$

with the distribution function $F_{X,\theta}$ like in (3.1). By Lemma 3.1, the estimators $\widehat{\psi}_{nk,\theta}$ are a.s. uniformly consistent and the processes $n^{\frac{1}{2}}(\widehat{\psi}_{nk,\theta} - \psi_{k,\theta})$ converge weakly to centered Gaussian process with bounded covariance functions. The function $m(\theta^T x)$ has the expansion

$$m_\theta(x) = \sum_{k\geq 0} b_{k,\theta}\psi_{k,\theta}(x) = \sum_{k\geq 0} b_{k,\theta}\psi_k(\theta^T x)$$

with the coefficients

$$b_{k,\theta} = \int_{\mathbb{R}} m(u)\psi_{k,\theta}(u)\, dF_{X,\theta}(u),$$

for $k \geq 1$, and $b_{0,\theta} = Em_\theta(X)$. For every integer $k \geq 1$, the coefficient $b_{k,\theta}$ is estimated by

$$\widehat{b}_{nk,\theta} = (n-1)^{-1} \sum_{i=1}^{n-1} Y_{(i)} \widehat{\psi}_{nki,\theta}(X_{(n:i)}), \tag{3.14}$$

where the larger observation $X_{(n:n)}$ is omitted, $Y_{(i)}$ is the response variable to $X_{(n:i)}$ and $\widehat{\psi}_{nki,\theta}$ is the estimator of ψ_k based on the estimator $\widehat{F}_{niX,\theta}$ of $F_{X,\theta}$ without the observation $X_{(n:i)}$. For every θ, the estimators $\widehat{b}_{nk,\theta}$ are a.s. uniformly consistent and, by Lemma 3.2, the variables $n^{\frac{1}{2}}(\widehat{b}_{nk,\theta} - b_{k,\theta})$ converge weakly to centered Gaussian variables with finite covariances.

At $\theta^T x$, the regression function m is approximated by the series

$$m_{n,\theta}(x) = \sum_{k=0}^{K_n} b_{k,\theta}\psi_{k,\theta}(x)$$

where $K_n = o(n)$, and it is a.s. consistently uniformly estimated by the process

$$\widehat{m}_{n,\theta}(x) = \sum_{k=0}^{K_n} \widehat{b}_{nk,\theta}\widehat{\psi}_{nk,\theta}(x). \tag{3.15}$$

By Lemma 3.3, for every θ

$$E_\theta\|\widehat{m}_{n,\theta} - m_{n,\theta}\|_2 = O(n^{-\frac{1}{2}}K_n).$$

The mean squared error of estimation in $L^2(\mathbb{R}, F_{X,\theta})$ is the process

$$l_n(\theta) = n^{-1}\sum_{i=1}^{n}\{Y_i - \widehat{m}_{n,\theta}(X_i)\}^2,$$

it converges a.s. uniformly under P_0 to

$$l(\theta) = \sigma_\varepsilon^2 + E\{m_\theta(\theta^T X) - m_0(\theta_0^T X)\}^2,$$

where the variance σ_ε^2 of the error ε is the limit as n tends to infinity of $n^{-1}\sum_{i=1}^n\{Y_i - m_0(\theta_0^T X_i)\}^2$. The function $l(\theta)$ is therefore minimum at θ_0 and the parameter is estimated by

$$\widehat{\theta}_n = \arg\inf_{\theta\in\Theta} l_n(\theta).$$

Lemma 3.4. *The estimator $\widehat{\theta_n^T}x)$ converges a.s. under P_0 to θ_0 and the estimator $\widehat{m}_n = \widehat{m}_{n,\widehat{\theta}_n}$ converges a.s. uniformly to m_0.*

Proof. The inequality

$$0 < l_n(\widehat{\theta}_n) \le \sup_{\theta\in\Theta}|l_n(\theta) - l(\theta)| + l(\theta)$$

for every θ implies $0 < l_n(\widehat{\theta}_n) \le l(\theta_0) + o_p(1)$ where $l(\theta_0) \le l(\widehat{\theta}_n)$ and the convergence of the process l_n implies the consistency of $\widehat{\theta}_n$. □

Proposition 3.6. *Under P_0, the variable $n^{\frac{1}{2}}K_n^{-1}(\widehat{\theta}_n - \theta_0)$ converges weakly to a centered Gaussian variable.*

Proof. The consistency of $\widehat{\theta}_n$ and $\widehat{m}_{n,\theta_0}$ provides the approximation

$$l'_n(\theta_0) = -(\widehat{\theta}_n - \theta_0)^T l''(\theta_0) + o_P(\|\theta - \theta_0\|),$$

where the derivative of l_n is

$$l'_n(\theta_0) = -n^{-1} \sum_{i=1}^{n} \{Y_i - \widehat{m}_{n,\theta_0}(X_i)\} \widehat{m}'_{n,\theta_0}(\theta_0^T X_i)\, X_i,$$

$$\widehat{m}'_{n,\theta_0}(u) = \sum_{k=0}^{K_n} \widehat{b}_{nk,\theta_0} \widehat{\psi}'_{nk}(u),$$

$$l''(\theta) = \int_{\mathcal{I}_X} \Big[\Big(x_j x_l \frac{\partial^2 m_\theta}{\partial\theta_j \theta_l} \Big)_{jl} (\theta_0^T x) \{m_\theta(\theta^T x) - m_0(\theta_0^T x)\} + \Big(x_j \frac{\partial m_\theta}{\partial \theta_j} \Big)_j^{\otimes 2} (\theta^T x) \Big] \, dF_X(x),$$

and $l'(\theta_0) = 0$. By Proposition 3.4, the process $\widehat{m}_{n,\theta_0} - m_0$ has the convergence rate $n^{-\frac{1}{2}} K_n$ and this is also the convergence rate of the variable $l'_n(\theta_0)$. It follows that the variable

$$n^{\frac{1}{2}} K_n^{-1} (\widehat{\theta}_n - \theta_0) = -l''^{-1}(\theta_0) n^{\frac{1}{2}} K_n^{-1} l'_n(\theta_0) + o(1)$$

converges weakly to a centered Gaussian variable with finite variance. □

Proposition 3.7. *Under the condition (3.7) and P_0, the MISE of $\widehat{m}_n$ has the order $O(n^{-\frac{1}{2}} K_n)$ and the process $K_n^{-1} n^{\frac{1}{2}} (\widehat{m}_n - m_0)$ converges weakly to a Gaussian process with finite expectation and variance functions.*

Proof. The order of the MISE under the conditions is a consequence of Proposition 3.4.

In a neighborhood V_0 of θ_0 in Θ, the functions $\psi_{k,\theta}$ are differentiable with respect to θ and it have the expansion

$$\psi_{k,\theta}(x) - \psi_{k,\theta_0}(x) = (\theta - \theta_0)^T \frac{\partial \psi_{k,\theta}(x)}{\partial \theta} + 0(\|\theta - \theta_0\|^2),$$

in the same way, the function $m_{n,\theta}(\theta^T x) = \sum_{k=0}^{K_n} \langle \psi_{k,\theta}, m_\theta > \psi_{k,\theta}(x)$ has an expansion depending on the expansion of the functions $\psi_{k,\theta}$ for $k = 0, \ldots, K_n$, the result is a consequence of the consistency Lemma 3.4 and Proposition 3.6. □

3.6 Multivariate nonparametric regressions

Let Y be a real variable and let $X = (X_1, \ldots, X_d)^T$ be a d dimensional vector of regressors with linearly independent components. The distribution function F_X of the variable X on a subset $\mathcal{I}_X$ of $\mathbb{R}^d$ has marginal distribution functions $F_1, \ldots, F_d$ on $\mathcal{I}_{X_j}$, $j = 1, \ldots, d$ respectively. We consider the nonparametric regression model for a n-sample of independent and identically distributed variables

$$Y_i = m(X_i) + \varepsilon_i,\ i = 1, \ldots, n,$$

with the multivariate regression function

$$E(Y_i \mid X_i) = m(X_i). \tag{3.16}$$

Let $\mathbb{X}_n = (X_{ij})_{i \le n, j \le d}$ be the observation matrix of dimension $n \times d$. The $d \times d$ dimensional symmetric matrix $\{\mathbb{X}_n - E\mathbb{X}_n\}^T\{\mathbb{X}_n - E\mathbb{X}_n\}$ is non singular, then it has d strictly positive eigenvalues λ_j and eigenvectors U_j of $\mathbb{R}^d$ such that $\{\mathbb{X}_n - E\mathbb{X}_n\}^T\{\mathbb{X}_n - E\mathbb{X}_n\}U_j = \lambda_j U_j$, for $j = 1, \ldots, d$. There exists a $d \times d$ dimensional matrix B such that the diagonal matrix Λ with diagonal values $\lambda_1, \ldots, \lambda_d$ satisfies

$$\{\mathbb{X}_n - E\mathbb{X}_n\}^T\{\mathbb{X}_n - E\mathbb{X}_n\} = B\Lambda B^{-1},$$

the matrix B defined an isomorphism u from $\mathcal{I}_X$ in $\mathbb{R}^d$ onto $\mathcal{I}_Z$ in $\mathbb{R}^d$ and the observation matrix $\mathbb{X}_n$ is mapped to the $n \times d$ dimensional matrix

$$\mathbb{Z}_n = \mathbb{X}_n B \tag{3.17}$$

of a n independent and identically distributed d dimensional vectors of independent variables such that

$$\{\mathbb{Z}_n - E\mathbb{Z}_n\}^T\{\mathbb{Z}_n - E\mathbb{Z}_n\}^T = B^{-1}\{\mathbb{X}_n - E\mathbb{X}_n\}^T\{\mathbb{X}_n - E\mathbb{X}_n\}B = \Lambda.$$

So there is a bijections between the vectors X_i and Z_i such that

$$X_{ij} = \sum_{l=1}^d \eta_{il} Z_{il}, \quad Z_{ij} = \sum_{l=1}^d \zeta_{il} X_{il}.$$

The regression model $m(X) = E(Y \mid X)$ is mapped from $\mathcal{I}_X$ onto a subset $\mathcal{I}_Z$ of $\mathbb{R}^d$ as

$$\mu(Z) = E(Y \mid Z) = \mu \circ u(X) = m(X). \tag{3.18}$$

The conditional expectation $\mu(z)$ of Y_i given $Z_i = z$ is expanded on the orthonormal basis of functions of $L^2(\mathbb{R}^d, F_Z)$ defined by (3.19) as a transform of the Laguerre orthonormal basis. By the independence of the components of Z,

$$1 - F_Z(z) = \prod_{j=1}^{d} \{1 - F_{Z_j}(z_j)\}$$

and the function $H_Z(z) = -\log\{1 - F_Z(z)\}$ is the sum

$$H_Z(z) = \sum_{j=1}^{d} H_{Z_j}(z_j).$$

The orthonormal basis is defined by the functions

$$\psi_k = L_k \circ H_Z, \; k \geq 0 \tag{3.19}$$

on $\mathbb{R}^d$. The equality $\langle L_0, L_k \rangle_{\mathcal{E}} = 0$, for the scalar product with respect to exponential density with parameter 1, implies

$$\int_{\mathcal{I}_Z} L_k \circ H_Z(z)\, dF_Z(z) = \langle \psi_0, \psi_k \rangle_{F_Z} = 0.$$

The marginal functions $\psi_k(z_j) = L_k \circ H_Z(z_j)$ are defined with components $z_l = 0$ for every $l \neq j$ hence $H_Z(z_j) = H_{Z_j}(z_j)$ like in (3.1). Furthermore

$$\int_{\mathcal{I}_Z} \psi_k(z_j)\psi_{k'}(z_l)\, dF_Z(z) = \int_{\mathcal{I}_{Z_j}} \int_{\mathcal{I}_{Z_l}} \psi_k(z_j)\psi_{k'}(z_l)\, dF_{Z_j}(z_j)\, dF_{Z_l}(z_l)$$

for all distinct integers j and l, this integral factorizes as the product of the integrals of functions of z_j and respectively z_l, it is therefore zero. For $l = j$, the integral $\int_{\mathcal{I}_Z} \psi_k(z_j)\psi_{k'}(z_j)\, dF_Z(z)$ reduces to $1_{\{k=k'\}}$, by the orthonormality of the functions ψ_k.

The transformed functions of the basis $\psi_k(z) = L_k \circ H_Z(z)$ satisfy

$$\begin{aligned}
&\int_{\mathcal{I}_Z} \mu(z) L_k \circ H_Z(z)\, dF_Z(z) \\
&= \sum_{l=0}^{\infty} b_l \int_{\mathbb{R}^d} L_l\Big(\sum_{j=1}^{d} s_j\Big) L_k\Big(\sum_{j=1}^{d} s_j\Big) e^{-\sum_{j=1}^{d} s_j}\, ds_1 \ldots, ds_d \\
&= b_k,
\end{aligned}$$

for every integer k and they are consistently estimated as

$$\widehat{\psi}_{nk}(z) = L_k \circ \widehat{H}_{nZ}(z),$$

with the estimator

$$\widehat{H}_{nZ}(z) = \sum_{j=1}^{d} \widehat{H}_{nZ_j}(z_j) \tag{3.20}$$

of the multivariate cumulative hazard function of Z, by the independence of the components of Z.

The real regression function $\mu(z)$ has the expansion $\sum_{k=0}^{\infty} b_k \psi_k(z)$ with coefficients $b_k = \int_{\mathbb{R}^d} \mu(z)\psi_k(z)\, dF_Z(z)$. The parameter

$$b_0 = \int_{\mathcal{I}_Z} \mu(z)\, dF_Z(z) = E\mu(Z) = EY$$

is estimated by $\widehat{b}_{0n} = n^{-1}\sum_{i=1}^{n} Y_i$ and, for $k \geq 1$, the components b_{kj} of the d dimensional parameter b_k is estimated by

$$\widehat{b}_{kjn} = (n-1)^{-1}\sum_{i=1}^{n-1} Y_{(i)} L_k \circ \widehat{H}_{niZ_j}(Z_{(n:i)j}) \tag{3.21}$$

$\widehat{H}_{niZ_j}(z_j)$ is defined by (3.3) using the empirical distribution function $\widehat{F}_{niZ_j}$ of the regressor Z_j, where the observation $Z_{(n:i)j}$ is omitted. In (3.20), the larger observations $Z_{(n:n)j}$ are omitted and $Y_{(i)}$ is the response variable to $Z_{(n:i)j}$.

Lemma 3.5. *For every integer $k \geq 1$, on every subset of $\mathcal{I}_Z$ where $\widehat{F}_{nZ}(z) < 1$, the estimator $\widehat{\psi}_{nk} = L_k \circ \widehat{H}_{nZ}$ defined by (3.20) is a.s. uniformly consistent and the process $n^{\frac{1}{2}}(\widehat{\psi}_{nk} - \psi_k)$ converges weakly to a centered Gaussian process with a bounded covariance function.*

The a.s. uniform convergence and the convergence rate are consequences of (3.20) and the convergence of the marginal estimators. Lemma 3.5 is proved like Lemma 3.1.

The variables $n^{\frac{1}{2}}(\widehat{b}_{nk} - b_k)$ converge weakly to centered Gaussian variables with variances

$$v_k^2 = E\{Y_i^2\widehat{\psi}_{nk}^2(Z_i)\} - b_k^2 = \sigma_\varepsilon^2 + \mathrm{Var}\{\mu(Z_i)\widehat{\psi}_{nk}(Z_i)\},$$

they are a $0(1)$ for every integer k. Let $K_n = o(n)$ be the number of functions used for the approximation of the series of the regression function μ by

$$\mu_n = \sum_{k=0}^{K_n} b_k \psi_k.$$

The function μ is estimated by the process

$$\widehat{\mu}_n(z) = \sum_{k=0}^{K_n} \widehat{b}_{nk}\widehat{\psi}_{nk}(z), \tag{3.22}$$

on every subset of $\mathcal{I}_Z$ in $\mathbb{R}^d$ where $\widehat{F}_{nZ}(z) < 1$. Like in Section 3.2 for the regression function m, the processes $n^{\frac{1}{2}}(\widehat{b}_{nk}\widehat{\psi}_{nk} - b_k\psi_k)$ converge weakly to centered Gaussian processes with a finite and non degenerated covariance functions, for every integer k.

The mean integrated squared error in $L^2(\mathcal{I}_Z, F_Z)$ of the regression function estimator is the sum

$$\mathrm{MISE}_n(\mu) = E\|\widehat{\mu}_n - \mu_n\|_2^2 + \|\mu - \mu_n\|_2^2$$

of the squared estimation error $E\|\widehat{\mu}_n - \mu_n\|_2^2$ and the squared error of approximation which is a $o(1)$. By the same proof as Lemma 3.3, as n tends to infinity

$$E\|\widehat{\mu}_n - \mu_n\|_2^2 = O(n^{-\frac{1}{2}}K_n),$$

it has the same rate as for a nonparametric regression function on a scalar regressor. The error is minimum as both terms have the same order therefore $K_n = o(n^{\frac{1}{2}})$ and we assume that the condition (3.7) is satisfied.

Lemma 3.6. *Let K_n tend to infinity as n tends to infinity with the rate $K_n^2 = o(n)$, then under the condition (3.7)*

$$\mathrm{MISE}_n(\mu) = O(n^{-1}K_n^2).$$

Proposition 3.8. *Under the condition (3.7), the process $n^{\frac{1}{2}}K_n^{-1}(\widehat{m}_n - m)$ converges weakly to a Gaussian process with finite expectation and variance functions.*

Proposition 3.9. *Under the condition (3.7), the estimation error $E\|\widehat{\mu}_n - \mu_n\|_2$ of the estimator $\widehat{\mu}_n$ defined by (3.22) has the order $O\Big(n^{-\frac{s}{2s+1}}\Big)$ for $K_n = O\Big(n^{\frac{1}{2(2s+1)}}\Big)$.*

The estimator of a regression function m belonging to the class W_{4s}, defined by (3.8), has the estimation error $O\Big(n^{-\frac{s}{2s+1}}\Big)$. Proposition 3.9 proves that the estimator (3.22) has a faster convergence rate than the kernel estimators for multidimensional regression variables.

3.7 Additive nonparametric regressions

Let Y be a real variable and let $X = (X_1, \ldots, X_d)^T$ be a d dimensional vector of regressors with linearly dependent components. The distribution function F_X on a set $\mathcal{I}_X$ has the marginal distribution functions $F_1, \ldots, F_d$ on $\mathcal{I}_{X_j}$, $j = 1, \ldots, d$ respectively. Let $(Y_i, X_{i1}, \ldots, X_{in})_{i=1,\ldots,n}$ be a sample of n independent and identically distributed observations of (Y, X) and let $(\varepsilon_i)_{i=1,\ldots,n}$ be a sample of errors with conditional mean $E(\varepsilon_i \mid X_i) = 0$ and variance σ_ε^2. The nonparametric regression model

$$Y_i = m(X_i) + \varepsilon_i,\ i = 1, \ldots, n,$$

is defined by a multivariate regression function $m(x) = E(Y \mid X = x)$, for $x = (x_1, \ldots, x_d)^T$ in a subset $\mathcal{I}_X$ of $\mathbb{R}^d$, and the expectation of Y is $EY = Em(X)$. In an additive nonparametric regression, the conditional expectation has the form

$$m(X_i) = m_0 + \sum_{j=1}^{d} m_j(X_{ij}) \tag{3.23}$$

with $m_0 = E\{m(X_i)\} = EY_i$ and with the real functions m_j defined on the real subsets $\mathcal{I}_{X_j}$ such that

$$E\{m_j(X_{ij})\} = 0.$$

For every j, let $F_{X^{(j)}}(x)$ be the distribution function of the sample $(X_i)_{i\le n}$ conditionally on $X_{ij} = x_j$, the univariate components of the additive model are identifiable as

$$\begin{aligned} m_j(x_j) &= E\{m(X_i) - m_0 \mid X_{ij} = x_j\} \\ &= \int_{\prod_{l\neq j} \mathcal{I}_{X_j}} m(x)\, dF_{X^{(j)}}(x) - m_0. \end{aligned}$$

Let $\widehat{m}_n$ be the nonparametric estimator of the multivariate regression function m defined in the previous section by (3.18) and (3.22), and let $\widehat{F}_n^{(j)}(x)$ be the empirical estimator of the distribution $F_{X^{(j)}}(x)$ conditionally on $X_{ij} = x_j$ for $i = 1, \ldots, n$. The constant m_0 has for estimator the empirical mean $\widehat{m}_{n0} = \bar{Y}_n$ of the sample $(Y_i)_i$ and the regression functions m_j have the consistent estimators

$$\widehat{m}_{nj}(x_j) = \int_{\prod_{l\neq j} \mathcal{I}_{X_j}} \widehat{m}_n(x)\, d\widehat{F}_n^{(j)}(x) - \widehat{m}_{n0}. \tag{3.24}$$

The asymptotic properties of the estimators (3.24) are deduced from the weak convergence of the empirical distribution function and from Lemma 3.6 and Proposition 3.8.

Proposition 3.10. *Under the condition (3.7), the processes $n^{\frac{1}{2}} K_n^{-1}(\widehat{m}_{nj} - m_j)$ converge weakly to a Gaussian process with finite expectation and variance functions.*

By the isomorphism (3.17), the observation matrix $\mathbb{X}_n$ is mapped to the $n \times d$ dimensional matrix $\mathbb{Z}_n = \mathbb{X}_n B$, an additive regression model (3.23) on the independent components of Z is written a

$$\mu(Z) = E(Y \mid Z) = E(Y) + \sum_{j=1}^{d} \mu_j(Z_j) \tag{3.25}$$

with the functions

$$\begin{aligned}\mu_j(z_j) &= E\{\mu(Z_i) \mid Z_{ij} = z_j\} - E(Y_i)\\ &= \int_{\prod_{l\neq j} \mathcal{I}_{Z_j}} \mu(z) \prod_{l\neq j} dF_{Z_l}(z_l) - m_0,\end{aligned}$$

and the independence of the components of Z_i entails the identifiability of the functions μ_j. They are estimated by projection on the orthonormal basis of functions $(\psi_{jk_j})_{k_j \geq 0}$ defined on $(L^2(\mathbb{R}, F_{Z_j})$, with the distribution function F_{Z_j} of the variable Z_j, for $j = 1, \ldots, d$. The orthonormal basis $(\psi_{jk_j})_{k_j \geq 0}$ is defined as the jth marginal of the basis (3.19) based on the Laguerre polynomials

$$\psi_{jk_j}(z_j) = L_k \circ H_{Z_j}(z_j), \tag{3.26}$$

$\psi_{jk_j}(z_j)$ is the value of $\psi_k(z) = L_k \circ H_Z(z)$ at z with values infinity except z_j for the jth component, it is also denoted $\psi_k(z_j)$.

For $j = 1, \ldots, d$, the function μ_j is approximated by the sum of K_{nj} functions of the basis

$$\mu_{nj}(z_j) = \sum_{k_j=1}^{K_{nj}} b_{jk_j} \psi_{jk_j}(z_j)$$

with the coefficients

$$b_{jk} = \int_{\mathcal{I}_{Z_j}} \mu_j(z_j) \psi_{jk}(z_j)\, dF_{Z_j}(z_j),\ k = 1, \ldots, K_{nj}.$$

The expectation $E\{Y_i \psi_{jk}(Z_{ij})\} = E\{\mu(Z_i)\psi_{jk}(Z_{ij})\}$ splits according to the additive regression model as

$$\begin{aligned}E\{\mu(Z_i)\psi_{jk}(Z_{ij})\} &= \sum_{l=1}^{d} E\{\mu_l(Z_{il})\psi_{jk}(Z_{ij})\}\\ &= \sum_{l=1}^{d} \sum_{k' \geq 0} b_{lk'} \int_{\mathcal{I}_{Z_j}} \int_{\mathcal{I}_{Z_l}} \psi_{lk'}(z_l)\psi_{jk}(z_j)\, dF_{jl}(z_j, z_l),\end{aligned}$$

where F_{jl} is the distribution function of (Z_j, Z_l). Therefore $E\{Y_i\psi_{jk}(Z_{ij})\}$ reduces to b_{jk}.

The function $\psi_k(z_j)$ depends on the marginal distribution function F_{Z_j} of the variable Z_j through the cumulative hazard function H_j of Z_j, it has the empirical estimator

$$\widehat{\psi}_{n,jk}(z_j) = L_k \circ \widehat{H}_{nj}(z_j) \tag{3.27}$$

where the estimator $\widehat{H}_{nj}(x)$ of the cumulative hazard function of the variables Z_{ij} is defined by (3.3). It satisfies

$$\begin{aligned} E\{\widehat{\psi}_{nk}(Z_i) \mid Z_{ij} = z_j)\} &- \psi_{jk}(z_j) \\ &= \{\widehat{H}_{nj}(z_j) - H_j(z_j)\}\{L'_k \circ H_j(z_j) + o_{as}(1)\}, \end{aligned} \tag{3.28}$$

the expansion (3.28) is the same as the expansion of $\widehat{\psi}_{n,jk}(z_j) - \psi_{jk}(z_j)$.

The uniform a.s. consistency of $\widehat{F}_{nj}$ on $\mathcal{I}_{Z_j}$ entails the uniform a.s. consistency of $\widehat{H}_{nj}$ on every sub-interval of $\mathcal{I}_{Z_j}$ where $\widehat{F}_{nj} < 1$, as n tends to infinity. Lemma 3.1 applies to the estimators $\widehat{\psi}_{n,jk}$.

Lemma 3.7. *For every integer $k \geq 1$, on every sub-interval of $\mathcal{I}_{Z_j}$ where $F_{Z_j}(z_j) < 1$, the estimator $\widehat{\psi}_{n,jk}$ is a.s. uniformly consistent and the process $n^{\frac{1}{2}}(\widehat{\psi}_{n,jk} - \psi_{jk})$ converges weakly to a centered Gaussian process with a bounded covariance function.*

For every integer $k \geq 1$, the coefficient b_{jk} is a.s. consistently estimated by the empirical mean

$$\widehat{b}_{n,jk} = (n-1)^{-1} \sum_{i=1}^{n-1} Y_{(i)} \widehat{\psi}_{ni,jk}(Z_{(n:i)j}) \tag{3.29}$$

where the larger observation $X_{(n:n)j}$ is omitted, $Y_{(i)}$ is the response variable to $Z_{(n:i)j}$ and $\widehat{\psi}_{ni,jk}$ is the estimator of $\psi_{jk} = \psi_k \circ H_j$ omitting the observation Z_{ij}. The variables $n^{\frac{1}{2}}(\widehat{b}_{n,jk} - b_{jk})$ converge weakly to centered Gaussian variables with variance

$$v_{jk}^2 = E\{Y_i^2 \widehat{\psi}_{ni,jk}^2(Z_{ij}\} - b_{jk}^2 = \sigma_\varepsilon^2 + \mathrm{Var}\{\mu(Z_{ij})\widehat{\psi}_{ni,jk}(Z_{ij})\} > 0.$$

The number K_{nj} of functions of the basis used in the approximation of the regression function μ_j by μ_{nj} tends to infinity with n and the squared error of approximation

$$\|\mu_j - \mu_{nj}\|_2^2 = \sum_{k > K_{nj}} b_{jk}^2$$

tends to zero as n tends to infinity. The function μ_j is a.s. consistently estimated by the sum of the estimators

$$\widehat{\mu}_{nj} = \sum_{k=1}^{K_{nj}} \widehat{b}_{n,jk} \widehat{\psi}_{n,jk}, \tag{3.30}$$

for $j = 1, \ldots, d$, and $\widehat{\mu}_n(z) = \sum_{j=1}^{d} \widehat{\mu}_{nj}(z_j)$.

The mean integrated squared errors of the regression functions μ_j in $L^2(\mathcal{I}_{Z_j}, F_{Z_j})$ have the same order as in Lemma 3.3. For the estimator of the regression functions μ, the mean integrated squared error in $L^2(\mathcal{I}_Z, F_Z)$ is expressed according to the sum of the squared errors for the functions μ_j

$$\begin{aligned}
\text{MISE}_n(\mu) &= E \left\| \sum_{j=1}^{d} (\widehat{\mu}_{nj} - \mu_{nj}) \right\|_2^2 + \left\| \sum_{j=1}^{d} (\mu_j - \mu_{nj}) \right\|_2^2 \\
&= E \sum_{j=1}^{d} \int_{\mathcal{I}_{Z_j}} (\widehat{\mu}_{nj} - \mu_{nj})^2 \, dF_{Z_j} + \sum_{j=1}^{d} \int_{\mathcal{I}_{Z_j}} (\mu_j - \mu_{nj})^2 \, dF_{Z_j},
\end{aligned}$$

by the independence of the components of the variable Z. It is the sum of the squared errors of estimation and approximation, they have the same orders as in Section 3.2 for a regression with a real variable X. The next lemma is similar to Lemma 3.3 and $\sum_{j=1}^{d} (\sum_{k=K_{nj}+1}^{\infty} b_{jk}^2)^{\frac{1}{2}} = o(1)$. The $\text{MISE}_n(\mu)$ is minimum as both errors have the same order, let

$$K_n^2 := \sum_{j=1}^{d} K_{nj}^2 = o(n),$$

for every $j = 1, \ldots, d$, we assume the condition

$$\sum_{j=1}^{d} \int_{\mathcal{I}_{Z_j}} (\mu_j - \mu_{nj})^2 \, dF_{Z_j} = O(n^{-\frac{1}{2}} K_n). \tag{3.31}$$

Lemma 3.8. *Let K_{nj} tend to infinity as n tends to infinity with the rate $K_{nj}^2 = o(n)$ for $j = 1, \ldots, d$, then*

$$E \left\| \sum_{j=1}^{d} (\widehat{\mu}_{nj} - \mu_{nj}) \right\|_2 = O\big((n^{-\frac{1}{2}} K_n^2)^{\frac{1}{2}}\big).$$

Under the condition (3.31), $\text{MISE}_n(\mu)$ *has the order* $(n^{-\frac{1}{2}} K_n^2)^{\frac{1}{2}}$.

Proposition 3.11. *In a nonparametric additive model with a regression function in C^s, $s \geq 2$, the convergence rate of the MISE error for an estimator defined by (3.29) and (3.30) under the conditions of Lemma 3.8 is a $O(n^{-\frac{s}{2s+1}})$ if $K_{nj} = O\Big(n^{\frac{1}{2(2s+1)}}\Big)$, for $j = 1, \ldots, d$.*

The proof is the same as for Proposition 3.3.

In the additive heteroscedastic model, the variance function $\sigma^2(z)$ is approximated on $\mathcal{I}_Z$ by a function $\sigma_n^2 = \sum_{k=0}^{L_n} c_k \psi_k$ where $L_n = o(n)$ and tends to infinity as n tends to infinity, with an approximation error $o(1)$. The coefficients c_k and the function σ^2 are still consistently estimated by (3.9) and (3.10) with the error

$$E\|\widehat{\sigma}_n^2 - \sigma^2\|_2 = O(n^{-\frac{1}{2}} L_n)$$

and the process $L_n^{-1} n^{\frac{1}{2}} (\widehat{\sigma}_n^2 - \sigma^2)$ converges weakly to a centered Gaussian process with a finite variance function.

3.8 Additive regression models with interactions

In a nonparametric regression model $Y_i = m(X_i) + \varepsilon_i$, $i = 1, \ldots, n$, for a n-sample of a real variable Y on a d dimensional regressor $X = (X_1, \ldots, X_d)^T$ belonging to a subset $\mathcal{I}_X$ of $\mathbb{R}^d$, we consider an ANOVA decomposition of the regression function m according to the interactions between the components of the regressor

$$\begin{aligned} m(x) = m_0 + \sum_{j=1}^{d} m_j(x_j) + \sum_{j \neq k=1}^{d} m_{jk}(x_j, x_k) + \cdots \\ + \sum_{j_1 \neq \cdots \neq j_d = 1}^{d} m_{j_1, \ldots, j_d}(x_{j_1}, \ldots, x_{j_d}), \end{aligned} \tag{3.32}$$

where $m_0 = EY_i$ is a constant, the regression functions m_j's are the main effects of the variables X_j, the functions m_{jk}'s are the two-way interactions of X_j and X_k, and the functions $m_{j_1,\ldots,j_d}$ are the interactions of order d for $X_{j_1}, \ldots, X_{j_d}$. They are the conditional expectations centered by the functions of the previous order

$$\begin{aligned} m_j(x_j) &= E\{m(X_i) \mid X_{ij} = x_j\} - E(Y_i) \\ m_{jk}(x_j, x_k) &= E\{m(X_i) \mid X_{ij} = x_j, X_{ik} = x_k\} - E\{m(X_i) \mid X_{ij} = x_j\} \\ &\quad - E\{m(X_i) \mid X_{ik} = x_k\} + E(Y_i). \end{aligned}$$

They are estimated recursively using the nonparametric estimator $\widehat{m}_n$ of the multivariate regression function m and the estimators $\widehat{m}_{nj}(x_j)$ of the functions m_j defined in Sections 3.6 and 3.7.

The isomorphism u of Section 3.7 maps the observation matrix $\mathbb{X}_n$ onto the matrix $\mathbb{Z}_n = \mathbb{X}_n B$ of a n-sample of p dimensional vectors of independent variables, by (3.17). An additive regression models with interactions is defined for the regression function $\mu(Z) = E(Y \mid Z)$, for z in a subset $\mathcal{I}_Z$ of $\mathbb{R}^d$, in the same form as (3.32)

$$\mu(z) = \mu_0 + \sum_{j=1}^{d} \mu_j(z_j) + \sum_{j \neq k=1}^{d} \mu_{jk}(z_j, z_k) + \cdots + \sum_{j_1 \neq \cdots \neq j_d = 1}^{d} \mu_{j_1,\ldots,j_d}(z_{j_1}, \ldots, z_{j_d}). \tag{3.33}$$

The identifiability of the interactions requires the orthogonality of the functional spaces describing the model (3.33), the function μ belongs to

$$\mathcal{M} = \{1\} \oplus \mathcal{M}_1 \oplus \cdots \oplus \mathcal{M}_d,$$

with functions μ_j in $\mathcal{M}_1$, μ_{jk} in $\mathcal{M}_2$, $\mu_{j_1,\ldots,j_d}(z_{j_1}, \ldots, z_{j_d})$ in $\mathcal{M}_d$.

In the model with first order interactions, we have

$$\mu_j(Z_j) = E(Y \mid Z_j) - \mu_0,$$
$$\mu_{jl}(Z_j, Z_l) = E(Y \mid Z_j, Z_l) - E(Y \mid Z_j) - E(Y \mid Z_l) + \mu_0,$$

and the interaction functions are centered variables.

In models (3.33), the parameter μ_0 is estimated by the empirical mean $\bar{Y}_n$ of the observations Y_i, $i = 1, \ldots, n$. Kernel estimators of the main effects and interaction functions are

$$\widehat{\mu}_{nh,j}(z_j) = \frac{\sum_{i=1}^n Y_i K_h(Z_{ij} - z_j)}{\sum_{i=1}^n K_h(Z_{ij} - z_j)} - \bar{Y}_n,$$
$$\widehat{\mu}_{nh,jl}(z_j, z_l) = \frac{\sum_{i=1}^n Y_i K_h(Z_{ij} - z_j) K_h(Z_{il} - z_l)}{\sum_{i=1}^n K_h(Z_{ij} - z_j) K_h(Z_{il} - z_l)} - \widehat{\mu}_{nh,j}(z_j) - \widehat{\mu}_{nh,l}(z_l) + \bar{Y}_n,$$

and so on, due to the independence of the components of the vector Z. The asymptotic properties of the multidimensional kernel estimators are well known, they are uniformly consistent and their convergence rates depends on the interaction order through the bandwidth h, the kernel estimators of

kth order interaction functions have the convergence rate $(nh^k)^{-\frac{1}{2}}$.

By projections on the orthonormal series $\psi_k = L_k \circ H_Z$ of Section 3.7 and under the identifiability constraints of orthogonality, the functions of models (3.33) are approximated by

$$\mu_{n,j}(z_j) = \sum_{k=0}^{K_{nj}} b_{jk}\psi_k(z_j) - \mu_0,$$
$$\mu_{n,jl}(Z_j, Z_l) = \sum_{k=1}^{K_{njl}} b_{jlk}\psi_k(z_j, z_l) - \mu_{n,j}(z_j) - \mu_{n,l}(z_l) + \mu_0,$$

etc., the approximation sizes K_{nj}, $K_{njl}, \ldots$, are $o(n)$ and they tend to infinity as n tends to infinity. The functions of the model are estimated by projections on the estimated functions $\widehat{\psi}_{n,jk}$ of the basis given by (3.27) for multivariate regressors, the estimators are

$$\widehat{\mu}_{n,j}(z_j) = \sum_{k=0}^{K_{nj}} \widehat{b}_{n,jk}\widehat{\psi}_{n,jk}(z_j) - \bar{Y}_n,$$
$$\widehat{\mu}_{n,jl}(z_j, z_l) = \sum_{k=0}^{K_{njl}} \widehat{b}_{n,jlk}\widehat{\psi}_{n,jlk}(z_j, z_l) - \widehat{\mu}_{n,j}(z_j) - \widehat{\mu}_{n,l}(z_l) + \bar{Y}_n,$$

with the estimated functions

$$\widehat{\psi}_{n,jlk}(z_j, z_l) = L_k \circ \widehat{H}_{n,jl}(z_j, z_l),$$
$$\widehat{H}_{n,jl}(z_j, z_l) = \widehat{H}_{nZ_j}(z_j) + \widehat{H}_{nZ_l}(z_l).$$

The estimators of the coefficients are similar to (3.29)

$$\widehat{b}_{n,jk} = (n-1)^{-1} \sum_{i=1}^{n-1} Y_{(i)}\widehat{\psi}_{ni,jk}(Z_{(n:i)j}) - \bar{Y}_n,$$

and under the constraints for the higher order terms

$$\widehat{b}_{n,jlk} = (n-1)^{-1} \sum_{i=1}^{n-1} Y_{(i)}\widehat{\psi}_{ni}(Z_{(n:i)j}, Z_{(n:i)l}) - \widehat{b}_{n,jk} - \widehat{b}_{n,lk} + \bar{Y}_n$$

where the estimators $\widehat{\psi}_{ni,jk}(Z_{ij})$ and $\widehat{\psi}_{ni}$ are defined omitting the ith observation, $Y_{(i)}$ is the response variable to $Z_{(n:i)}$. The estimators of the basis satisfy the properties of Lemma 3.7, those of the coefficients are a.s. consistent and have the convergence rate $n^{-\frac{1}{2}}$, with asymptotically Gaussian limits.

The mean integrated squared errors of the regression function in $L^2(\mathcal{I}_Z, F_Z)$ is expressed according to the sum of the errors of the estimators of the main and interaction functions of the model, they have the same order as in Lemma 3.3. Like in the additive regression model, in the first order interaction model, the orthogonality of the functions and the components of Z entail

$$\begin{aligned}
\text{MISE}_n(\mu) &= E\|\widehat{\mu}_n - \mu_n\|_2^2 + \|\mu_n - \mu\|_2^2 \\
&= \sum_{j=1}^d \Big\{ E\int_{\mathcal{I}_{Z_j}} (\widehat{\mu}_{n,j} - \mu_{n,j})^2 \, dF_{Z_j} + \int_{\mathcal{I}_{Z_j}} (\mu_j - \mu_{nj})^2 \, dF_{Z_j} \Big\} \\
&\quad + \sum_{j\neq l=1}^d \Big\{ E\int_{\mathcal{I}_{Z_j,Z_l}} (\widehat{\mu}_{n,jl} - \mu_{n,jl})^2 \, dF_{Z_j,Z_l} \\
&\qquad + \int_{\mathcal{I}_{Z_j,Z_l}} (\mu_{jl} - \mu_{n,jl})^2 \, dF_{Z_j,Z_l} \Big\}.
\end{aligned}$$

For all integers j and l, the mean integrated squared estimation errors have the orders $\text{MISE}_n(\mu_j) = O(n^{-1}K_{nj}^2)$ and $\text{MISE}_n(\mu_{jl}) = O(\big(n^{-1}K_{njl}^2)$, the approximation errors $\|\mu_{n,j} - \mu_j\|_2$ and $\|\mu_{n,jl} - \mu_{jl}\|_2$ converge to zero, as n tends to infinity. The error $\text{MISE}_n(\mu)$ is minimum as all errors have the same order therefore $\sum_{j=1}^d K_{n,j}^2$ and $\sum_{j\neq l=1}^d K_{n,jl}^2$ are $o(n)$ and we assume that the following conditions which extend (3.31) are fulfilled, for all j, l

$$\begin{aligned}
K_{nj}^2 &= o(n), \quad \int_{\mathcal{I}_{Z_j}} (\mu_j - \mu_{nj})^2 \, dF_{Z_j} = O(n^{-\frac{1}{2}} K_{nj}), \\
K_{njl}^2 &= o(n), \quad \int_{\mathcal{I}_{Z_j,Z_l}} (\mu_{jl} - \mu_{n,jl})^2 \, dF_{Z_j,Z_l} = O(n^{-\frac{1}{2}} K_{njl}). \quad (3.34)
\end{aligned}$$

Lemma 3.9. *Let K_{nj} and K_{njl} tend to infinity as n tends to infinity for $j, l = 1, \ldots, d$, then under the conditions (3.34)*

$$\begin{aligned}
E\|\sum_{j=1}^d (\widehat{\mu}_{n,j} - \mu_j)\|_2 &= O\Big(n^{-\frac{1}{2}} \Big(\sum_{j=1}^d K_{nj}^2\Big)^{\frac{1}{2}}\Big), \\
E\|\sum_{j\neq l=1}^d (\widehat{\mu}_{n,jl} - \mu_{jl})\|_2 &= O\Big(n^{-\frac{1}{2}} \Big(\sum_{j\neq l=1}^d K_{n,jl}^2\Big)^{\frac{1}{2}}\Big).
\end{aligned}$$

Proposition 3.3 extends to models (3.32).

Proposition 3.12. *In model (3.32) with regression functions in W_{4s}, $s \geq 1$, the optimal convergence rate of the MISE error for the estimators*

$\widehat{\mu}_n$ is a $O\Big(n^{-\frac{s}{2s+1}}\Big)$ as the approximations sizes $K_{nj}, K_{n,j_1j_2}, \ldots, K_{n,j_1\cdots j_d}$ are $O\Big(n^{\frac{1}{2(2s+1)}}\Big)$, for $j_1, \ldots, j_d = 1, \ldots, d$.

The weak convergence of Proposition 3.4 extends to the estimators of the functions μ_j and the interaction functions, according to the approximations sizes.

We consider now a model where the multivariate regression functions $m_{j_1,\ldots,j_l}(x_{j_1}, \ldots, x_{j_l})$ of the model (3.32) factorize as products $\prod_{k=1}^{l} m_{j_k}(x_{j_k})$, for $l = 1, \ldots, d$, with univariate regressions functions m_{j_k} of $L^2(F_{X_k})$ having respectively a convergent expansion on an orthonormal basis of functions. The product of these expansions leads to replace the sums

$$\sum_{k=0}^{K_{j_1,\ldots,j_l}} b_{k,j_1,\ldots,j_l}\psi_k = \sum_{k_1=0}^{K_{j_1}} \cdots \sum_{k_l=0}^{K_{j_l}} b_{k_1,\ldots,k_l,j_1,\ldots,j_l}\psi_{k_1}\cdots\psi_{k_l}$$

by an expansion where the products $b_{k_1,j_1}\cdots b_{k_l,j_l}$ replace the coefficients $b_{k_1,\ldots,k_l,j_1,\ldots,j_l}$ in the expression of $\mu_{j_1,\ldots,j_l}$. The coefficients b_{k_ν,j_ν} are estimated by

$$\widehat{b}_{n,k_\nu,j_\nu} = (n-1)^{-1}\sum_{i=1}^{n-1} Y_{(i)}\widehat{\psi}_{ni,k_\nu,j_\nu}(z_{j_\nu})$$

and the properties of the estimated regression functions are the same are in the additive model.

3.9 Semi-parametric models

The linear model generalizes to parametric regressions $Y_i = m_\theta(X_i) + \varepsilon_i$ with a twice continuously differentiable function $\theta \mapsto m_\theta(x)$, uniformly on x, and under integrability conditions. The variable Y belongs to $L^2(\mathbb{R})$ and the error is centered and has a variance σ^2, its density f may be known or unknown. Let Θ be an open bounded subset of $\mathbb{R}^d$, let X be a real random regressor and let g be a function from $\Theta \times \mathbb{R}$ into a subset of $\mathbb{R}$, twice continuously differentiable on Θ uniformly in x in $\mathcal{I}_X$.

A nonparametric regression of Y_i on a parametric transformation of the regressors X_i is expressed as

$$Y_i = m \circ g_\theta(X_i) + \varepsilon_i,\ i = 1, \ldots, n \tag{3.35}$$

with an unknown function m and a parametric function g_θ such as a change of location and scale of the regressors. We assume that the function $g_\theta(x)$ belongs to $C^2(\Theta)$ uniformly on x, m belongs to $C^2(\mathbb{R})$ and Y is $L^2(\mathbb{R})$. Under the probability P_0 of the sample $(X_i, Y_i)_{i=1,\ldots,n}$, the parameter has the value θ_0 belonging to Θ, it is estimated with the functions m in a two step procedure. At an arbitrary θ, let

$$Z(\theta) = g_\theta(X)$$

with distribution function G_θ and the empirical estimator $\widehat{G}_{n\theta}$, and let $(\varphi_{k\theta})_{k\geq 0}$ be an orthonormal basis of functions of $L^2(G_\theta)$, it is estimated by $(\widehat{\varphi}_{n,k\theta})_{k\geq 0}$ defined like in Section 3.2 with $\widehat{G}_{n\theta}$. An expansion of the function m on the basis $(\varphi_{k\theta})_{k\geq 0}$ is approximated by $m_K = \sum_{k=0}^K b_{k\theta}\varphi_{k\theta}$, the coefficients are estimated by

$$\widehat{b}_{n,k\theta} = (n-1)^{-1} \sum_{i=1}^{n-1} Y_{(i)}\, \widehat{\varphi}_{ni,k\theta}(Z_{(n:i)}(\theta)),\ k = 0, \ldots, K$$

and $m \circ g_\theta$ is estimated by

$$\widehat{m}_{n,K\theta}(z) = \sum_{k=0}^{K} \widehat{b}_{n,k\theta} \widehat{\varphi}_{n,k\theta}(z),$$

with the notations of Section 3.2. The empirical error of estimation is

$$\sigma^2_{nK}(\theta) = n^{-1} \sum_{i=1}^{n} \{Y_i - \widehat{m}_{n,K\theta} \circ g_\theta(X_i)\}^2,$$

it is minimum at

$$\widetilde{\theta}_{nK} = \arg\min_{\theta\in\Theta} \sigma^2_{nK}(\theta).$$

Under P_0, by the a.s. consistency of the estimated basis and the differentiability condition, $\sigma^2_{nK}(\theta)$ converges a.s. uniformly on Θ to the function

$$\sigma^2_K(\theta) = \sigma^2 + E\{m \circ g_{\theta_0}(X_i) - m \circ g_\theta(X_i)\}^2$$

which is minimum at θ_0. By the same proof as Lemma 3.4, the estimator $\widetilde{\theta}_{nK}$ is therefore a.s. consistent.

If the density f of the error is known, the likelihood of the sample is

$$L_n(\theta, m) = \prod_{i=1}^{n} f(Y_i - m \circ g_\theta(X_i)),$$

it is maximum at

$$\widehat{\theta}_{nK} = \arg\min_{\theta\in\Theta} L_n(\theta, \widehat{m}_{n,K\theta}).$$

Let $l_n(\theta, m) = \log L_n(\theta, m) = \sum_{i=1}^n \log f(Y_i - m \circ g_\theta(X_i))$, the maximum likelihood estimator $\widehat{\theta}_n$ of the parameter is solution of the score equation

$$\dot{l}_{n,\theta}(\theta, \widehat{m}_{n,K\theta}) = -\sum_{i=1}^n \frac{f'}{f}(Y_i - m \circ g_\theta(X_i)) m' \circ g_\theta(X_i) \dot{g}_\theta(X_i) = 0$$

where the derivative of g is with respect to the parameter θ. The properties of the maximum likelihood entail the consistency of $\widehat{\theta}_n$ under P_0.

Proposition 3.6 generalizes to the minimum variance estimator $\widetilde{\theta}_n$ and to the maximum likelihood estimator $\widehat{\theta}_n$ which have the convergence rate $n^{-\frac{1}{2}} K_n$, with $K_n = o((n)$ and K_n tending to infinity as n tends to infinity. The variables $n^{\frac{1}{2}} K_n^{-1}(\widetilde{\theta}_n - \theta_0)$ and $n^{\frac{1}{2}} K_n^{-1}(\widehat{\theta}_n - \theta_0)$ converge weakly to centered Gaussian variables with finite variances. Finally, the regression function m is estimated by

$$\widehat{m}_n = \widehat{m}_{n,K\widehat{\theta}_n},$$

with the maximum likelihood estimator $\widehat{\theta}_n$ or by $\widetilde{m}_n = \widehat{m}_{n,K\widetilde{\theta}_n}$ with the minimum variance estimator $\widetilde{\theta}_n$. Let $m_0 = m \circ g_{\theta_0}$.

Proposition 3.13. *The processes $n^{\frac{1}{2}} K_n^{-1}(\widehat{m}_n - m_0)$ and $n^{\frac{1}{2}} K_n^{-1}(\widetilde{m}_n - m_0)$ converge weakly to Gaussian processes with finite variances, uniformly on every bounded subsets of $\mathcal{I}_X$.*

The proof is similar to the proof of Proposition 3.7, with a twice continuously differentiable function $m \circ g_\theta$.

The linear model has also a generalization as a linear regression on unknown functions of the regressors $Y_i = m_\theta(X_i) + \varepsilon_i$, with

$$m_\theta(X_i) = \sum_{j=1}^d m_j(X_{ij})\theta_j,\ i = 1, \ldots, n, \tag{3.36}$$

with d dimensional vectors of unknown coefficients $\theta = (\theta_1, \ldots, \theta_d)^T$ and functions $m = (m_1, \ldots, m_d)^T$. The observations are samples of linearly independent regressors $X_i = (X_{i1}, \ldots, X_{id})^T$ and Y_i, the error variables ε_i are independent such that $E(\varepsilon_i \mid X_i) = 0$ and $\mathrm{Var}(\varepsilon_i \mid X_i) = \sigma^2$ is strictly positive. Model (3.36) extends the additive nonparametric regression of Section 3.7 and it has been estimated by kernel smoothing and cubic splines in iterative algorithms, here we define an orthonormal series estimator of m.

Let $\mathbb{X}_n = (X_{ij})_{i \leq n, j \leq d}$ be the $n \times d$ dimensional matrix of the regressors, it is mapped by (3.17) to a $n \times d$ dimensional matrix

$$\mathbb{Z}_n = \{\mathbb{X}_n - E\mathbb{X}_n\}B$$

such that $\mathbb{Z}_n^T \mathbb{Z}_n$ is a diagonal matrix Λ with strictly positive eigenvalues and the regression model $m(X) = E(Y \mid X)$ is mapped onto a subset $\mathcal{I}_Z$ of $\mathbb{R}^d$ as $m(X) = \mu(XB) = \mu(Z)$.

The n dimensional vector $\mathbb{Y}_n = (Y_i)_{i \leq n}$ and the $n \times d$ dimensional matrices $\mathbb{Z}_n = (Z_{ij})_{i=1,\dots,n, j=1,\dots,d}$ and $M(\mathbb{Z}_n) = \mu_j(Z_{ij})_{i=1,\dots,n, j=1,\dots,d}$ reparametrize the model (3.36) as

$$E(\mathbb{Y}_n \mid \mathbb{Z}_n) = M(\mathbb{Z}_n)\theta.$$

Let $(\varphi_{jk})_{k \geq 0}$ be an orthogonal basis of functions of $L^2(\mathbb{R}, F_{Z_j})$ and let $\mu_j(z_j) = \sum_{k \geq 0} a_{jk}\varphi_{jk}(z_j)$, the orthogonality of the components of Z_i and of the functions of the basis implies

$$E\{Y_i \varphi_{jk}(Z_{ij})\} = \theta_j E\{\mu_j(Z_{ij})\varphi_{jk}(Z_{ij})\} = a_{jk}\theta_j.$$

The approximation of $\mu_j(z_j)$ by

$$\mu_{nj}(z_j) = \sum_{k=0}^{K_{nj}} a_{jk}\varphi_{jk}(z_j),$$

where $K_{nj} = o(n)$ tends to infinity as n tends to infinity, provides a consistent estimator of $a_{jk}\theta_j$

$$\widehat{a_{jk}\theta_j}_n = (n-1)^{-1} \sum_{i=1}^{n-1} Y_{(i)}\, \widehat{\varphi}_{n,jk}(Z_{(n:i)j}), \tag{3.37}$$

with an estimation of the basis like in (3.2). Furthermore, the best linear estimator of the parameter θ minimizes the mean squared error

$$\widehat{\sigma}_n^2 = n^{-1} \sum_{i=1}^{n} \{Y_i - m^T(X_i)\theta\}^2,$$

it has the form

$$\widehat{\theta}_n = \{\widehat{M}_n^T(\mathbb{Z}_n)\widehat{M}_n(\mathbb{Z}_n)\}^{-1}\widehat{M}_n^T(\mathbb{Z}_n)\mathbb{Y}_n \tag{3.38}$$

with nonparametric estimators of the functions μ_j, or using a generalized inverse if the components of the estimator $\widehat{M}_n^T(\mathbb{Z}_n)$ are not linearly independent.

A consistent iterative algorithm starting from the estimators $\widehat{M}_n^{(0)}$ of the additive nonparametric regression (3.16) determines an initial estimator $\widehat{\theta}_n^{(0)}$ of the parameter θ by (3.38). For the kth step of the algorithm with initial estimator $\widehat{\theta}_n^{(k-1)}$, the estimators $\widehat{a}_{n,jk}^{(k-1)}$ and $\widehat{M}_n^{(k-1)}$ are defined from (3.37), then (3.38) yields $\widehat{\theta}_n^{(k)}$ and $\widehat{M}_n^{(k)}$. As the number of iterations increases and n tends to infinity, $\widehat{\theta}_n^{(k)}$, and respectively $\widehat{M}_n^{(k)}$, converge to θ_0, and respectively M_0.

3.10 Nonparametric varying-coefficient linear regression

In a linear model with random coefficients, let $(Y_i, X_i, U_i)_{i=1,\dots,n}$ be a sample of independent and identically distributed random variables such that

$$Y_i = m(U_i)X_i + \varepsilon_i, \tag{3.39}$$

where $(\varepsilon_i)_i$ is a sequence of independent and identically distributed variables with conditional densities $f_{\varepsilon|U}$ given U, its conditional expectation is zero and $E(\varepsilon_i^2 \mid X_i, U_i) = \sigma^2(U_i)$. If the variables U_i are unobserved in (3.39), the expectation of Y_i conditionally on X_i is

$$E(Y_i \mid X_i) = E\{m(U_i) \mid X_i\}\, X_i$$

and (3.39) is confounded with a nonparametric regression of Y_i on X_i. With independent variables U_i and X_i, the model reduces to a linear regression of Y_i and X_i.

The expectation and the variance of Y_i conditionally on (U_i, X_i) are

$$\begin{aligned} E\{Y_i \mid U_i, X_i\} &= m(U_i)\{X_i - E(X_i \mid U_i)\} + r(U_i), \qquad (3.40)\\ r(U_i) &= m(U_i)\, E(X_i \mid U_i),\\ \operatorname{Var}(Y_i \mid U_i, X_i) &= \sigma^2(U_i), \end{aligned}$$

where $m(U_i)$ and $\widetilde{X}_i = X_i - E(X_i \mid U_i)$ are independent, then

$$r(U_i) = E\{Y_i \mid U_i\}.$$

For a real variable U, the real functions $r(u)$ and $h(u) = E\{X \mid U = u\}$ are estimated by projections on an orthonormal basis of functions $(\psi_k)_{k\geq 0}$ of $L^2(I_U, G)$, like in Section 3.2 then an estimator of the functions $m(u)$ is deduced as their ratio

$$\widehat{m}_n(u) = \frac{\widehat{h}_n(u)}{\widehat{r}_n(u)} 1_{\{\widehat{r}_n(u)>0\}}.$$

The estimators are consistent and their convergence rate is $n^{-\frac{1}{2}}K_n$ where K_n is the number of functions of the basis in the approximation of the functions r and h. Lemma 3.3 provides the estimation errors of $\widehat{h}_n$ and $\widehat{r}_n$ and this is also the estimation error of $\widehat{m}_n$.

Under the conditions (3.7), Proposition 3.4 is still valid for the estimator $\widehat{m}_n$ which converges weakly to a Gaussian process.

Model (3.39) extends to models with several regression functions

$$\begin{aligned} Y_i &= \sum_{j=1}^{p} m_j(U_{ij})X_{ij} + \varepsilon_i\\ &= M^T(U_i)X_i + \varepsilon_i \end{aligned}$$

with the vectors $X_i = (X_{i1}, \ldots, X_{ip})^T$, $U_i = (U_{i1}, \ldots, U_{ip})^T$ and $M(U_i) = (m_1(U_{i1}, \ldots, m_p(U_{ip}))^T$. The observations are the sample $(Y_i, X_i, U_i)_{i=1,\ldots,n}$. We assume that the components of the variables U_i are linearly independent so they are mapped to vectors $\widetilde{U}_i$ with independent components and $m_j(U_{ij})$ is mapped to a function $\mu_j(\widetilde{U}_{ij})$ of $L^2(F_{\widetilde{U}_i})$, for all i and j. By (3.40), the conditional expectation of Y_i is expressed linearly according to the conditional expectations $h_j(u_j) = E(X_{ij} \mid \widetilde{U}_{ij} = u_j)$

$$r(u) = E\{Y_i \mid \widetilde{U}_i = u\} = \sum_{j=1}^{p} \mu_j(u_j) h_j(u_j) := h^T(u)\mu(u).$$

The functions r and h_j are estimated by projections on an orthonormal basis of functions of $L^2(F_{\widetilde{U}})$ and the vector of functions $\mu = (\mu_1, \ldots, \mu_p)^T$ is estimated by

$$\widehat{\mu}_n(u) = \{\widehat{h}_n \widehat{h}_n^T\}^{-1} \widehat{r}_n \widehat{h}_n^T$$

where the matrix $\widehat{h}_n \widehat{h}_n^T$ has the dimensions $p \times p$. The estimators $\widehat{r}_n$, $\widehat{h}_n$ and therefore $\widehat{\mu}_n$ are a.s. consistent and their convergence rate is $n^{-\frac{1}{2}} K_n$ where K_n is the number of functions of the basis in the approximation of the functions r and h. The estimation error estimation error of $\widehat{\mu}_n$ has still the same order as the errors of $\widehat{h}_n$ and $\widehat{r}_n$. Under the conditions (3.7), Proposition 3.4 provides the weak convergence of the estimator $\widehat{\mu}_n$ to a Gaussian process.

3.11 Comparison of regression curves

The comparison of two regression curves on a bounded interval $\mathcal{I}_X$ of their support and for independent variables sets (X_1, Y_1) and (X_2, Y_2) relies on tests of the hypothesis

$$H_0 : m_1(x) = E(Y_1 | X_1 = x) = E(Y_2 | X_2 = x) = m_2(x)$$

for every x in $\mathcal{I}_X$. The general alternative is the existence of a sub-interval $\mathcal{J}_X$ of $\mathcal{I}_X$ where $E(Y_1 | X_1 = x)$ and $E(Y_2 | X_2 = x)$ are distinct for every x of the sub-interval. Let $(X_{ij}, Y_{ij})_{j=1,\ldots,n_i, i=1,2}$ be a sample of independent observations of the variables and let $n = n_1 + n_2$ be the total sample size such that $n^{-1} n_1$ converge to p in $]0, 1[$ as n tends to infinity, so n_1 and n_2 are $O(n)$.

For each sub-sample, let $\widehat{m}_{in}$ be the estimators defined by (3.6) of the regression functions m_i and let

$$B_{in} = K_{n_i}^{-1} n_i^{\frac{1}{2}} (\widehat{m}_{in} - m_i), i = 1, 2$$

be the normalized processes. Under conditions (3.7), the processes B_{in} converge weakly to processes B_i, $i = 1, 2$, determined in Proposition 3.4. Under the conditions on n_1 and n_2, one can choose K_{n_1} and K_{n_2} of the same order and such that $K_{n_i} K_n^{-1}$ converges to a strictly positive finite limit ρ_i, for $i = 1, 2$. The approximation errors are such that $\mu_{n_i} = K_{n_i}^{-1} n_i^{\frac{1}{2}} (m_{in} - m)$, $i = 1, 2$, converge to limits μ_i, and the processes $K_{n_i}^{-1} n_i^{\frac{1}{2}} (\widehat{m}_{in} - m_{in})$ converge weakly to centered Gaussian processes B_i with variances V_i, for $i = 1, 2$.

A statistic for a Kolmogorov–Smirnov test of H_0 relies on the difference of the estimators of the regression functions

$$T_n = \sup_{x \in \mathcal{I}_X} \left| \frac{(n_1 n_2)^{\frac{1}{2}} K_n}{n^{\frac{1}{2}} K_{n_1} K_{n_2}} \{\widehat{m}_{1n}(x) - \widehat{m}_{2n}(x)\} \right|. \tag{3.41}$$

Proposition 3.14. *Under Conditions, the statistic T_n converges weakly under H_0 to the supremum of a Gaussian process*

$$T_0 = \sup_{\mathcal{I}_X} |(1-p)^{\frac{1}{2}} \rho_2^{-1} B_1 - p^{\frac{1}{2}} \rho_1^{-1} B_2|.$$

Under a fixed alternative, it tends to infinity.

Proof. Under H_0, $m_1 = m_2$ and the statistic is expressed as

$$T_n = \sup_{x \in \mathcal{I}_X} \left| \frac{n_2^{\frac{1}{2}} K_n}{n^{\frac{1}{2}} K_{n_2}} B_{1n}(x) - \frac{n_1^{\frac{1}{2}} K_n}{n^{\frac{1}{2}} K_{n_1}} B_{2n}(x) \right|.$$

Under H_0 and the conditions, B_{1n} and B_{2n} converge weakly to a Gaussian processes and T_n converges weakly to T_0, this limit is bounded in probability on a bounded interval $\mathcal{I}_X$. Under an alternative K of distinct functions m_1 and m_2, the statistic is written as

$$\begin{aligned} T_n = \sup_{x \in \mathcal{I}_X} \Bigg| & \frac{n_2^{\frac{1}{2}} K_n}{n^{\frac{1}{2}} K_{n_2}} B_{1n}(x) - \frac{n_1^{\frac{1}{2}} K_n}{n^{\frac{1}{2}} K_{n_1}} B_{2n}(x) \\ & + \frac{(n_1 n_2)^{\frac{1}{2}} K_n}{n^{\frac{1}{2}} K_{n_1} K_{n_2}} \{m_1(x) - m_2(x)\} \Bigg| \end{aligned}$$

and it tends to infinity due to the normalization of the difference $m_1 - m_2$ of the regression functions. □

Let $(A_n)_{n \geq 1}$ be a sequence of local alternatives defined by regression functions

$$m_{jn} = m + n_j^{-\frac{1}{2}} K_{n_j} r_{jn_j} \tag{3.42}$$

such that $(r_{jn_j})_{n_j \geq 1}$ converges uniformly on I_X to a function r_j as n tends to infinity, for $j = 1, 2$, and the function $(1-p)^{\frac{1}{2}}\rho_2 r_1)$ is different from $p^{\frac{1}{2}}\rho_1 r_2(x)$.

Proposition 3.15. *The statistic T_n converges weakly under the local alternatives A_n to*

$$T = \sup_{x \in \mathcal{I}_X} |(1-p)^{\frac{1}{2}}\rho_2(B_1 + r_1)(x) - p^{\frac{1}{2}}\rho_1(B_2 + r_2)(x)|.$$

Proof. Under A_n the difference of the regression functions is

$$m_{1n} - m_{2n} = n_1^{-\frac{1}{2}} K_{n_1} r_{1n_1} - n_2^{-\frac{1}{2}} K_{n_2} r_{2n_2}$$

and, in the expression of T_n

$$\frac{(n_1 n_2)^{\frac{1}{2}} K_n}{n^{\frac{1}{2}} K_{n_1} K_{n_2}}\{m_{1n}(x) - m_{2n}(x)\} = \frac{n_2^{\frac{1}{2}} K_n}{n^{\frac{1}{2}} K_{n_2}} r_{1n_1} - \frac{n_1^{\frac{1}{2}} K_n}{n^{\frac{1}{2}} K_{n_2}} r_{2n_2}$$

converges to $(1-p)^{\frac{1}{2}}\rho_2^{-1} r_1 - p^{\frac{1}{2}}\rho_1^{-1} r_2$. □

The variable T is the supremum of a process shifted of the function $(1-p)^{\frac{1}{2}}\rho_2 r_1(x) - p^{\frac{1}{2}}\rho_1 r_2(x)$ as compared to T_0, the test against the local alternatives is therefore consistent.

A Cramer-von Mises statistic is an integrated squared difference between the estimators of m_1 and m_2

$$T_{2n} = \frac{n_1 n_2 K_n^2}{n K_{n_1}^2 K_{n_2}^2} \int_{\mathcal{I}_X} \{\widehat{m}_{1n}(x) - \widehat{m}_{2n}(x)\}^2 w_n(x)\, d\widehat{F}_{X,n}(x),$$

where the sequence of weighting functions w_n converge uniformly to a function w integrable with respect to F_X, one can choose it as the inverse of the estimator of the variances of the estimator $\widehat{m}_n$ of the regression function under H_0.

Proposition 3.16. *Under Conditions, the statistic T_{2n} converges weakly under H_0 to the variable*

$$T_{02} = \int_{\mathcal{I}_X} \{(1-p)^{\frac{1}{2}}\rho_2^{-1} B_1(x) - p^{\frac{1}{2}}\rho_1^{-1} B_2(x)\}^2 w(x)\, dF_X(x).$$

Under a fixed alternative, it tends to infinity. Under the local alternatives (3.42), T_{2n} converges weakly to

$$\int_{\mathcal{I}_X} \{(1-p)^{\frac{1}{2}}\rho_2^{-1}(B_1 + r_1) - p^{\frac{1}{2}}\rho_1^{-1}(B_2 + r_2)\}^2 w(x)\, dF_X.$$

Tests of sub-models in the multivariate models of Sections 3.6 and 3.8 are performed with the statistics T_n or T_{2n} depending on the terms omitted in the regression under the hypothesis, like in the parametric analysis of variance.

3.12 Goodness-of-fit tests in regression

Let $\mathcal{M}_\Theta = \{m = m_\theta \in C^2(\Theta), \theta \in \Theta\}$ be a parametric regression model defined on a bounded interval $\mathcal{I}_X$, for an open and bounded subset of $\mathbb{R}^d$, $d \geq 1$. A goodness-of-fit test for the family of models $\mathcal{M}_\Theta$ is performed with a statistic which compares a nonparametric estimator of a regression function to its estimator in $\mathcal{M}_\Theta$. Let $(X_j, Y_j)_{j=1,\dots,n}$ be a sample of independent observations such that $E(Y_j \mid X_j) = m(X_j)$, we consider a partition of the sample as two independent sub-samples of sizes n_1 and n_2 of order $O(n)$ and such that $n^{-1}n_1$ converges to p in $]0,1[$ as n tends to infinity. We assume that the conditional expectations in each samples are

$$E(Y_{1j} \mid X_{1j}) = m(X_{1j}),$$
$$E(Y_{2j} \mid X_{2j}) = m_\theta(X_{2j}),$$

in $\mathcal{I}_X$, with m_θ in $\mathcal{M}_\Theta$. A test of comparison of the regression curves in $\mathcal{I}_X$ provides a goodness-of-fit test for the family of models $\mathcal{M}_\Theta$.

In $\mathcal{M}_\Theta$, the least squares estimator and the maximum likelihood estimators $\widehat{\theta}_{n_2}$ of the parameter θ obtained from the second sub-sample are consistent with the convergence rate $n_2^{-\frac{1}{2}}$, this is also the convergence rate of $m_{\widehat{\theta}_{n_2}}$ and the process $n_2^{\frac{1}{2}}(m_{\widehat{\theta}_{n_2}} - m)$ converges weakly under H_0 to a centered Gaussian process with variance function $v(x)$. Let $\widehat{m}_{n_1}$ be the nonparametric estimator of the first sub-sample based on a projection on K_{n_1} functions of an estimated basis and let

$$T_n = n_1^{\frac{1}{2}} K_{n_1}^{-1} \sup_{x \in \mathcal{I}_X} |\widehat{m}_{n_1}(x) - m_{\widehat{\theta}_{n_2}}(x)|. \tag{3.43}$$

Proposition 3.17. *Under Conditions, the statistic T_n converges weakly under H_0 to the supremum of the Gaussian process B_1. Under the alternative, it tends to infinity.*

Proof. Under H_0, the convergence rate of the nonparametric estimator implies $n_1^{\frac{1}{2}} K_{n_1}^{-1}\{\widehat{m}_{n_1}(x) - m_{\widehat{\theta}_{n_2}}(x)\} = n_1^{\frac{1}{2}} K_{n_1}^{-1}\{\widehat{m}_{n_1}(x) - m(x)\} + o_p(1)$ and the asymptotic behavior of the statistic T_n is the same as for a test of a known function in $\mathcal{M}_\Theta$. The limit of T_n follows, where B_1 is defined like in Proposition 3.14. Under an alternative of two distinct regression functions m_1 and m_2, we have

$$\begin{aligned} n_1^{\frac{1}{2}} K_{n_1}^{-1}\{\widehat{m}_{n_1}(x) - m_{\widehat{\theta}_{n_2}}(x)\} &= n_1^{\frac{1}{2}} K_{n_1}^{-1}\{\widehat{m}_{n_1}(x) - m_1(x)\} \\ &\quad - n_1^{\frac{1}{2}} K_{n_1}^{-1}\{m_{\widehat{\theta}_{n_2}}(x) - m_2(x)\} \\ &\quad + n_1^{\frac{1}{2}} K_{n_1}^{-1}\{m_1(x) - m_2(x)\}, \end{aligned}$$

the first term converges weakly to B_1, the second term converges to zero and the last term diverges, as n tends to infinity. □

Let $(A_n)_{n\geq 1}$ be a sequence of local alternatives defined by regression functions $m_{1n} = m + n_1^{-\frac{1}{2}} K_{n_1} r_{1n}$ and $\theta_n = \theta + n_2^{-\frac{1}{2}} r_{2n}$ such that $m_\theta = m$ and there exist limits r_j for which $\|r_{jn_j} - r_j\|_\infty$ converges to zero as n tends to infinity, for $j = 1, 2$.

Proposition 3.18. *The statistic T_n converges weakly under A_n to*

$$T = \sup_{\mathcal{I}_X} |(1-p)^{\frac{1}{2}} \rho_2(B_1 + r_1)|.$$

Proof. Under A_n the difference of the regression functions satisfies

$$n_1^{\frac{1}{2}} K_{n_1}^{-1} \{m_{1n} - m_{2n}\} = r_{1n_1} + o(1)$$

and, in the expression of T_n

$$n_1^{\frac{1}{2}} K_{n_1}^{-1} \{\widehat{m}_{n_1}(x) - m_{\widehat{\theta}_{n_2}}(x)\} = n_1^{\frac{1}{2}} K_{n_1}^{-1} \{\widehat{m}_{n_1}(x) - m_1(x)\} + r_{1n_1} + o_p(1),$$

it converges weakly to the supremum of the weighted sum of r_1 and B_1. □

Let $\mathcal{H} = \{h = h_{\theta,m}, \theta \in \Theta, m \in \mathcal{M}, h \in C^2(\Theta)\}$ be a semi-parametric regression model defined on a bounded interval $\mathcal{I}_X$ by a parameter θ and an unknown function m in $\mathcal{M}$, like the projection pursuit regression model (3.13) and the regression model with transformed explanatory variables (3.35). Let $(X_j, Y_j)_{j=1,\dots,n}$ be a sample of independent observations partitioned as two independent sub-samples of sizes n_1 and n_2 of order $O(n)$ such that $n^{-1} n_1$ converges to p in $]0, 1[$ as n tends to infinity, such that

$$E(Y_{1j} \mid X_{1j}) = m(X_{1j}),\ j = 1, \dots, n_1,$$
$$E(Y_{2j} \mid X_{2j}) = h_{\theta,m}(X_{2j}), j = 1, \dots, n_2,$$

on $\mathcal{I}_X$, with $h_{\theta,m}$ in $\mathcal{H}$. In the semi-parametric models (3.13) and (3.35), the estimator of the function $m_{\theta,g}$ based on a projection on K_{n_2} functions of an estimated basis has the convergence rate $n_2^{-\frac{1}{2}} K_{n_2}^{-1}$ like in the nonparametric model.

A goodness-of-fit test for the family of models $\mathcal{M}$ is performed with a statistic T_n defined by (3.41) which compares the nonparametric estimator of a regression function and its estimator in $\mathcal{M}$. Its asymptotic behavior under H_0 and alternatives are given by Propositions 3.14 and 3.15.

3.13 Estimation under bias sampling

Let $(X_{ij}, Y_{ij})_{i=1,\ldots,I, j=1,\ldots,n_i}$ be a hierarchical sample of a regression model such that for $i = 1, \ldots, I$, the variables $(X_{ij}, Y_{ij})_{j=1,\ldots,n_i}$ have densities

$$f_{ni}(x, y) = \frac{w_i(x, y)}{W_i} g(x, y) \tag{3.44}$$

with known weight functions $w_i(x, y)$ and an unknown density g, the densities are weighted by the scalars $W_i = \int w_i \, dG$. The density of the whole sample is a weighted sum of the densities of the sub-samples, with weights $\lambda_{ni} = n^{-1} n_i$ depending on the sub-sample sizes and $n = \sum_{i=1}^I n_i$

$$f_n(x, y) = \sum_{i=1}^I \lambda_{ni} f_{ni}(x, y) = \Big\{ \sum_{i=1}^I \frac{\lambda_{ni} w_i(x, y)}{W_i} \Big\} g(x, y). \tag{3.45}$$

The estimation of the density G and the scalar W_i has been studied with iterative procedures by Vardy (1982), Gill et al. (1988), Bickel et al. (1993).

Example 4.1. In regression models, the densities f_{ni} of variables (X_i, Y_i) may be biased according restrictions of the values of the regressors X_i in fixed or random intervals I_l, $l = 1, \ldots, L$, with the weighting functions $w_l(x) = 1_{I_l}(x)$, or by right-censoring of X_i by an independent observed variable C_i, then $w_i(X_i) = 1_{\{X_i \le C_i\}}$. The weights in (3.45) are $W_l = G_X(I_l)$, and respectively $W = \int P(C > x) \, dG_X(x)$, they are unknown. The density of Y conditionally on $X = x$ in I_l is $f_{Y|X}(y; x) = g_{Y|X}(y; x) w_l(x) W_l^{-1}$. In both cases the estimator of G relies on the inversion of the expression in brackets in (3.45), on the set where it is strictly positive.

Example 4.2. Let Y be a binary variable depending on a covariate X in logistic model with $P(Y = 1 \mid X) = p(X)$ and let f be the unknown density of the regressor X, then $p = P(Y = 1) = \int p(x) f(x) \, dx$ and the densities f_i of X conditionally on $Y = i$, for $i = 0, 1$, are

$$f_1(x) = \frac{p(x)}{p} f(x), \qquad f_2(x) = \frac{1 - p(x)}{1 - p} f(x).$$

Let n_i be the number of observations with $Y_j^* = i$ and let $n = n_0 + n_1$, the ratio $\lambda_n = n_0^{-1} n_1$ is an estimator of $\lambda = (1 - p)^{-1} p$ and p is estimated by $p_n = n^{-1} n_1$. The distribution function $F = pF_1 + (1 - p)F_2$ of X is estimated by the mixture of the sub-distribution functions.

For a real variable X with a distribution function F with density (3.45) under biased sampling, the estimators of the normalizing constant and the distribution function G with density g in equation (3.44) are solutions of the equations

$$G(x) = \int_{\{y \le x\}} \frac{J_n(y) dF_n(y)}{\sum_{i=1}^I W_i^{-1} \lambda_{ni} w_i(y)}, \tag{3.46}$$

with $J_n(y) = 1_{\{\sum_{i=1}^I W_i^{-1} \lambda_{ni} w_i(y) > 0\}}$ and the convention $\frac{0}{0} = 0$, then

$$W_i = \int_{\mathbb{R}} w_i(y)\, dG(y) = \int_{\mathbb{R}} \frac{w_i(y)}{\sum_{j=1}^I \lambda_{nj} w_j(y) W_j^{-1}}\, dF_n(y).$$

Vardi (1982) proposed iterative algorithms for the estimation of the parameters and the estimation of G follows from the empirical estimator of the distribution function F_n of the sample with density (3.45).

Assuming that the weight functions in (3.45) depend only on the variable x, the regression function of the sample

$$m(x) = E_{F_n}\{Y \mid X = x\}$$

reduces to $E_G\{Y \mid X = x\}$ which can be estimated like in sections 3.2, 3.7, 3.6 or 3.8, according to the regression model.

For a sample $(X_i, Y_i)_{i=1,\dots,n}$ with a single density f defined by (3.44), the expression of the regression function does no depend on the normalization constant $W = E_G\{w(X, Y)\}$, it reduces to

$$\begin{aligned} m(x) &= \frac{\int_{\mathbb{R}} y w(x, y) g(x, y)\, dy}{\int_{\mathbb{R}} w(x, y) g(x, y)\, dy} \\ &= E_G\Big\{\frac{Y w(x, Y)}{E_G\{w(x, Y) \mid X = x\}} \mid X = x\Big\}, \end{aligned}$$

for every x such that $E_G\{w(x, Y) \mid X = x\} > 0$. It is estimable from the empirical distribution $\widehat{F}_n$ of the sample, with a known weight function.

For the sample under bias sampling with density (3.45) and for every strictly positive function w_i, the inversion of (3.44) provides an estimator of the constant W_i

$$\widehat{W}_{ni} = \Big\{n_i^{-1} \sum_{j=1}^{n_i} w_i^{-1}(X_{ij}, Y_{ij}) 1_{\{w_i(X_{ij}, Y_{ij}) > 0\}}\Big\}^{-1},$$

with the convention $\frac{0}{0} = 0$.

Proposition 3.19. *The variables $\widehat{W}_{ni}$ are a.s. consistent and $n_i^{\frac{1}{2}}(\widehat{W}_{ni} - W_i)$ converges weakly to a centered Gaussian variable with a finite variance.*

In Example 4.1, for right-censored regression variables and with $w_i(x) = 1_{\{x \le C_i\}}$, the estimator of W_i is

$$\widehat{W}_{ni} = \Big\{ n_i^{-1} \sum_{j=1}^{n_i} 1_{\{X_{ij} \le C_i\}} \Big\}^{-1}$$

and it is strictly positive if all observations of the regression variables are not censored in the i-th sub-sample. For regressors censored by intervals I_l, the estimators of the constants W_l are

$$\widehat{W}_{nil} = \Big\{ n_i^{-1} \sum_{j=1}^{n_i} 1_{\{X_{ij} \in I_l\}} \Big\}^{-1}.$$

With positive weight functions, the equality $W_i = 0$ is equivalent to $w_i = 0$, G-almost surely and (3.45) reduces to

$$f_n(x,y) = \sum_{i \in \mathcal{I}'} \lambda_{ni} f_{ni}(x,y)$$

with a sum on indices in the set $\mathcal{I}'$ where the constants W_i are strictly positive. Reducing the estimator of the distribution function F_n to the set $\mathcal{I}'_n$ of significantly strictly positive constants $\widehat{W}_{ni}$, the distribution function G is estimable from (3.45) by plugging

$$\begin{aligned}
\widehat{G}_n(x,y) &= n^{-1} \sum_{i \in \mathcal{I}'_n} \sum_{j=1}^{n_i} \frac{1_{\{X_{ij} \le x, Y_{ij} \le y\}}}{\sum_{i' \in \mathcal{I}'_n} \lambda_{ni'} w_{i'}(X_{ij}, Y_{ij}) \widehat{W}_{ni'}^{-1}} \\
&= \sum_{i \in \mathcal{I}'_n} \lambda_{ni} \int_{-\infty}^{x} \int_{-\infty}^{y} \frac{\widehat{F}_{i,n_i}(ds,dt)}{\sum_{j \in \mathcal{I}'_n} \lambda_{nj} w_j(s,t) \widehat{W}_{nj}^{-1}},
\end{aligned}$$

where $\widehat{F}_{i,n_i}$ is the empirical distribution function of the i-th sub-sample $(X_{ij}, Y_{ij})_{j=1,\ldots,n_i}$, for i in $\mathcal{I}'_n$. The estimator $\widehat{G}_n$ is a.s. consistent and its convergence rate is $n^{-\frac{1}{2}}$.

Let λ_i be the limit of λ_{ni} as n tends to infinity, the denominator $D_n(s,t)$ of the integral with respect to $\widehat{F}_{i,n_i}(s,t)$ converges uniformly a.s. to $D(s,t) = \sum_{j \in \mathcal{I}'} \lambda_j w_j(s,t) W_j^{-1}$. Let B_0 be the standard Brownian bridge and, for every i in $\mathcal{I}'_n$, let B_i be the Gaussian limit of the process $n^{\frac{1}{2}}(\lambda_{ni}^{\frac{1}{2}} D_n^{-1} - \lambda_i^{\frac{1}{2}} D^{-1})$,

Proposition 3.20. *The process* $Z_n(x,y) = n^{\frac{1}{2}}\{\widehat{G}_n(x,y) - G(x,y)\}$ *converges weakly to a centered Gaussian process*

$$Z(x,y) = \sum_{i \in \mathcal{I}'_n} \lambda_i^{-\frac{1}{2}} \int D^{-1}\, dB_0 \circ F_i + \sum_{i \in \mathcal{I}'_n} \lambda_i^{\frac{1}{2}} \int B_i\, dF_i.$$

Proof. The process Z_n is the sum

$$Z_n(x,y) = \sum_{i\in\mathcal{I}'_n} \lambda_{ni}^{-\frac{1}{2}} \int D_n^{-1} n_i^{\frac{1}{2}} \, d(\widehat{F}_{i,n_i} - F_i)$$
$$+ \sum_{i\in\mathcal{I}'_n} \int n^{\frac{1}{2}} (\lambda_{ni} D_n^{-1} - \lambda_{ni}^{\frac{1}{2}} \lambda_i^{\frac{1}{2}} D^{-1}) \, dF_i$$

and its weak convergence to the centered Gaussian process Z is a consequence of the convergence of the variables $n_i^{\frac{1}{2}}(\widehat{W}_{ni} - W_i)$ and of the marginal empirical processes $n_i^{\frac{1}{2}}(\widehat{F}_{i,n_i} - F_i)$. □

The regression function $m(x)$ is estimable as previously from the sub-samples of $\mathcal{I}'_n$. Wu (2000) studied a kernel estimator and a local polynomial estimator of the regression function under identifiability conditions. The marginal density $g_X(x)$ of X has a series estimator

$$\widehat{g}_{nX}(x) = \sum_{k=0}^{K_n} \int_{\mathcal{I}_X \times \mathcal{I}_Y} \widehat{a}_{kn} \psi_k(s) \, \widehat{G}_n(ds, dy)$$

with the estimator of the coefficients $\widehat{a}_{kn} = \int_{\mathcal{I}_X \times \mathcal{I}_Y} \psi_k(x) \, \widehat{G}_n(dx, dy)$ defined like (2.17), in the same way the conditional probability function $p(y;x) = P_G(Y \le y, X = x)$ and the function

$$\pi(y) = \int_{\mathcal{I}_X} \psi_k(x) \, G(dx, y)$$

have the estimators

$$\widehat{p}_n(y;x) = \sum_{k=0}^{K_n} \widehat{\pi}_{kn}(y) \psi_k(x),$$
$$\widehat{\pi}_{kn}(y) = \int_{\mathcal{I}_X} \psi_k(x) \widehat{G}_n(dx, y),$$

with K_n satisfying the conditions (3.7). The estimator $\widehat{p}_n$ is a.s. consistent and it has the same convergence rate as the estimator of the nonparametric regression function in Section 3.2.

An a.s. consistent estimator of the conditional expectation of Y under the distribution function G is deduced from the estimator of the conditional probability function $p(y;x)$.

$$\widehat{m}_n(x) = \int_{\mathcal{I}_Y} y \widehat{p}_n(dy;x),$$

its convergence rate is the same as estimator $\widehat{p}_n$, as a consequence of Proposition 3.20. The proof is the same as for Proposition 3.4.

Proposition 3.21. *Under the condition (3.7), the processes* $K_n^{-1}n^{\frac{1}{2}}(\widehat{p}_n - p)$ *and* $K_n^{-1}n^{\frac{1}{2}}(\widehat{m}_n - m)$ *converge weakly to Gaussian processes with finite expectation and variance functions.*

The optimal convergence rate of the estimators $\widehat{p}_n$ $\widehat{m}_n$ for a conditional probability p and a regression function m in W_{4s} is $n^{\frac{s}{2s+1}}$, as the optimal number of functions of the basis is $K_n = O(n^{\frac{1}{2(2s+1)}})$.

Chapter 4

Nonparametric generalized linear models

4.1 Nonparametric GLM models

In generalized linear models a variable Y is linked with a predicted value depending linearly on a vector X of explanatory variables, in the form $EY = g(\beta^T X)$ with a known function g. The variance of a variable Y is determined by its distribution, its overdispersion is modeled with random effect which modifies its variance. For a binomial variable Y with a random distribution $B(n, p)$, the probability $p = P(Y = 1)$ is a variable with expectation $\mu =$ and variance σ^2 which entails

$$\begin{aligned} EY &= n\mu, \\ \operatorname{Var} Y &= \operatorname{Var}\{E(Y \mid p)\} + E\{\operatorname{Var}(Y \mid p)\} = n^2\sigma^2 + nE\{p(1-p)\} \\ &= n\mu(1-\mu) + n(n-1)\sigma^2, \end{aligned}$$

the variance of Y is therefore larger than in the binomial model $B(n, \mu)$. The parameters μ and σ^2 are estimated from the empirical mean and variance of a sample of the variable Y.

For a stochastic multinomial variable Y with random probabilities $p_1, \ldots, p_J$, let μ_j in $]0, 1[$ be the expectation of p_j, let $\sigma_j^2 = \sigma_{jj}^2$ be its variance and let $\sigma_{jj'}^2$ be the covariance of p_j and $p_{j'}$, for distinct j and $j' = 1, \ldots, J$. The expectation and the variance of Y are

$$\begin{aligned} EY &= n \sum_{j=1}^{J} \mu_j := n\mu, \\ \operatorname{Var} Y &= \operatorname{Var}\{E(Y \mid p_1, \ldots, p_J)\} + E\{\operatorname{Var}(Y \mid p_1, \ldots, p_J)\} \\ &= n(n-1) \sum_{j,j'=1}^{J} \sigma_{jj'}^2 + n\mu(1-\mu). \end{aligned}$$

The empirical ratio $\widehat{\mu}_{nj} = n^{-1}\sum_{i=1}^{n} 1_{\{Y_i=j\}}$ of the variables Y with value j in a n-sample is an estimator of μ_j and the covariance of $\widehat{\mu}_{nj}$ and $\widehat{\mu}_{nj}$ converges to $\sigma^2_{jj'}$, for all j and j'. An empirical estimator of the covariance $\sigma^2_{jj'}$ is deduced from the empirical variance of Y.

The logistic model is defined for a Bernoulli variable Y with strictly positive probabilities depending on a regression variable X of $\mathbb{R}^d$ through a nonparametric model

$$p(X) = P(Y=1 \mid X) = \frac{e^{m(X)}}{1+e^{m(X)}}, \tag{4.1}$$
$$1-p(X) = P(Y=0 \mid X) = \frac{1}{1+e^{m(X)}}.$$

The regression function m is deduced from their ratio as

$$m(x) = \log \frac{P(Y=1 \mid X)}{P(Y=0 \mid X)}.$$

Let $(X_i, Y_i)_{i=1,\ldots,n}$ be a sample of independent and identically distributed observations of the variables and let $\mathbb{Z}_n = (Z_{i1}, \ldots, Z_{id})_{i=1,\ldots,n}$ be the sample of independent variables defined as $\mathbb{Z}_n = \mathbb{X}_n B$ by diagonalization of the empirical covariance matrix of the regressors, like in Section 3.7. By the change of variable (3.17), the probability function p is mapped to $\pi(z) = p(x)$.

Let $(\psi_k)_{k\geq 0}$ be the orthonormal basis of polynomials in $L^2(\mathbb{R}^d, F_Z)$ defined by (3.1) from Laguerre's polynomials L_k and the distribution function F_Z of the explanatory vector Z. The function π has the expansion of Section 3.6

$$\pi(z) = \sum_{k\geq 0} b_k \psi_k(z),$$

with $b_k = \int \pi(z)\psi_k(z)\, dF_Z(z)$. The independence of the components of Z imply that its survival function $\bar{F}_Z(z) = 1 - F_Z(z)$ is the product of its marginals F_{Z_j} and the function $H_Z(z) = -\log\{1 - F_Z(z)\}$ is the sum of its marginals $H_Z(z) = \sum_{j=1}^{d} H_{Z_j}(z_j)$. Let $\widehat{H}_{nZ}$ be the empirical estimator (3.20) of H_Z, the functions ψ_k have the empirical estimators of Section 3.6

$$\widehat{\psi}_{nk} = L_k \circ \widehat{H}_{nZ}.$$

Let K_n be the number of functions of the series for the approximation of π by $\pi_n(z) = \sum_{k=0}^{K_n} b_k\psi_k(z)$, a nonparametric estimator $\widehat{\pi}_n$ of the probability function π is defined as

$$\widehat{\pi}_n(z) = \sum_{k=0}^{K_n} \widehat{b}_{nk}\widehat{\psi}_{nk}(z)$$

where the coefficients have the estimators

$$\widehat{b}_{nk} = (n-1)^{-1} \sum_{i=1}^{n-1} Y_{(i)}\widehat{\psi}_{nik}(Z_{(n:i)}), \tag{4.2}$$

for $k = 0, \ldots, K_n$, with the notations of (3.4) and $K_n = o(n)$ tending to infinity as n tends to infinity. The mean integrated squared error in $L^2(F_Z)$ of the estimator is the sum

$$\text{MISE}_{K_n,n}(\pi) = E\|\widehat{\pi}_n - \pi_n\|_2^2 + \|\pi - \pi_n\|_2^2$$

of the squared estimation error $E\|\widehat{\pi}_n - \pi_n\|_2^2$ and the squared error of approximation which is a $o(1)$. By the same proof as Lemma 3.3, as n tends to infinity

$$E\|\widehat{\pi}_n - \pi_n\|_2^2 = O(n^{-\frac{1}{2}}K_n).$$

The error is minimum as both terms have the same order therefore we set $K_n = o(n^{\frac{1}{2}})$, the process $n^{\frac{1}{2}}K_n^{-1}(\widehat{\pi}_n - \pi)$ converges weakly to a Gaussian process and Proposition 3.9 applies.

The regression function $\mu(z) = m(x)$ is consistently estimated by

$$\widehat{\mu}_n(z) = \log \frac{\widehat{\pi}_n(z)}{1 - \widehat{\pi}_n(z)}$$

and it has the same convergence rate as $\widehat{\pi}_n$. The convergence properties of $\widehat{\pi}_n$ extend to $\widehat{\mu}_n$. The additive regression model and the models with interactions of Sections 3.7 and 3.8 apply to the Bernoulli variables with a logistic transformation and Propositions 3.11 and 3.12 are still valid.

A multinomial variable Y with values in $\{1, \ldots, J\}$ and with strictly positive probabilities $p_1, \ldots, p_J$ has the probability density

$$L(Y) = \prod_{j=1}^{J} p_j^{\delta_{Y,j}},$$

under the constraints $\sum_{j=1}^{J} p_j = 1$, p_j in $]0,1[$ for every j, and with the indicator variables $\delta_{Y,j} = 1_{\{Y=j\}}$. The empirical estimators of the probabilities are defined from a n-sample $(Y_i)_{i=1,\ldots,n}$ as $\widehat{p}_{nj} = n^{-1}\sum_{i=1}^{n} 1_{Y_i=j}$, for $j = 1, \ldots, J$. They have the variance $n^{-1}p_j(1-p_j)$ and the covariances $-n^{-1}p_jp_{j'}$, for all distinct indices j and j' in $\{1, \ldots, J\}$.

Multinomial variables depending on regressors in a logistic model have the conditional probabilities

$$p_j(X) = P(Y = j \mid X) = \frac{e^{m_j(X)}}{1 + e^{m_j(X)}}. \tag{4.3}$$

Let $(X_i, Y_i)_{i=1,\dots,n}$ be a sample of independent and identically distributed observations of a multinomial variable Y and its explanatory variable X in $\mathbb{R}^d$, and let $\mathbb{Z}_n = (Z_{i1}, \dots, Z_{id})_{i=1,\dots,n}$ be the sample of independent variables defined by the change of variable (3.17), the probability functions p_j are mapped to $\pi_j(z) = p_j(x)$ and $\mu_j(Z) = m_j(X)$ such that

$$\pi_j(Z) = \frac{e^{\mu_j(Z)}}{1 + e^{\mu_j(Z)}}.$$

The functions π_j are estimated by projection on the orthonormal basis of functions $(\psi_k)_{k\geq 0}$ of $L^2(\mathbb{R}^d, F_Z)$ where $\psi_k = L_k \circ H_Z$ is defined by (3.1) from Laguerre's polynomials L_k and the distribution function of the regressor Z, for $k \geq 0$. They have the empirical estimators $\widehat{\psi}_{nk} = L_k \circ \widehat{H}_{nZ}$. Let $\pi_j(z) = \sum_{k\geq 0} b_{kj}\psi_k(z)$, the constraints $\sum_{j=1}^J p_j(x) = 1$ for every x is equivalent to $\sum_{j=1}^J \pi_j(x) = 1$ for every x and by the expansion of the functions π_j, it is equivalent to

$$\sum_{j=1}^J b_{0j} = 1, \quad \sum_{j=1}^J b_{kj} = 0, \; k \geq 1.$$

The constraint is satisfied for coefficients $b_{0j} = \int_{\mathcal{I}_Z} \pi_j(z)\, dF_Z(z)$ as a consequence of the constraint on the probabilities π_j and for the coefficients

$$b_{kj} = \int_{\mathcal{I}_Z} \pi_j(z)\psi_k(z)\, dF_Z(z)$$

as a consequence of the orthogonality of ψ_0 and ψ_k.

The coefficients have the estimators

$$\widehat{b}_{n,jk} = (n-1)^{-1} \sum_{i=1}^{n-1} 1_{\{Y_{(i)}=j\}} \widehat{\psi}_{nik}(Z_{(n:i)}),$$

for $j = 1, \dots, J$ and $k = 0, \dots, K_n$, with $K_n = o(n)$ tending to infinity as n tends to infinity. They have the variances

$$\operatorname{Var} \widehat{b}_{n,jk} = n^{-1} E[p_j(Z_i)\{1 - p_j(Z_i)\}\psi_k^2(Z_i)] - n^{-1} b_{kj}^2$$

and the covariances

$$\operatorname{Cov}(\widehat{b}_{n,jk}, \widehat{b}_{n,j'k'}) = -n^{-1} b_{kj} b_{k'j'}, \text{ for } j \neq j',$$

and

$$\mathrm{Cov}(\widehat{b}_{n,jk}, \widehat{b}_{n,jk'}) = -n^{-1}E[p_j(Z_i)\{1 - p_j(Z_i)\}\psi_k(Z_i)\psi_{k'}(Z_i)] \\ -n^{-1}b_{kj}b_{k'j}, \text{ for } k \neq k'.$$

Let $\pi_{nj}(z) = \sum_{k=0}^{K_n} b_{jk}\psi_k(z)$, the estimators of the conditional probabilities π_j are

$$\widehat{\pi}_{nj}(z) = \sum_{k=0}^{K_n} \widehat{b}_{n,jk}\widehat{\psi}_{nk}(z), \; j = 1, \ldots, m,$$

on every interval where the distribution functions of the regressors is strictly lower than one. Their variances and covariances are deduced from those of the estimated coefficients and of the estimated functions of the basis. The regression function $\mu_j(z) = m_j(xB)$ are consistently estimated by

$$\widehat{\mu}_{nj}(z) = \log \frac{\widehat{\pi}_{nj}(z)}{1 - \widehat{\pi}_{nj}(z)}$$

and they have the same convergence rate as $\widehat{\pi}_{nj}$, given in Proposition 3.9.

The multinomial models are special cases of the nonparametric generalized regression models. Let Y be a discrete variable with values in a set $\{1, \ldots, J\}$ and let X be a d dimensional vector of regressors with a distribution function F_X on a set $\mathcal{I}_X$ of $\mathbb{R}^d$. In a nonparametric generalized linear model, the conditional probabilities of a n-sample $(X_i, Y_i)_{i=1,\ldots,n}$ of (X, Y) are

$$p_j(X_i) = P(Y_i = j \mid X_i) = g \circ m_j(X_i), \tag{4.4}$$

with a known monotone function g of $C^1(\mathbb{R})$, having an inverse g^{-1}, and with unknown functions m_j from $\mathcal{I}_X$ to a $\mathbb{R}$, under the constraint $\sum_{j=1}^{m} p_j(x) = 1$ for every x of $\mathcal{I}_X$. The conditional expectation of Y is deduced as

$$E(Y_i \mid X_i) = \sum_{j=1}^{J} j p_j(X_i). \tag{4.5}$$

The conditional probabilities p_j of Y may be estimated by projection on a basis of orthonormal functions of $l^2(F_X)$ which requires the orthogonalization of the regression vector like in Section 3.7. Nonparametric estimators $(\widehat{m}_{1n}, \ldots, \widehat{m}_{Jn})$ of the regression functions $m_j(x) = g^{-1} \circ p_j(x)$ are deduced from the inversion of (4.4), they have the same convergence rate as the estimator of the functions $p_j(x)$. The projection pursuit model, the additive and ANOVA models of Sections 3.5 3.7, 3.6 and 3.8 apply to the regression functions $m_j(x) = g^{-1} \circ p_j(x)$ with a multivariate regressor. Their asymptotic properties of the estimators of the functions p_j and m_j are the same as in the regression models.

In a nonparametric generalized linear model for a variable Y with values in a real set and a d dimensional vector of regressors X with a distribution function F_X on a set $\mathcal{I}_X$ of $\mathbb{R}^d$, the conditional expectation of a n-sample $(X_i, Y_i)_{i=1,\ldots,n}$ of (X, Y) is

$$r(X_i) = E(Y_i \mid X_i) = g \circ m(X_i), \tag{4.6}$$

with a known monotone function g of $C^1(\mathbb{R}^d)$ and a function m defined from $\mathcal{I}_X$ to $\mathbb{R}$. The conditional expectation r is estimated like in Section 3.6 and an estimator of the regression function $m(x) = g^{-1} \circ r(x)$ is deduced. The models of Chapter 3 and asymptotic properties of their estimators are extended to the conditional functions r and m.

The projection pursuit model of Section 3.5 extends the GLM models, the conditional expectation of Y has the form

$$E(Y \mid X) = g \circ m(\theta^T X) = r(\theta^T X), \tag{4.7}$$

it is defined by a known monotone function g of $C^1(\mathbb{R})$, a univariate regression function m and a parameter θ in an open subset Θ of $\mathbb{R}^d$. The estimators of the function r and θ are defined by projections like in Section 3.5 in model (4.7) and a nonparametric estimator of the regression function $m = g^{-1} \circ r$ follows. The estimators are consistent and asymptotically Gaussian, with the same convergence rate as in Section 3.5.

The model (4.7) is generalized to a model for (X_i, Y_i) with conditional expectation $E(Y_i \mid X_i) = g \circ \{\sum_{j=1}^d \theta_j m_j(X_{ij})\}$.

4.2 Biased sampling for Bernoulli variables

In case-control studies, individuals are not uniformly sampled in the population: for rare events, they are sampled so that the cases of interest (individuals with $Y_i = 1$) are sufficiently represented in the sample but the proportion of cases in the sample differs from its proportion in the general population. Let S_i be the sampling indicator of individual i in the global population, it is independent of the covariate X_i and

$$P(S_i = 1 | Y_i = 1) = \lambda_1, \; P(S_i = 1 | Y_i = 0) = \lambda_0$$

with $\lambda_1 > \lambda_0$. The distribution function of (S_i, Y_i) conditionally on $X_i = x$ is defined as

$$\begin{aligned}
P(S_i = 1, Y_i = 1 | X_i = x) &= P(S_i = 1 | Y_i = 1) P(Y_i = 1 | X_i = x) \\
&= \lambda_1 p(x), \\
P(S_i = 1, Y_i = 0 | X_i = x) &= P(S_i = 1 | Y_i = 0) P(Y_i = 0 | X_i = x) \\
&= \lambda_0 \{1 - p(x)\}, \\
P(S_i = 1 | X_i = x) &= P(S_i = 1, Y_i = 1 | X_i = x) \\
&\quad + P(S_i = 1, Y_i = 0 | X_i = x) \\
&= \lambda_1 p(x) + \lambda_0 \{1 - p(x)\}.
\end{aligned}$$

Let

$$\theta = \frac{\lambda_0}{\lambda_1}, \quad \alpha(x) = \theta \frac{1 - p(x)}{p(x)}.$$

The variable (X_i, Y_i) is observed conditionally on $S_i = 1$ and the distribution function of Y_i conditionally on $S_i = 1, X_i = x$ is

$$\begin{aligned}
\pi(x) = P(Y_i = 1 | S_i = 1, X_i = x) &= \frac{\lambda_1 p(x)}{\lambda_1 p(x) + \lambda_0 \{1 - p(x)\}} \\
= \frac{p(x)}{p(x) + \theta\{1 - p(x)\}} &= \frac{1}{1 + \alpha(x)},
\end{aligned}$$

reversely

$$\alpha(x) = \frac{1 - \pi(x)}{\pi(x)}.$$

The probability $p(x)$ is deduced from θ and $\pi(x)$ by the relation

$$p(x) = \frac{\theta \pi(x)}{1 - (1 - \theta)\pi(x)} = \frac{\theta}{\theta + \alpha(x)} \tag{4.8}$$

and the conditional bias of the probability functions

$$\pi(x) - p(x) = \frac{(1 - \theta)\pi(x)(1 - \pi(x))}{1 - (1 - \theta)\pi(x)}$$

is strictly positive. The logistic regression model is stable under bias sampling up to a constant, the function $\psi(x) = \log[p(x)\{1 - p(x)\}^{-1}]$ is replaced by

$$\log \frac{\pi(x)}{1 - \pi(x)} = \psi(x) - \log \theta.$$

The function α is identifiable and θ must be known or estimated from a preliminary study before the estimation of the probability function p.
Let γ be the relative inverse of the proportion of cases in the population

$$\gamma = \frac{P(Y=0)}{P(Y=1)} = \frac{1 - \int p(x)\,dF_X(x)}{\int p(x)\,dF_X(x)}. \tag{4.9}$$

Under the biased sampling, we have

$$P(Y_i = 1|S_i = 1) = \frac{1}{1+\theta\gamma},$$
$$P(Y_i = 0|S_i = 1) = \frac{\theta\gamma}{1+\theta\gamma},$$

and γ is modified as

$$P(Y=0|S=1)P^{-1}(Y=1|S=1) = \theta\gamma.$$

The product $\theta\gamma$ may be directly estimated from the observed Bernoulli variables Y_i by maximisation of the likelihood

$$L_n = \prod_{i=1}^n [\{P(Y_i = 1|S_i = 1)\}^{Y_i}\{P(Y_i = 0|S_i = 1)\}^{1-Y_i}]^{1_{\{S_i=1\}}},$$
$$\theta\widehat{\gamma}_n = \frac{\sum_{i=1}^n (1-Y_i)1_{\{S_i=1\}}}{\sum_{i=1}^n Y_i 1_{\{S_i=1\}}}.$$

For random observations of the variable X, the function $\pi(x)$ is estimated by projection on an orthonormal basis of functions $(\psi_k)_{k\geq 0}$ of $L^2(\mathcal{I}_X, F_X)$, defined as (3.2). Let $\pi(x) = \sum_{k\geq 0} \rho_k\psi_k(x)$, the coefficients p_k, $k = 0, \ldots, K_n$ and the probability function $p(x)$ have the estimators

$$\widehat{\rho}_{nk} = \frac{\sum_{i=1}^n Y_i S_i \widehat{\psi}_{nk}(X_i)}{\sum_{i=1}^n Y_i S_i \widehat{\psi}_{nk}(X_i) + \sum_{i=1}^n (1-Y_i) S_i \widehat{\psi}_{nk}(X_i)},$$
$$\widehat{\pi}_n(x) = \sum_{k=0}^{K_n} \widehat{\rho}_{nk}\widehat{\psi}_{nk}(x),$$

where $K_n = o(n)$ tends to infinity as n tends to infinity. By Lemma 3.1, the estimators $\widehat{\rho}_{nk}$ are a.s. consistent and the variables $n^{\frac{1}{2}}(\widehat{\varepsilon}_{nk} - \rho_k)$ converge weakly to centered Gaussian variables. The weak convergence of the process $\widehat{\pi}_n$ is deduced from its MISE and Proposition 3.4.

Proposition 4.1. *Under the condition (3.7), the process $K_n^{-1}n^{\frac{1}{2}}(\widehat{\pi}_n - \pi)$ converges weakly to a Gaussian process.*

The estimator of the probability function $p(x)$ is deduced from $\widehat{\pi}_n$ and an estimator $\widehat{\theta}_n$ of θ by the relation (4.8), it has the convergence rate $K_n n^{-\frac{1}{2}}$ and asymptotic properties follow from their convergence.

In the logistic model (4.1) for a Bernoulli variable, the regression function $m(x) = \log p(x) - \log\{1 - p(x)\}$ is estimated using the estimator $\widehat{p}_n(x)$ of the probability function $p(x)$.

4.3 Biased sampling in a multinomial model

Under biased sampling in a multinomial model with covariate X, a variable Y with values in a discrete set $\mathcal{Y} = \{1, \ldots, d\}$ and random probabilities $p_j(X)$ is sampled with more importance for selected values of Y in a subset $\mathcal{Y}_s$ of $\mathcal{Y}$. Let S_i be the sampling indicator of individual i in the global population, it is assumed independent of the covariate X_i and with $\lambda_j > \lambda_k$ for all j in $\mathcal{Y}_s$ and k in $\bar{\mathcal{Y}}_s = \mathcal{Y} \setminus \mathcal{Y}_s$. The distribution function of (S_i, Y_i) conditionally on $X_i = x$ is defined by the probabilities

$$\lambda_j = P(S_i = 1 | Y_i = j), \qquad p_j(x) = P(Y_i = j | X_i = x)$$

such that $\sum_{j=1}^d p_j(x) = 1$, and for every j

$$P(S_i = 1, Y_i = j | X_i = x) = \lambda_j p_j(x), \; j = 1, \ldots, d,$$

$$P(S_i = 1 | X_i = x) = \sum_{j=1}^d P(S_i = 1, Y_i = j | X_i = x) = \sum_{j=1}^d \lambda_j p_j(x).$$

For j and k in $\mathcal{Y}$, the ratio $\theta_{jk} = \lambda_k \lambda_j^{-1}$ is known or estimated from preliminary observations, let

$$\alpha_{jk}(x) = \frac{\lambda_k p_k(x)}{\lambda_j p_j(x)} = \theta_{jk} \frac{p_k(x)}{p_j(x)},$$

and let

$$\alpha_j(x) = \sum_{k \neq j, k=1}^d \alpha_{jk}(x) = \sum_{k \neq j, k=1}^d \theta_{jk} \frac{p_k(x)}{p_j(x)}.$$

The variable (X_i, Y_i) is observed conditionally on $S_i = 1$ and the distribution function of Y_i conditionally on $S_i = 1, X_i = x$ is

$$\pi_j(x) = P(Y_i = j | S_i = 1, X_i = x) = \frac{\lambda_j p_j(x)}{\sum_{k=1}^d \lambda_k p_k(x)} = \frac{1}{1 + \alpha_j(x)},$$

so $\alpha_j(x) = \{1 - \pi_j(x)\} \pi_j^{-1}(x)$. The conditional bias of the probability functions is strictly positive

$$\pi_j(x) - p_j(x) = p_j(x) \Big\{ \frac{1}{\sum_{k \neq j, k=1}^d \theta_{jk} p_k(x)} - 1 \Big\}.$$

For random observations of the variable X, the logarithm of the likelihood of the observations Y_i conditionally on $S_i = 1$ and X_i is

$$l_n = \sum_{i=1}^n \sum_{j=1}^d S_i 1_{\{Y_i=j\}} \log \pi_j(X_i).$$

The functions $\pi_j(x)$ have expansions on an orthonormal basis of functions $(\psi_k)_{k\geq 0}$ of $L^2(\mathcal{I}_X, F_X)$, defined by (3.2). Let $\pi_j(x) = \sum_{k\geq 0} \rho_{jk}\psi_k(x)$, with the coefficients $\rho_{jk} = E\{\pi_j(X)\psi_k(X)\}$, they are estimated by maximization of the likelihood as

$$\widehat{\rho}_{njk} = \frac{\sum_{i=1}^n 1_{\{Y_i=j\}} S_i \widehat{\psi}_{nk}(X_i)}{\sum_{i=1}^n \sum_{\ell=1}^d 1_{\{Y_i=\ell\}} S_i \widehat{\psi}_{nk}(X_i)},$$

$$\widehat{\pi}_{nj}(x) = \sum_{k=0}^{K_n} \widehat{\rho}_{njk} \widehat{\psi}_{nk}(x),$$

where $K_n = o(n)$ tends to infinity as n tends to infinity. By Lemma 3.1, the estimators $\widehat{\rho}_{njk}$ are a.s. consistent and $n^{\frac{1}{2}}(\widehat{\rho}_{njk} - \rho_{jk})$ converges weakly to Gaussian variables. Let π, and respectively $\widehat{\pi}_n$, be the vector with components π_j, and respectively $\widehat{\pi}_{nj}$, for $j = 1, \ldots, d$, the weak convergence of the process $\widehat{\pi}_n$ is deduced from its MISE and Proposition 3.4.

Proposition 4.2. *Under the condition (3.7), the process $K_n^{-1} n^{\frac{1}{2}}(\widehat{\pi}_n - \pi)$ converges weakly to a Gaussian process.*

The estimators of the probability functions $p_j(x)$ are deduced from $\widehat{\pi}_n$, with estimators $\widehat{\theta}_{njk}$ of the parameters θ_{jk}, their asymptotic properties follow from Proposition 4.2.

4.4 Additive GLM model

Let $(Y_i, X_i)_{i=1,\ldots,n}$ be a sample of regression variables with a d dimensional vector of linearly independent explanatory variables $X_i = (X_{ij})_{j=1,\ldots,d}$ satisfying an additive GLM model

$$E(Y_i \mid X_i) = g \circ \Big(\eta_0 + \sum_{j=1}^d \eta_j(X_{ij})\Big), \tag{4.10}$$

for $i = 1, \ldots, n$ and $j = 1, \ldots, d$, with a known function g. The multivariate regression function $m(x) = E(Y_i \mid X_i)$ has a nonparametric estimator $\widehat{m}_n(x)$ defined in Section 3.6.

Let h be the inverse of the function g, the process $\widehat{l}_n(x) = h \circ \widehat{m}_n(x)$ of the linear model $l(x) = h \circ m(x) = \eta_0 + \sum_{j=1}^d \eta_j(x_j)$ has the expansion

$$h \circ \widehat{m}_n(x) - h \circ m(x) = \frac{\widehat{m}_n(x) - m(x)}{g' \circ h \circ m(x)} \{1 + o_{as}(1)\},$$

its asymptotic properties follow from Lemma 3.6 and Propositions 3.8.

Lemma 4.1. *Under the condition (3.7), the process $n^{\frac{1}{2}} K_n^{-1}\{\widehat{l}_n - l\}$ converges weakly to a Gaussian process with finite expectation and variance functions.*

The constant c is consistently estimated by

$$\widehat{\eta}_{n0} = n^{-1} \sum_{i=1}^n h \circ \widehat{m}_n(X_i)$$

and, by Lemma 4.1, it is asymptotically Gaussian, with the convergence rate $n^{-\frac{1}{2}} K_n$. The components η_j of the additive regression (4.10) are defined by integration of the linear function l as

$$\eta_j(x_j) = E\{l(X_i) \mid X_{ij} = x_j\} - \eta_0$$

and their estimation relies on the integration of the estimator $\widehat{l}_n$ of the function $l(x)$. Let $X_i^{(j)}(x)$ be the d dimensional vector with components x_j and the components of X_i except X_{ij}, then

$$\widehat{\eta}_{nj}(x_j) = n^{-1} \sum_{i=1}^n \widehat{l}_n(X_i^{(j)}(x)) - \widehat{\eta}_{n0}.$$

The weak convergence of the estimators $\widehat{\eta}_{nj}$ is deduced from Proposition 3.8 for the weak convergence of the process $n^{\frac{1}{2}} K_n^{-1}(\widehat{m}_n - m)$.

Proposition 4.3. *Under the condition (3.7) and for a function g of $C^1(\mathbb{R})$, the processes $n^{\frac{1}{2}} K_n^{-1}(\widehat{\eta}_{nj} - \eta_j)$ converge weakly to a Gaussian process with finite expectation and variance functions.*

The additive GLM model is extended to models with interactions, the conditional regression function has the form

$$m(X_i) = E(Y_i \mid X_i) = g \circ \Big\{\mu_0 + \sum_{j=1}^d m_j(X_{ij}) + \sum_{j' \neq j=1}^d m_{jj'}(X_{ijj'}) + \cdots\Big\}$$

with functions $m_j(x_j)$, $m_{jj'}(x_{jj'})$, etc. satisfying the orthogonality constraints of the additive regression model (3.32) with interactions. They are expressed as orthogonal conditional expectations given $X_{ij} = x_j$, respectively $X_{ijj'} = x_{jj'}$, and we can estimate them by the differences of the corresponding empirical conditional expectations of the estimator $h \circ \widehat{m}_n$.

4.5 Time varying effects in GLM models

A stationary model for binary longitudinal data $(Y_{ij})_{i=1,\ldots,I,j=1,\ldots,n_i}$ depending on a d-dimensional vector of covariates Z_i was defined by Zeger, Liang and Self (1985) as a logistic model with a conditional expectation function of (Y_{ij-1}, Z_i) through a linear model. The conditional probabilities were defined as

$$\begin{aligned} p_{ij} &= P(Y_{ij} = 1 \mid Y_{ij-1}, Z_i) = p(Z_i) + \{Y_{ij-1} - p(Z_i)\}\rho \\ &= \rho Y_{ij-1} + (1-\rho)p(Z_i), \end{aligned} \tag{4.11}$$

with a stationary correlation $\rho = \mathrm{corr}(Y_{ij}, Y_{ij-1} \mid Z_i)$, it follows that for every $j = 1, \ldots, n_i$

$$E(p_{ij} \mid Z_i) = P(Y_{ij} = 1 \mid Z_i) = \rho^j E(Y_{i0} \mid Z_i) + (1-\rho^j)p(Z_i).$$

Assuming that $E(Y_{i0} \mid Z_i) = p(Z_i)$, (4.11) entails

$$E(Y_{ij} \mid Z_i) = E(p_{ij} \mid Z_i) = p(Z_i). \tag{4.12}$$

Under the assumption of a linear model for logit $\pi(Z_i)$, the estimators of the parameters are consistent and asymptotically Gaussian.

In a nonparametric logit model with explanatory variables Z_i in a bounded subset $\mathcal{I}_Z$ of $\mathbb{R}^d$, the probability function $p(Z_i)$ of (4.11) is mapped to $\pi(\widetilde{Z}_i)$ where the components of $\widetilde{Z}$ are independent, by the change of variable (3.17) used in Section 4.1. The function $\pi(\widetilde{z})$ has an expansion on an orthonormal basis $(\psi_k)_k$ of $L^2(F_{\widetilde{Z}})$ and it is estimated by projection on the estimated basis $(\widehat{\psi}_{nk})_k = L_k \circ \widehat{H}_{n\widetilde{Z}}$ defined with the empirical estimator (3.20). The function π has the expansion $\pi(\widetilde{z}) = \sum_{k=0}^{\infty} b_k \psi_k$ and it is approximated by $\pi_n(\widetilde{z}) = \sum_{k=0}^{K_n} b_k \psi_k$ where $K_n = o(n)$ tends to infinity as $n = \sum_{i=1}^{I} n_i$ tends to infinity.

Under the assumption $E(Y_{i0} \mid Z_i) = p(Z_i)$ and by (4.12), the coefficients b_k have the estimators

$$\widehat{b}_{nk} = I^{-1} \sum_{i=1}^{I} (n_i - 1)^{-1} \sum_{j=1}^{n_i - 1} Y_{i(j)} \widehat{\psi}_{nik}(Z_{(n:i)}),$$

for $k = 0, \ldots, K_n$, with the notations of (3.4). Let $K_n = o(n^{\frac{1}{2}})$, the MISE error of the estimator $\widehat{\pi}_n(\widetilde{z}) = \sum_{k=0}^{K_n} \widehat{b}_{nk} \widehat{\psi}_{nk}$ of $\pi(\widetilde{z})$ is

$$E\|\widehat{\pi}_n - \pi\|_2^2 = O(n^{-1}K_n^2)$$

as the estimation error and the approximation error have the same order, and the process $n^{\frac{1}{2}} K_n^{-1}(\widehat{\pi}_n - \pi)$ converges weakly to a Gaussian process.

In a logit model (4.1) for $p(z)$, the equation (4.11) entails

$$m(Z_i) = \log \frac{E(Y_{ij} = 1 \mid Z_i)}{P(Y_{ij} = 0 \mid Z_i)} = \log \frac{p(Z_i)}{1 - p(Z_i)}$$

and the function $\mu(\widetilde{z}) = m(z)$ is estimated using the estimator $\widehat{\pi}_n$, its asymptotic properties follow.

Let Y be a real variable and let X be a d dimensional vector of regressors and let $(X_{ij}, Y_{ij})_{i=1,\dots,I,j=1,\dots,n_i}$ be a n-sample of the regression variables in I sub-populations, with $n = \sum_{i=1}^{I} n_i$. We assume that for every i, $n^{-1}n_i$ converge to a strictly positive limit p_i, as n tends to infinity, and that the covariates are not linearly dependent. The projection pursuit model (4.7) is extended by considering distinct parameters θ_i in $\mathbb{R}^d$ for each sub-sample. The conditional model for the expectation of the variables Y_{ij} is

$$r_{\theta_i}(X_{ij}) = E(Y_{ij} \mid X_{ij}) = g \circ m(\theta_i^T X_{ij}) \tag{4.13}$$

with a regression function m of $C^2(\mathbb{R})$ and parameter vectors θ_i in subsets Θ_i of $\mathbb{R}^d$. Let θ_{0i} be the parameter values under the probability P_0 of the sample. At θ_i, the variable $W_{\theta,i} = \theta_i^T X_i$ has the distribution function $F_{W_{\theta,i}}$ on a subset $\mathcal{I}_{W_{\theta,i}}$ of $\mathbb{R}$ and its empirical estimator is denoted $\widehat{F}_{n,W_{\theta,i}}$, for $i = 1, \dots, I$.

The function $r(w_i) = g \circ r(w_i)$ is estimated by projection on the orthonormal basis of functions $(\psi_k)_{k \geq 0}$ defined on $(L^2(\mathcal{I}_{W_{\theta,i}}, F_{W_{\theta,i}})$ like in Section 3.2. The orthonormal basis of $(L^2(\mathcal{I}_{W_{\theta,i}}, F_{W_{\theta,i}})$ is defined like (3.1), with the Laguerre polynomials as

$$\psi_{ik} = L_k \circ H_{W_{\theta,i}}, \ k \geq 0, \tag{4.14}$$

with the cumulative hazard function $H_{W_{\theta,i}} = -\log(1 - F_{W_{\theta,i}})$. For every i, the function $r(w_i)$ is approximated by the sum of K_{ni} functions of the basis

$$r_{n\theta i}(w_i) = \sum_{k=0}^{K_{ni}} b_{ik}\psi_{ik}(w_i)$$

with the coefficients

$$b_{ik} = \int_{\mathcal{I}_{w_i}} r(w_i)\psi_{ik}(w_i)\, dF_{W_{\theta,i}}(w_i), \ k = 0, \dots, K_{ni}.$$

The functions ψ_{ik} have the empirical estimators

$$\widehat{\psi}_{nik}(w_i) = L_k \circ \widehat{H}_{n\theta,i}(w_i) \tag{4.15}$$

where $\widehat{H}_{n\theta,i}(w_i)$ is the empirical estimator of the cumulative hazard function of the variables $W_{\theta,i}$, they satisfy Lemma 3.1. The coefficients b_{ik} are estimated by

$$\widehat{b}_{nik} = (n_i - 1)^{-1} \sum_{j=1}^{n_i-1} Y_{i(j)} \widehat{\psi}_{nik}(W_{\theta,i(n_i:j)}), \tag{4.16}$$

where the larger observation $W_{\theta,i(n_i:n_i)}$ is omitted so the variables $\widehat{\psi}_{nik}(W_{\theta,i(n_i:j)})$, $j = 1,\ldots,n_i - 1$, are bounded as n_i tends to infinity, and $Y_{i(j)}$ is the response variable to the regressors $W_{\theta,i,(n_i:j)}$. The estimators $\widehat{b}_{nik}$ are consistent and asymptotically Gaussian, by Lemma 3.2.

The regression function r has the estimators

$$\widehat{r}_{n,\theta i}(w_i) = \sum_{k=0}^{K_{ni}} \widehat{b}_{nik} \widehat{\psi}_{nik}(w_i).$$

The mean squared estimation error of $r_{\theta i}$ is

$$l_n(\theta) = n^{-1} \sum_{i=1}^{I} \sum_{i=1}^{n_i} \{Y_{ij} - \widehat{r}_{n,\theta i}(W_{\theta,ij})\}^2,$$

and the parameter θ_i is estimated by $\widehat{\theta}_{ni}$ which minimizes the function

$$l_{ni}(\theta_i) = \sum_{i=1}^{n_i} \{Y_{ij} - \widehat{r}_{n,\theta_i i}(W_{\theta_i,ij})\}^2.$$

Finally, the functions r and m are estimated by

$$\widehat{r}_n(w) = I^{-1} \sum_{i=1}^{I} \widehat{r}_{n,\widehat{\theta}_{ni} i}(w),$$

$$\widehat{m}_n = g^{-1} \circ \widehat{r}_n.$$

At the true parameter values θ_{0i}, the variables $l_{ni}(\theta_{0i})$ converges a.s. under P_0 to the variance σ_i^2 of the variable Y_i and this is the minimum value of the limit of l_{ni}. It follows that the estimators $\widehat{\theta}_{ni}$ are a.s. consistent and the estimators $\widehat{r}_n$ converges a.s. uniformly to the true function $r(\theta_0^T x)$.

Proposition 4.4. *Under P_0, the variables $n_i^{\frac{1}{2}} K_{ni}^{-1}(\widehat{\theta}_{ni} - \theta_{0i})$ converge weakly to centered Gaussian variable, for $i = 1,\ldots,I$.*

The proof is similar to the proof of Proposition 3.6 for the projection pursuit model, by the minimization of $l_{ni}(\theta_i)$ for every i. Let $K_n^2 = \sum_{i=1}^{I} n n_i^{-1} K_{ni}^2$, then $K_n^{-1} K_{nj}$ converges to a strictly positive limit as n tends to infinity.

The mean integrated squared error of $\widehat{r}_n$ is

$$\begin{aligned} \mathrm{MISE}_n(r) &= \sum_{i=1}^I E \int_{\mathcal{I}_{W_{\theta,i}}} \{\widehat{r}_n(w_i) - r(w_i)\}^2 \, dF_{\theta_{0i}^T X_i}(w_i) \\ &= \sum_{i=1}^I E \int_{\mathcal{I}_{W_{\theta,i}}} \{\widehat{r}_n(w_i) - r_n(w_i)\}^2 \, dF_{\theta_{0i}^T X_i}(w_i) \\ &\quad + \sum_{i=1}^I \int_{\mathcal{I}_{W_{\theta,i}}} \{r_n(w_i) - r(w_i)\}^2 \, dF_{\theta_{0i}^T X_i}(w_i), \end{aligned}$$

where the first term of $\mathrm{MISE}_n(r)$ is the estimation error, its order is $O(I^{-2}\sum_{i=1}^I n_i^{-1}K_{ni}^2) = O(n^{-1}K_n^2)$ by Lemma 6.1 and with I independent sub-samples, the second term is a $o(1)$. As n tends to infinity, the number K_{nj} of the functions approximating r are supposed to satisfy the conditions

$$K_{nj} = o(n), \quad \|r_0 - r_{0K_n}\|_2 = O((n^{-\frac{1}{2}}K_n^2)^{\frac{1}{2}}), \tag{4.17}$$

then the MISE is a $O(n^{-1}K_n^2)$, as n tends to infinity.

Proposition 4.5. *Under P_0 and the conditions (4.17), the processes $n^{\frac{1}{2}}K_n^{-1}(\widehat{r}_n - r_0)$ and $n^{\frac{1}{2}}K_n^{-1}(\widehat{m}_n - m_0)$ converge weakly to Gaussian processes with finite expectation and variance functions.*

The weak convergence is a consequence of Proposition 4.4, it is proved like Proposition 3.7 in the regression model (3.13).

4.6 GLM with random effects

We consider a parametric logistic model with a random effects for a sample $(Y_i)_{i=1,\ldots,n}$ of a Bernoulli variable with random probabilities following a model depending on independent samples of regressors $(X_i)_{i=1,\ldots,n}$ and $(A_i)_{i=1,\ldots,n}$,

$$p_\theta(X_i, A_i) = P_\theta(Y_i = 1 \mid X_i, A_i)$$

where A_i are unobserved random variables with density f. Let

$$p_\theta(X_i, A_i) = \frac{\exp\{\theta^T X_i + h(A_i)\}}{1 + \exp\{\theta^T X_i + h(A_i)\}}$$

with a known function h, the log-likelihood of the sample $(Y_i)_{i=1,\ldots,n}$ conditionally on $(X_i)_{i=1,\ldots,n}$ is

$$\begin{aligned} l_n(\theta) &= \sum_{i=1}^n \log \int p_\theta^{Y_i}(X_i, a_i)\{1 - p_\theta(X_i, a_i)\}^{1-Y_i} f(a_i) \, da_i \\ &= \sum_{i=1}^n \log \int \frac{\exp\{Y_i(\theta^T X_i + h(a_i))\}}{1 + \exp\{\theta^T X_i + h(a_i)\}} f(a_i) \, da_i. \end{aligned}$$

For every parametric density, its parameter and the parameter θ are estimated by maximization of the log-likelihood using an EM algorithm. Assuming that the variables A_i have a multinomial distribution with probabilities p_k, for $k = 1, \ldots, K$, $l_n(\theta)$ is the log-likelihood of a finite discrete mixture with

$$\begin{aligned}\int p_\theta(X_i, a_i)^{Y_i}\{1 - p_\theta(X_i, a_i)\}^{1-Y_i} f(a_i)\, da_i \\ = \sum_{k=1}^{K} p_k p_\theta(X_i, h_k)^{Y_i}\{1 - p_\theta(X_i, h_k)\}^{1-Y_i} \\ = \sum_{k=1}^{K} p_k \frac{\exp\{Y_i(\theta^T X_i + h_k)\}}{1 + \exp\{\theta^T X_i + h_k\}}.\end{aligned}$$

The logistic model with a random effects extends to a hierarchical sampling for a sample $(Y_{ij})_{i=1,\ldots,I,j=1,\ldots,n_i}$ of a Bernoulli variable with random probabilities

$$p_\theta(X_{ij}, A_i) = P_\theta(Y_{ij} = 1 \mid X_{ij}, A_i)$$

conditionally on independent samples of regressors $(X_{ij})_{i=1,\ldots,I,j=1,\ldots,n_i}$ and $(A_i)_{i=1,\ldots,I}$, such that A_i are unobserved variables with density f. Let

$$p_\theta(X_{ij}, A_i) = \frac{\exp\{\theta^T X_{ij} + h(A_i)\}}{1 + \exp\{\theta^T X_{ij} + h(A_i)\}}$$

with a known function h, the log-likelihood of the sample $(Y_{ij})_{i,j}$ conditionally on $(X_{ij})_{i,j}$ is

$$\begin{aligned} l_n(\theta) &= \sum_{i=1}^{I} \log\Big\{\int \prod_{j=1}^{n_i} p_\theta^{Y_{ij}}(X_{ij}, a_i)\{1 - p_\theta(X_{ij}, a_i)\}^{1-Y_{ij}} f(a_i)\, da_i\Big\} \\ &= \sum_{i=1}^{I} \log \int \prod_{j=1}^{n_i} \frac{\exp\{Y_{ij}(\theta^T X_{ij} + h(a_i))\}}{1 + \exp\{\theta^T X_{ij} + h(a_i)\}} f(a_i)\, da_i.\end{aligned}$$

In a GLM where $E(Y_{ij} \mid X_{ij}, A_i) = g(\theta^T X_{ij} + h(A_i))$ and the variable Y_{ij} has a density $\phi(Y_{ij}, g(\theta^T X_{ij} + h(A_i))$, the log-likelihood of the sample $(Y_{ij})_{i,j}$ conditionally on $(X_{ij})_{i,j}$ is

$$l_n(\theta) = \sum_{i=1}^{I} \log \int \prod_{j=1}^{n_i} \phi(Y_{ij}, g(\theta^T X_{ij} + h(A_i)) f(a_i)\, da_i,$$

in every model the parameters of f, g and θ are estimated by maximization of the log-likelihood, with an EM algorithm.

4.7 Estimation of a conditional Poisson process

Parametric models of aggregate observations for dependent individuals following a Poisson variable with a random parameter

$$\lambda_i = \mu_i e^{r_\theta(X_i)+\varepsilon_i}, \quad i = 1, \ldots, n,$$

have been defined by linear regressions on a vector X_i of random risk factors and a Gaussian error ε_i due to unobserved factors such as geographical variations. Integrating with respect to the distribution $\mathcal{N}(0, \sigma_\varepsilon^2)$ of the error yields the intensity

$$\lambda_i = \mu_i e^{r_\theta(X_i)-\frac{1}{2}\sigma_\varepsilon^2},$$

the model extends to nonparametric regressions.

Let Y be a Poisson process with a conditional expectation function defined on a subset $\mathcal{I}_X$ of $\mathbb{R}$, support of a regression variable X having a distribution function F_X

$$E(Y \mid X = x) = \lambda(x),$$

its conditional probabilities are

$$P(Y = k \mid X = x) = e^{-\lambda(x)} \frac{\lambda^k(x)}{k!}, \; k \geq 0. \tag{4.18}$$

We assume that $\mu = E\{\lambda(X)\}$ and $\text{Var}\{\lambda(X)\}$ are finite. The probabilities of the variable Y are integrated as

$$P(Y = k) = \int_{\mathcal{I}_X} e^{-\lambda(x)} \frac{\lambda^k(x)}{k!} \, dF_X(x).$$

Let $(Y_i)_{i=1,\ldots,n}$ be a sample of independent and identically distributed observations of the variable Y, the expectation μ of $\lambda(X)$ has the empirical estimator

$$\widehat{\mu}_n = n^{-1} \sum_{i=1}^n e^{-\widehat{\lambda}_n(X_i)} \frac{\{\widehat{\lambda}_n(X_i)\}^{Y_i}}{(Y_i - 1)!} \tag{4.19}$$

depending on an estimator $\widehat{\lambda}_n$ of the function λ. The intensity function λ is estimated by projection on the orthonormal basis of functions $(\psi_k)_{k\geq 0}$ of $L^2(\mathcal{I}_X, F_X)$ defined like in Section 3.1 as

$$\psi_j(x) = L_j \circ H_X(x), \; j \geq 0,$$

they are estimated by the processes

$$\widehat{\psi}_{nj}(x) = L_j \circ \widehat{H}_n(x)$$

on every sub-interval of $\mathcal{I}_X$ where $F_X(x) < 1$ and the estimators satisfy Lemma 3.1. Let $\lambda = \sum_{j\geq 0} b_j\psi_j$ with the coefficients

$$b_j = \int_{\mathcal{I}_X} \lambda(x)\psi_j(x)\,dF_X(x),$$

for $j \geq 1$, and $b_0 = E\lambda(X)$. We have

$$E\{Y_i\psi_j(X_i)\} = \int_{\mathcal{I}_X} \lambda(x)\psi_j(x)\,dF_X(x) = b_j$$

so the coefficient b_j is consistently estimated by the empirical mean

$$\widehat{b}_{nj} = (n-1)^{-1}\sum_{i=1}^{n-1} Y_{(i)}\widehat{\psi}_{nij}(X_{(n:i)}),\ j \geq 1, \tag{4.20}$$

where $\widehat{\psi}_{nij}$ is the estimator where the ith observation $X_{(n:i)}$ is omitted. By Lemma 3.1, the estimator (4.20) is such that

$$\widehat{b}_{nj} = (n-1)^{-1}\sum_{i=1}^{n-1} Y_{(i)}\psi_j(X_{(n:i)}) + O_p(n^{-\frac{1}{2}}).$$

Furthermore $E(Y^2 \mid X) = \lambda(X) + \lambda^2(X)$, the variance of $Y\psi_j$ is

$$\sigma_j^2 = E\{Y^2\psi_j^2(X)\} - b_j^2 = E\{\lambda(X)\psi_j^2(X)\} + \mathrm{Var}\{\lambda(X)\psi_j(X)\}$$

and the covariance of $Y\psi_j$ and $Y\psi_{j'}$ is

$$\begin{aligned}\sigma_{jj'}^2 = E\{Y^2\psi_j(X)\psi_{j'}(X)\} - b_jb_{j'} &= E\{\lambda(X)\psi_j(X)\psi_{j'}(X)\}\\ &\quad + \mathrm{Cov}\{\lambda(X)\psi_j(X), \lambda(X)\psi_{j'}(X)\},\end{aligned}$$

using these notations, the variance of $\widehat{b}_{nj}$ is asymptotically equivalent to $n^{-1}\sigma_j^2$ and the covariance of $\widehat{b}_{nj}$ and $\widehat{b}_{nj'}$ is asymptotically equivalent to $n^{-1}\sigma_{jj'}^2$. Under the condition that $\sigma_{jj'}^2$ is finite for all j and j', it follows that the variables $n^{\frac{1}{2}}(\widehat{b}_{nk} - b_k)$ converge weakly to centered Gaussian variables with strictly positive variances σ_k^2, for every k.

Let $N_n = Y_{(n:n)}$ be the largest value observed in a sample of n independent observations $Y_1, \ldots, Y_n$ of the random Poisson process, the function λ has the estimator

$$\widehat{\lambda}_n(x) = \sum_{j=0}^{N_n} \widehat{b}_{nj}\widehat{\psi}_{nj}(x).$$

By Lemma 3.1, the processes $n^{\frac{1}{2}}(\widehat{b}_{nk}\widehat{\psi}_{nk} - b_k\psi_k)$ converge weakly to centered Gaussian processes with finite and strictly positive variance functions, for every interger k. Conditionally on N_n, the expectation of $\widehat{\lambda}_n$ conditionally on N_n is a.s. equivalent to

$$\lambda_n(x) = \sum_{j=0}^{N_n} b_j \psi_j(x)$$

and it converges to $\lambda(x)$ as N_n tends to infinity with n. The variance of $n^{\frac{1}{2}}\{\widehat{\lambda}_n(x) - \lambda_n(x)\}$ conditionally on N_n is a sum

$$v_n(x) = \sum_{j=0}^{N_n} v_{n,j}(x) + \sum_{j' \neq j=0}^{N_n} v_{n,jj'}(x)$$

where $v_{n,jj'}(x)$ is the covariance of $\widehat{b}_{nj}\widehat{\psi}_{nj}(x)$ and $\widehat{b}_{nj'}\widehat{\psi}_{nj'}(x)$, and $v_{n,j}(x)$ is the variance of the jth term. Let

$$V_n = \int_{\mathcal{I}_X} v_n(x)\, dx.$$

Conditionally on N_n, the mean integrated squared error of the estimator in $L^2(\mathbb{R}, F_X)$ is the sum

$$\text{MISE}_n(\lambda) = E(\|\widehat{\lambda}_n - \lambda_n\|_2^2 \mid N_n) + \|\lambda - \lambda_n\|_2^2$$

of the conditional squared estimation error $E(\|\widehat{\lambda}_n - \lambda_n\|_2^2 \mid N_n)$ and the square error of the approximation of λ by λ_n, $\|\lambda - \lambda_n\|_2^2 = \sum_{j=N_n+1}^{\infty} b_j^2$, which converges to zero as n and N_n tend to infinity.

Proposition 4.6. *Conditionally on N_n, the estimator $\widehat{\lambda}_n$ of the function λ satisfies*

$$\text{MISE}_n(\lambda) = O(n^{-1}N_n^2) + o(1) = o(1).$$

Proof. The squared conditional estimation error is the integral of $n^{-1}v_n$

$$n^{-1}V_n = n^{-1}\sum_{j=0}^{N_n} \int_{\mathcal{I}_X} v_{n,j}(x)\, dx + n^{-1}\sum_{j=0}^{N_n}\sum_{j' \neq j=0}^{N_n} v_{n,jj'}(x)\, dx,$$

where the variance and covariance are $v_{n,jj'} = n^{-1}v_{jj'} + o(n^{-1})$ with finite $v_{jj'}$ depending on the covariance of the estimated functions of the basis and on $\sigma_{jj'}^2$, by Lemmas 3.1 and 3.3. The squared estimation error of $\widehat{\lambda}_n(x)$ conditionally on N_n is $n^{-1}V_n + \sum_{j=N_n+1}^{\infty} b_j^2$ and it is a $O(n^{-1}N_n^2)$. For every $\varepsilon > 0$, we have $P(Y^2 > n\varepsilon) \leq \mu(n\varepsilon)^{-\frac{1}{2}}$ and

$$P(n^{-1}N_n^2 > \varepsilon) = 1 - P^n(Y^2 < n\varepsilon) = 1 - \left\{1 - \frac{\mu}{(n\varepsilon)^{\frac{1}{2}}}\right\}^n = o(1),$$

it follows that $N_n = o_p(n^{\frac{1}{2}})$ as n tends to infinity. □

Proposition 4.7. *On every compact subset of $\mathcal{I}_X$, the process $N_n^{-1} n^{\frac{1}{2}} \{\widehat{\lambda}_n(X) - \lambda_n(X)\}$ converges weakly to a centered Gaussian variable with a finite variance function.*

Proof. On a compact set the function $\max_{j\geq 0} \psi_j$ is uniformly bounded and $\max_{j\geq 0} \int_{\mathcal{I}_X} \sigma_{jj'}^2(x)\, dF_X(x)$ is bounded for all j and j', it follows that $nN_n^{-2} \int_{\mathcal{I}_X} v_n(x)\, dF_X(x)$ converges to a finite limit as n tends to infinity. The expectation and the variance of the variable $n^{\frac{1}{2}}\{\widehat{\lambda}_n(X) - \lambda_n(X)\}$ converge to finite limits, its weak convergence is a consequence of the weak convergence of the processes $n^{\frac{1}{2}}(\widehat{b}_{nk}\widehat{\psi}_{nk} - b_k\psi_k)$. The proof is similar to the proof of Proposition 2.9 for a compound Poisson process. □

The model (4.18) generalizes to a regression vector X and the nonparametric intensity is estimated after orthogonalization of the components of the regressor, like in Section 3.7. The multivariate models of Sections 3.5 to 3.8 apply to the nonparametric intensity $\lambda(X)$ and the convergence of its estimators are similar to the convergence in the nonparametric regression models.

4.8 Tests in nonparametric GLM models

In the generalized nonparametric regression models, the expectation of the variable Y conditionally on X depends on a function g without error. Let F_{XY} be the distribution function of (X, Y) and let $F_{Y|X}$ be the conditional distribution function of Y, the model for the expectation of Y has the form

$$E(Y \mid X = x) = \int y\, dF_{Y|X=x}(y) = g \circ m(x)$$

and the variance of Y conditionally on $X = x$ is

$$\text{Var}(Y \mid X = x) = \int y^2\, dF_{Y|X=x}(y) - g^2 \circ m(x) = \sigma_X^2(x).$$

In the multinomial models, the conditional expectation and variance of the response variable Y is a function of the probabilities $p_j(X) = g \circ m(X)$

$$E(Y \mid X = x) = \sum_{j=1}^{d} j p_j(x),$$

$$\text{Var}(Y \mid X = x) = \sum_{j=1}^{d} j^2 p_j(x) - \Big\{\sum_{j=1}^{d} j p_j(x)\Big\}.$$

In conditional Poisson processes, the conditional variance is the expectation of the process. The empirical distribution of the variable Y determines the model.

Let $(X_{1i}, Y_{1i})_{i=1,\dots,n_1}$ and $(X_{2i}, Y_{2i})_{i=1,\dots,n_2}$ be independent samples of discrete variables with conditional probabilities

$$p_{jk}(X_{ji_j}) = P(Y_{ji_j} = k \mid X_{ji_j}) = g \circ m_{jk}(X_{ji_j})$$

with a monotone function g, for all integers k in the support of the variables Y_1 and Y_2, for $i_j = 1, \dots, n_j$ and $j = 1, 2$. In the nonparametric GLM model (4.4), the conditional probabilities $p_{jk}(X_{ji_j})$ of Y_{ji_j} given X_{ji_j} have estimators $\widehat{p}_{njk}$ obtained by projection on a basis of orthonormal functions of $L^2(F_{X_j})$ after the orthogonalization of the regression vector like in Section 3.7. Nonparametric estimators of the functions m_{jk} are deduced by inversion of the function g as $\widehat{m}_{njk} = g^{-1}(\widehat{p}_{njk})$, for $j = 1, 2$. The estimators $\widehat{p}_{njk}$ and $\widehat{m}_{njk}$ have the same convergence rate studied in Chapter 3.

A test for the comparison of the distribution functions of the samples is a test of comparison of their conditional probabilities $p_{1k}(x)$ and $p_{2k}(x)$ in an interval $\mathcal{I}_X$ of the common support, with the hypothesis $H_0 : p_{1k} = p_{2k}$ in $\mathcal{I}_X$, for every integer k in the support of the variables Y_j, it is equivalent to a test of the hypothesis $m_{1k} = m_{2k}$ in $\mathcal{I}_X$, for every k in the support of the variables Y_j, $j = 1, 2$. The tests performed in Section 3.11 are extended their statistics are the means of the statistics previously defined for every integer k in the support of the variables Y_1 and Y_2.

Tests of sub-models in the multivariate models 3.6 and 3.8 for the conditional probabilities are performed in the same way with means of statistics depending on the terms omitted in the regression under the hypothesis, like in the parametric analysis of variance.

Let $(X_i, Y_i)_{i=1,\dots,n}$ be a sample of independent observations such that $p_j(X_i) = P(Y_i = j \mid X_i) = m_j(X_i)$ for every integer $j = 1, \dots, J$ in the support of the variables Y_i. A goodness of fit tests for a parametric GLM models $\mathcal{M}_\Theta = \{m = m_\theta \in C^2(\Theta), \theta \in \Theta\}$ on an interval $\mathcal{I}_X$ in the support of the regressor is performed with a statistic that compares a nonparametric estimators of a regression functions to its estimator in $\mathcal{M}_\Theta$. We consider a partition of the sample as two independent sub-samples of sizes n_1 and respectively n_2, of order $O(n)$ and such that $n^{-1}n_1$ converges to p in $]0, 1[$

as n tends to infinity. We assume

$$p_j(X_{1i}) = P(Y_{1i} = j \mid X_{1i}) = g \circ m_j(X_{1i}),$$
$$p_{\theta_j}(X_{2i}) = P(Y_{2i} = j \mid X_{2i}) = g \circ m_{\theta_j}(X_{2i}), \tag{4.21}$$

with m_{θ_j} in $\mathcal{M}_\Theta$, for very $j = 1, \ldots, J$. A test of comparison of the J regression functions m_j and m_{θ_j} in $\mathcal{I}_X$ provides a a goodness-of-fit test for the family of models $\mathcal{M}_\Theta$.

In $\mathcal{M}_\Theta$, the least squares estimators and the maximum likelihood estimators $\widehat{\theta}_{nj}$ of the parameters θ_j obtained from the second sub-sample are consistent and their convergence rate is $n_2^{-\frac{1}{2}}$, this is also the convergence rate of $m_{\widehat{\theta}_{nj}}$, let $v(x)$ be the limiting variance of $n_2^{\frac{1}{2}}(m_{\widehat{\theta}_{nj}} - m_j)$, for every $j = 1, \ldots, J$, under H_0. Let $\widehat{m}_{nj}$ be the nonparametric estimators of the function m_j based on the first sub-sample based on a projection on K_{n_1} functions of an estimated basis and let

$$B_{nj} = K_{n_1}^{-1} n_1^{\frac{1}{2}}(\widehat{m}_{nj} - m_j),\ j = 1, \ldots, J$$

be the normalized processes. Under conditions (3.7), the processes B_{nj} converge weakly to the processes B_j determined in Proposition 3.4. If J is finite, let

$$T_n = n_1^{\frac{1}{2}} K_{n_1}^{-1} J^{-1} \sum_{j=1}^{J} \sup_{x \in \mathcal{I}_X} |\widehat{m}_{nj}(x) - m_{\widehat{\theta}_{nj}}|, \tag{4.22}$$

if J is infinite, let J_n converge to infinite as n tends to infinite and let

$$T_n = n_1^{\frac{1}{2}} K_{n_1}^{-1} J_n^{-1} \sum_{j=1}^{J_n} \sup_{x \in \mathcal{I}_X} |\widehat{m}_{nj}(x) - m_{\widehat{\theta}_{nj}}|. \tag{4.23}$$

The proofs of the following propositions are consequences of Propositions 3.17 and 3.18.

Proposition 4.8. *Under Conditions, the statistic T_n converges weakly under H_0 to the supremum of the mean of Gaussian processes B_j. Under fixed alternatives, it tends to infinity.*

Let $(A_n)_{n \geq 1}$ be a sequence of local alternatives defined by regression functions $m_{nj} = m_j + n_1^{-\frac{1}{2}} K_{n_1} r_{nj}$ and $\theta_{nj} = \theta_j + n_2^{-\frac{1}{2}} a_{nj}$ such that $m_{\theta_j} = m_j$ and the sequences $(r_{nj})_{n_j \geq 1}$, and respectively $(a_{nj})_{n_j \geq 1}$, converge uniformly on I_X to functions r_j, and respectively to a limit a_j, as n tends to infinity, for $j = 1, \ldots, J$.

Proposition 4.9. *The statistic T_n converges weakly under A_n to*

$$T = \lim_n J_n^{-1} \sum_{j=1}^{J_n} \sup_{\mathcal{I}_X} |(1-p)^{\frac{1}{2}} \rho_2(B_j + r_j)|.$$

If J is finite, T is replaced by $J^{-1} \sum_{j=1}^{J} \sup_{\mathcal{I}_X} |(1-p)^{\frac{1}{2}} \rho_2(B_j + r_j)|$ in Proposition 4.9.

The semi-parametric GLM model (4.21) generalizes to a GLM model where the function m is semi-parametric. Let

$$\mathcal{H} = \{h = h_{\theta,m}, \theta \in \Theta, m \in \mathcal{M}, h \in C^2(\Theta) \cap C^2(\mathcal{M})\}$$

be a semi-parametric regression model defined on an interval $\mathcal{I}_X$ by a parameter θ and an unknown function m in $\mathcal{M}$, like the projection pursuit regression model (3.13) and the regression model with transformed explanatory variables (3.35) where $h_{\theta,m}(x) = m \circ \eta_\theta(x)$ with an unknown function m and a parametric function η_θ such that the map $\theta \mapsto \eta_\theta(x)$ is $C^2(\Theta)$ uniformly in x.

Let $(X_j, Y_j)_{j=1,\ldots,n}$ be a sample of independent observations partitioned as two independent sub-samples of sizes n_1 and n_2 of order $O(n)$ such that $n^{-1}n_1$ converges to p in $]0,1[$ as n tends to infinity, and let

$$\begin{aligned} p_j(X_{1i}) &= P(Y_{1i} = j \mid X_{1i}) = g \circ m_j(X_{1i}),\ i = 1, \ldots, n_1, \\ p_{\theta_j,m_j}(X_{2i}) &= P(Y_{2i} = j \mid X_{2i}) = g \circ h_{\theta_j,m_j}(X_{2i}),\ i = 1, \ldots, n_2, \end{aligned} \quad (4.24)$$

on $\mathcal{I}_X$, with $h_{\theta,m}$ in $\mathcal{H}$. In the semi-parametric models (3.13) and (3.35), the estimator of the function $h_{\theta,m}$ based on a projection on K_{n_2} functions of an estimated basis of functions has the convergence rate $n_2^{-\frac{1}{2}} K_{n_2}^{-1}$ like in the nonparametric regression model.

A goodness-of-fit test for the family of models $\mathcal{H}$ is performed with the mean statistics T_n defined by (4.22) or (4.23) which compares the nonparametric estimators of the regression functions m_j and their estimators $h_{\widehat{\theta}_{nj},\widehat{m}_{nj}}$ in $\mathcal{H}$. Its asymptotic behavior under H_0 and local alternatives are given by Propositions 4.8 and 4.9.

A test of homogeneity in the projection pursuit GLM model (4.13) has the hypothesis $H_0 : E(Y_{ij} \mid X_{ij}) = g \circ m(\theta^T X_{ij})$ for all $i = 1, \ldots, I$ and $j = 1, \ldots, n_i$, against the alternative of a model with distinct coefficients $\theta_1, \ldots, \theta_I$, for each sub-samples. The test of the hypothesis is performed by a comparison of the estimators of $r(\theta_i^T X_{ij})$ and $r(\theta^T X_{ij})$, for $i = 1, \ldots, I$, with the statistic

$$T_{nI} = n^{\frac{1}{2}} K_n^{-1} \sum_{i=1}^{I} \sup_{\mathbb{R}} |\widehat{r}_n - \widehat{r}_{n,i}|, \tag{4.25}$$

where $\widehat{r}_{n,i} = \widehat{r}_{n,\widehat{\theta}_{ni}i}$ is defined in Section 4.5. Under the hypothesis H_0, the processes $n^{\frac{1}{2}} K_n^{-1}(\widehat{r}_n - r_0)$, and respectively $n_i^{\frac{1}{2}} K_{ni}^{-1}(\widehat{r}_{ni} - r_0)$, converges to a Gaussian process W_0, and respectively W_{0i}, depending on $r_0(x) = r(\theta_0^T x)$, the common expectation of Y_{ij} conditionally on X_{ij} in each sub-sample. Let ρ_i be the limit of $n^{\frac{1}{2}} n_i^{-\frac{1}{2}} K_n^{-1} K_{ni}$ as n tends to infinity.

Proposition 4.10. *Under the hypothesis H_0, the statistic T_{nI} converges weakly to the variable $T_{0I} = \sum_{i=1}^{I} \sup_{\mathbb{R}} |W_0 - \rho_i W_{0i}|$. Under fixed alternatives, T_n diverges.*

Let $(A_n)_{n\geq 1}$ be a sequence of local alternatives such that $\theta_{ni} = \theta_0 + n_i^{-\frac{1}{2}} K_{ni} h_{ni}$, for $i = 1, \ldots, I$, and $\theta_n = \theta_0 + n^{-\frac{1}{2}} K_n h_n$, where the sequences $(h_{ni})_n$, and respectively $(h_n)_n$, converge as n tends to infinity to non zero limits h_j, and respectively h, with $\rho_i h_i$ different from h for every $i = 1, \ldots, I$.

Proposition 4.11. *The statistic T_{nI} converges weakly under A_n to the variable $T_I = \sum_{i=1}^{I} \sup_{\mathbb{R}} |W_0 - \rho_i W_{0i} + h - \rho_i h_i|$.*

Let $(Y_i)_{i=1,\ldots,n}$ be a sample of a Poisson process with a conditional parameter $\lambda(X)$, defined by model (4.18). A goodness of fit test for the hypothesis H_0 of a Poisson variable with a scaler parameter λ against the alternative of a conditional intensity $\lambda(X)$ is performed with the empirical log-likelihood ratio variable

$$\begin{aligned} l_n &= \sum_{i=1}^{n} \log \frac{e^{-\widehat{\lambda}_n(X_i)} \{\widehat{\lambda}_n(X_i)\}^{Y_i}}{e^{-\widehat{\lambda}_{0n}} \widehat{\lambda}_{0n}^{Y_i}} \\ &= \sum_{i=1}^{n} Y_i \log \frac{\widehat{\lambda}_n(X_i)}{\widehat{\lambda}_{0n}} - \sum_{i=1}^{n} \{\widehat{\lambda}_n(X_i) - \widehat{\lambda}_{0n}\} \end{aligned}$$

where $\widehat{\lambda}_n$ is the estimator of the function λ in model (4.18) and $\widehat{\lambda}_{0n} = \bar{Y}_n$ is the estimator of the parameter λ in a Poisson model. By Proposition 4.7 the estimator $\widehat{\lambda}_n$ of the conditional function λ satisfies

$$\widehat{\lambda}_n(x) = \lambda(x) + U_n(x)$$

where $U_n = \alpha_n^{\frac{1}{2}} B_n$, $\alpha_n = n^{-1} N_n^2$ and the process $B_n(x) = \alpha_n^{-\frac{1}{2}}\{\widehat{\lambda}_n(x) - \lambda(x)\}$ converges weakly to a Gaussian process B with a finite variance function, under the alternative. Under H_0, let λ_0 be the parameter value, we have $B_n(x) = \alpha_n^{-\frac{1}{2}}\{\widehat{\lambda}_n(x) - \lambda_0\}$ and we obtain the approximation

$$\alpha_n^{-1}\{\widehat{\lambda}_n(X_i) - \widehat{\lambda}_{0n}\} = B_n(x) + o_p(1).$$

According to the convergence rate of the process U_n, a test of the hypothesis H_0 relies on the normalized statistic $T_n = N_n^{-2} l_n$.

Proposition 4.12. *Under the hypothesis H_0, the statistic T_n converges weakly to the variable $T_0 = \frac{1}{2\lambda_0^2}\int y B^2(x)\, F(dx, dy)$. Under fixed alternatives, T_n diverges.*

Proof. A second order expansion of the logarithm in the expression of l_n entail

$$\begin{aligned}
n^{-1} l_n &= n^{-1}\sum_{i=1}^n Y_i\Big[\frac{\widehat{\lambda}_n(X_i) - \widehat{\lambda}_{0n}}{\widehat{\lambda}_{0n}} - \frac{\{\widehat{\lambda}_n(X_i) - \widehat{\lambda}_{0n}\}^2}{2\widehat{\lambda}_{0n}^2}\Big] \\
&\quad - n^{-1}\sum_{i=1}^n \{\widehat{\lambda}_n(X_i) - \widehat{\lambda}_{0n}\} + o_p(1) \\
&= n^{-1}\sum_{i=1}^n Y_i\Big[2\frac{\widehat{\lambda}_n(X_i)}{\widehat{\lambda}_{0n}} - \frac{\widehat{\lambda}_n^2(X_i)}{2\widehat{\lambda}_{0n}^2}\Big] - \widehat{\lambda}_n(X_i) - \frac{\widehat{\lambda}_{0n}}{2} + o_p(1),
\end{aligned}$$

it converges in probability to zero. Arguing like in Proposition 2.5, under H_0 we have

$$\begin{aligned}
N_n^{-2}\sum_{i=1}^n Y_i \log\frac{\widehat{\lambda}_n(X_i)}{\widehat{\lambda}_{0n}} &= N_n^{-2}\sum_{i=1}^n Y_i\frac{U_n(X_i)}{\lambda_0} - N_n^{-2}\sum_{i=1}^n Y_i\frac{U_n^2(X_i)}{2\lambda_0^2} + o(1) \\
&= \alpha_n^{\frac{1}{2}}\sum_{i=1}^n Y_i\frac{B_n(X_i)}{\lambda_0} - n^{-1}\sum_{i=1}^n Y_i\frac{B_n^2(X_i)}{2\lambda_0^2} + o_p(1),
\end{aligned}$$

$$\begin{aligned}
N_n^{-2}\sum_{i=1}^n \{\widehat{\lambda}_n(X_i) - \widehat{\lambda}_{0n}\} &= N_n^{-1} n^{-\frac{1}{2}}\sum_{i=1}^n B_n(X_i) + o_p(1) \\
&= N_n^{-1}\int B_n\, d\nu_n + o_p(1),
\end{aligned}$$

where ν_n is the empirical process of the sample. It follows that the statistic T_n is asymptotically equivalent to

$$\frac{1}{2\lambda_0^2}\int y B_n^2(x)\,\widehat{F}_n(dx, dy).$$

Under the alternative, $\alpha_n^{-1}\{\widehat{\lambda}_n(X_i) - \widehat{\lambda}_{0n}\} = B_n(x) + \alpha_n^{-1}\{\lambda(x) - \lambda_0\} + o_p(1)$ and the second term diverges as n tends to infinity. □

Let $(A_n)_{n\geq 1}$ be a sequence of local alternatives defined by hazard functions $\lambda_{1n}(x) = \lambda_0 + n^{-\frac{1}{2}} N_n h_{1n}(x)$ and for a scalar intensity $\lambda_{2n} = \lambda_0 + n^{-\frac{1}{2}} h_{2n}$ such that there exist non zero limits h_j of h_{jn}, as n tends to infinity, for $j = 1, 2$.

Proposition 4.13. *The statistic T_n converges weakly under A_n to the variable $T = \frac{1}{2\lambda_0^2} \int y(B + h_1)^2(x)\, F(dx, dy)$.*

Proof. Under A_n the difference of the hazard functions satisfies

$$\alpha_n^{-\frac{1}{2}} \{\lambda_{1n} - \lambda_{2n}\} = h_{1n_1} + o(1)$$

and $\alpha_n^{-\frac{1}{2}} \{\widehat{\lambda}_n(x) - \widehat{\lambda}_{0n}\} = B_n(x) + h_{1n_1}(x) + o(1)$. In the expression of T_n, the second order term of the logarithmic expansion becomes

$$n^{-1} \sum_{i=1}^{n} Y_i \frac{\{B_n(X_i) + h_{1n}(X_i)\}^2}{2\lambda_0^2} + o_p(1),$$

it converges weakly to the integral of $(B + h_1)^2$. □

Chapter 5

Deconvolution and inverse problems

5.1 Estimation of the distribution of errors

In an additive model with error $Y = X + \varepsilon$, the variable Y is observed and ε is a centered error variable independent of X, with density known up to a variance parameter. Estimators of the density of X and of the variance of the error have been defined in Section 1.3. The Laplace and Fourier transforms of Y are multiplicative, the transforms for X is then the ratio of the transforms for Y and ε.

Hall and Caroll (1988) proposed to estimate the density of the variable X by smoothing the inversion of the estimated Fourier transform of a n-sample $(Y_i)_{i=1,\ldots,n}$, under the assumption that the error has a known density. Let $\widehat{\phi}_{nY}(t) = n^{-1}\sum_{i=1}^{n} e^{-itY_i}$ be the empirical Fourier transform of the variable Y and let ϕ_ε be the Fourier transform of the error ε, then the Fourier transform of the density of X is estimated by

$$\widehat{\phi}_{nX}(t) = \widehat{\phi}_{nY}(t)\phi_K(ht)\phi_\varepsilon^{-1}(t)$$

using a kernel K with bandwidth h. The estimator of the density has the form

$$\widehat{f}_{n,X}(x) = n^{-1}\sum_{i=1}^{n} K_{h,\varepsilon}(x - Y_i).$$

Fan and Truong (1993) extended this kernel method to the estimation of a nonparametric regression function in a model where the regressor is observed with an independent error, their estimator has the same form as the kernel estimator for a regression function with an observed regressor, using the kernel K_ε. They established an optimal convergence rate $(\log n)^{-\frac{k}{\beta}}$ for a regression function of C^k and an error with density in C^β.

Here we define estimators of the nonparametric regression function on regressors observed with error, based on the Laplace transform and series estimation. Its optimal convergence rate is lower than the rate of estimator for a regression function on observed covariates, it remains a power of n so it is better than the logarithmic rate obtained by Fan and Truong.

Let $Y = m(X) + \varepsilon$ be the regression model of a real variable Y on a variable X belonging to a subset $\mathcal{I}_X$ of $\mathbb{R}^d$, with an unknown function m and an error variable such that $E(\varepsilon \mid X) = 0$ and $\sigma^2 = \mathrm{Var}\, \varepsilon$ is finite. We first consider the estimation of the regression function m and the distribution of the error variable from a sample $(X_i, Y_i)_{i=1,\dots,n}$. The Laplace transform of the variable Y is the product of the transform

$$L_{m(X)}(t) = \int_0^\infty e^{-tm(x)}\, dF_X(x)$$

for $m(X)$ and the Laplace transform of the error variable ε

$$L_Y(t) = L_{m(X)}(t)\, L_\varepsilon(t),$$

for every t in $\mathbb{R}_+$. Let $\widehat{m}_n$ be a series estimator of the regression function m, based on the projection of the sample $(Y_i)_i$ on K_n functions of the basis $(\psi_k)_k$ defined by (3.19), the Laplace transforms L_Y and $L_{m(X)}$ have the empirical estimators

$$\widehat{L}_{n,Y}(t) = n^{-1} \sum_{i=1}^n e^{-tY_i} = \int e^{-ty}\, d\widehat{F}_{nY}(y),$$

$$\widehat{L}_{n,\widehat{m}_n(X)}(t) = n^{-1} \sum_{i=1}^n e^{-t\widehat{m}_n(X_i)}.$$

The Laplace transform of the error variable is then estimated by their ratio

$$\widehat{L}_{n,\varepsilon}(t) = \frac{\sum_{i=1}^n e^{-tY_i}}{\sum_{i=1}^n e^{-t\widehat{m}_n(X_i)}},$$

for t in $\mathbb{R}_+$ and it tends to zero as t tends to infinity. An estimator of the variance of the error variable ε is obtained as $\widehat{\sigma}_n^2 = n^{-1} \sum_{i=1}^n \{Y_i - \widehat{m}_n(X_i)\}$.

The asymptotic properties of the estimated regression function imply the consistency of the estimator $\widehat{L}_{n,\widehat{m}_n(X)}$.

Proposition 5.1. *If $\int_0^\infty m(x)e^{-tm(x)}\, dF_X(x)$ if finite, the estimators $\widehat{L}_{n,\widehat{m}_n(X)}$ and $\widehat{L}_{n,\varepsilon}$ are consistent, they have the same convergence rate $n^{-\frac{1}{2}}K_n$ as $\widehat{m}_n$.*

Proof. The convergence rate of $\widehat{L}_{n,\widehat{m}_n(X)}$ is deduced from the convergence rate of $\widehat{m}_n$ by the expansion

$$\begin{aligned}\widehat{L}_{n,\widehat{m}_n(X)}(t) - L_{m(X)}(t) &= \int_0^\infty \{e^{-t\widehat{m}_n(x)} - e^{-tm(x)}\}\,dF_X(x)\\ &\quad + \int_0^\infty -e^{-t\widehat{m}_n(x)}\,d\{\widehat{F}_n(x) - F_X(x)\}\\ &= \Big[\int_0^\infty \{e^{-t\widehat{m}_n(x)} - e^{-m(x)}\}\,dF_X(x)\Big]\{1+o_p(1)\}.\end{aligned}$$

The consistency of $\widehat{m}_n$ and a first order differentiation entail

$$\begin{aligned}&\int_0^\infty \{e^{-t\widehat{m}_n(x)} - e^{-tm(x)}\}\,dF_X(x)\\ &= \Big[\int_0^\infty \{\widehat{m}_n(x) - m(x)\}\{1 - tm(x)\}e^{-tm(x)}\,dF_X(x)\Big]\{1+o_p(1)\},\end{aligned}$$

under the condition and by a converging expansion $m(x) = \sum_{k\geq 0} b_k\varphi_k(x)$ on a basis of $L^2(\mathcal{I}_X, F_X)$, the series $\sum_{k\geq 0} b_k \int_0^\infty \varphi_k(x)e^{-m(x)}\,dF_X(x)$ converges and the series $\sum_{k\geq 0} \widehat{b}_{nk} \int_0^\infty \varphi_k(x)e^{-tm(x)}\,dF_X(x)$ is finite by the convergence properties of the estimators. It follows that $\widehat{L}_{n,\widehat{m}_n(X)}$ converges to $L_{m(X)}$ with the same convergence rate as $\widehat{m}_n$. The estimator $\widehat{L}_{n,Y}$ is consistent with the convergence rate $n^{-\frac{1}{2}}$ hence $\widehat{L}_{n,\varepsilon}$ has the same rate of convergence as $\widehat{m}_n$, given in Proposition 3.9. □

By inversion of this estimated Laplace transform, the distribution function G_ε of the error variable is estimated by the deconvolution formula (1.1) as

$$\widehat{G}_{n,\varepsilon}(s) = \lim_{t\to\infty} \sum_{k\leq st} \frac{(-1)^k}{k!} t^k \widehat{L}_{n\varepsilon}^{(k)}(t). \tag{5.1}$$

If the regression function m is bounded, the proof of Proposition 5.1 implies that the derivatives $\widehat{L}_{n\varepsilon}^{(k)}$, $k \geq 1$, of the estimator $\widehat{L}_{n\varepsilon}$ have the same convergence rate as $\widehat{L}_{n\varepsilon}$ and this is also the rate of convergence for the estimator $\widehat{G}_{n,\varepsilon}$ of the distribution function G_ε.

The density g of the error variable ε is estimated by projection on the estimated basis $(\widehat{\psi}_{\varepsilon,nk})_k$ of $(\psi_{\varepsilon,k})_{k\geq 0}$ in $L^2(\mathbb{R}, G_\varepsilon)$, like in Section 3.6

$$\widehat{g}_n(s) = \sum_{k=0}^{k_n} \widehat{c}_{nk}\widehat{\psi}_{\varepsilon,nk}(s)$$

with the estimators $\widehat{c}_{nk} = \int_{\mathbb{R}} \widehat{\psi}_{\varepsilon,nk}\,d\widehat{G}_{n,\varepsilon}$ of the coefficients $c_k = \int_{\mathbb{R}} \psi_{\varepsilon,k}\,dG$ of the series $g = \sum_{k\geq 0} c_k\psi_{\varepsilon,k}$ for the density of G_ε. The variables $\widehat{c}_{nk}$ have the convergence rate of the estimator $\widehat{G}_{n,\varepsilon}$, the MISE of $\widehat{g}_n$ is therefore

$$\mathrm{MISE}_n(g) = O(n^{-1}K_n^2k_n^2) + \|g - g_n\|^2$$

where the squared approximation error is $\|g-g_n\|^2 = \sum_{k>k_n} c_k^2$, an optimal size k_n is obtained under the conditions

$$k_n = o(n^{\frac{1}{2}} K_n^{-1}), \quad \|g - g_n\|_2 = O(n^{-\frac{1}{2}} K_n k_n). \tag{5.2}$$

The squared estimation error $\|\widehat{g}_n - g_n\|^2$ depends on the integrated variance of the estimators $\widehat{c}_{nk}$ and $\widehat{\psi}_{nk}$. The bias $g_n - g$ depends on the approximation error and it converges to zero if $\|m - m_n\|_2 = o(n^{-\frac{1}{2}} K_n k_n)$.

Proposition 5.2. *Under the condition (5.2), the error $E\|\widehat{g}_n - g\|_2$ for an estimator of a density g of W_{4s}, $s > 1$, is $n^{-\frac{4s^2}{(2s+1)})^2}$.*

Proof. For a density g of the space W_{4s}, $s > 1$, we have $\|g - g_n\|_2^2 = O(k_n^{-4s})$ and for the optimal numbers K_n for the estimation of a regression function m in W_{4s}, and k_n for the estimation of the density g, we have $k_n^{-4s} = 0(n^{-\frac{2s}{2s+1}} k_n^2)$ which yields $k_n = 0(n^{\frac{s}{(2s+1)})^2}$ and

$$\|g - g_n\|_2^2 = O(k_n^{-4s}) = 0(n^{-\frac{4s^2}{(2s+1)})^2}).$$

□

This order is the optimal order of the MISE error of the series estimator $\widehat{g}_n$ defined by the deconvolution formula for the Laplace transform, with the estimator $\widehat{m}_n$ of the regression function.

Proposition 5.3. *Under the conditions (5.2), the process $n^{\frac{1}{2}} K_n^{-1} k_n^{-1}(\widehat{g}_n - g)$ converges weakly to a Gaussian process with finite expectation and variance functions.*

The error-in-variables regression model is defined as

$$Y_i = m(X_i) + \varepsilon_i,$$
$$X_i = W_i + U_i,$$

for an observed sample $(W_i, Y_i)_{i=1,\dots,n}$, with mutually independent samples $(W_i)_{i=1,\dots,n}$, $(U_i)_{i=1,\dots,n}$ and $(\varepsilon_i)_{i=1,\dots,n}$. The variables X_i are independent and unobserved, with an unknown density f_X, the variables U_i are independent with a known symmetric density f_U and the variables ε_i are independent, centered and have a finite variance σ^2, for $i = 1, \dots, n$. If the density of the error ε is known, the Laplace transform of Y is the product

$$L_Y(t) = L_{m(X)}(t) L_\varepsilon(t)$$

and $L_{m(X)}$ is estimated by the ratio

$$\widehat{L}_{n,m(X)}(t) = L_\varepsilon^{-1}(t) \widehat{L}_{n,Y}(t)$$

and the convergence rate of this estimator is $n^{-\frac{1}{2}}$. The empirical distribution $\widehat{F}_{n,m(X)}$ of $(m(X_i))_{i\le n}$ is deduced by (5.1) and $m(X)$ has the estimator

$$\widehat{m}_n(X) = \int_{\mathbb{R}} y\, d\widehat{F}_{n,m(X)}(y),$$

the convergence rate of $\widehat{F}_{n,m(X)}$ and $\widehat{m}_n(X)$ is still $n^{-\frac{1}{2}}$ and the empirical variable $n^{\frac{1}{2}}\{\widehat{m}_n(X) - Em(X)\}$ converges weakly to centered Gaussian variable. Let $Z = m(X)$ in $m(\mathcal{I}_X)$ and, for all z in $m(\mathcal{I}_X)$ and $\delta > 0$, let $V_\delta(z)$ be an interval of length δ centered at z, the function $m(x)$ is a.s. consistently estimated as

$$\widehat{m}_n(x) = \lim_{\delta_n \to 0} \frac{\int_{V_{\delta_n}(m(x))} z\, d\widehat{F}_{n,m(X)}(z)}{\widehat{F}_{n,m(X)}(V_{\delta_n}(m(x)))}. \tag{5.3}$$

The variances of $\delta_n^{-1}\widehat{F}_{n,m(X)}(V_{\delta_n}(m(x)))$ and $\delta_n^{-1}\int_{V_{\delta_n}(m(x))} z\, d\widehat{F}_{n,m(X)}(z)$ are $O(\delta_n^{-1}n^{-1})$, this is also the order of the variance of $\widehat{m}_n(x)$, for every x.

Proposition 5.4. *Let $\delta_n^{-1} = o(n)$, the process $(n\delta_n)^{\frac{1}{2}}(\widehat{m}_n - m)$ converges weakly to a centered Gaussian process.*

If the density f_ε is unknown, the density $f_X = f_W \star f_U$ of X is estimated as the convolution

$$\widehat{f}_{nX}(x) = \widehat{f}_{nW} \star f_U(x),$$

with a series estimator $\widehat{f}_{nW}$ of the density f_W using K_n functions of the estimated basis $(\widehat{\psi}_{W,nk})_{k\ge 0}$ of $(\psi_{W,k})_{k\ge 0}$ in $L^2(F_W)$ defined like in Section 3.6. Its convergence rate under the conditions (3.7) is $n^{-\frac{1}{2}}K_n$.

The Laplace transform of $m(X)$ is deduced as

$$\widehat{L}_{n,m(X)}(t) = \int_{\mathcal{I}_X} e^{-m(x)}\widehat{f}_{nX}(x)\, dx$$

and by the inversion formula (1.1), the distribution function $F_{m(X)}$ of $m(X)$ has the estimator

$$\widehat{F}_{n,m(X)}(y) = \lim_{t\to\infty} \sum_{k\le yt} \frac{(-1)^k}{k!} t^k \widehat{L}^{(k)}_{n,m(X)}(t). \tag{5.4}$$

The derivatives $\widehat{L}^{(k)}_{n,m(X)}$, $k \ge 0$, of the estimator $\widehat{L}_{n,m(X)}$ and the estimator $\widehat{F}_{n,m(X)}$ have the same convergence rate $n^{-\frac{1}{2}}K_n$ as the estimator $\widehat{f}_{nX}$ of the density f_X, under the conditions (3.7). An estimator of the function m is deduced as previously by (5.3), with $\delta_n = o(1)$. The variance of $\widehat{m}_n(x)$

is now a $O(n^{-1}\delta_n^{-1}K_n^2) = o(1)$ for every x, its convergence rate is therefore $(n\delta_n)^{-\frac{1}{2}}K_n$.

Proposition 5.5. *Under the conditions (3.7) for $\widehat{f}_{nW}$ and with $n^{-1}\delta_n^{-1}K_n^2 = o(1)$, the process $(n\delta_n)^{\frac{1}{2}}K_n^{-1}(\widehat{m}_n - m)$ converges weakly to a Gaussian process.*

An estimator of the Laplace transform of ε is deduced from the inversion

$$\widehat{L}_{n\varepsilon}(t) = \frac{\widehat{L}_{nY}(t)}{L_{\widehat{m}_n(X)}(t)}$$

and the distribution function G_ε of the error variable is estimated by deconvolution as (5.1).

In the generalized nonparametric regression models, the expectation of the variable Y conditionally on X depends on a function g without error. Let F_{XY} be the distribution function of (X, Y) and let $F_{Y|X}$ be the conditional distribution function of Y, the model for the expectation of Y has the form

$$E(Y \mid X = x) = \int y\, dF_{Y|X=x}(y) = g \circ m(x) := r(x),$$

its variance conditionally on $X = x$ is $\sigma_X^2(x) = \int y^2\, dF_{Y|X=x}(y) - r^2(x)$.

GLM models with error in covariates are defined as above by the distribution of Y, its expectation and its variance conditionally on an unobserved covariate X, with

$$X_i = W_i + U_i$$

where U has a known density f_U on a real interval $\mathcal{I}_U$. If U has a symmetric distribution, the density of X is $f_X = f_W \star f_U(w)$ and it is estimated from a n-sample $(W_i)_{i=1,\dots,n}$ as $\widehat{f}_{nX}$ by convolution of $f_U(w)$ and a series estimator of f_W. A discrete variable Y with probabilities $p_k(X)$ conditionally on X has the conditional probability $\pi_k(W)$ given W, such that

$$p_k(x) = \int_{\mathcal{I}_U} \pi_k(x-u)\, dF_U(u) = \pi_k \star f_U(x). \tag{5.5}$$

Estimators of the probabilities $\pi_k(W)$ are defined by projections on an orthonormal basis of functions $(\psi_k)_{k\geq 0}$ of $L^2(\mathcal{I}_W, F_W)$, estimated by $(\widehat{\psi}_{nk})_{k\geq 0}$ defined as (3.2) from $(W_i)_{i=1,\dots,n}$. Let $\pi_k(w) = \sum_{j\geq 0} q_{kj}\psi_j(w)$

be a convergent series with the coefficients $q_{kj} = \int_{\mathcal{I}_W} \pi_k(w)\psi_j(w)\,dw$, they have the consistent estimators

$$\widehat{q}_{nkj} = n^{-1} \sum_{i=1}^{n} 1_{\{Y_i=k\}} \widehat{\psi}_{nj}(W_i),$$

$$\widehat{\pi}_{nk}(w) = \sum_{j=0}^{J_n} \widehat{q}_{nkj} \widehat{\psi}_{nj}(w).$$

Their asymptotic behavior is deduced from Lemma 3.2 and Proposition 3.4.

Proposition 5.6. *Under the condition (3.7), the processes $K_n^{-1} n^{\frac{1}{2}} (\widehat{\pi}_{nk} - \pi_k)$ converge weakly to Gaussian processes on $\mathcal{I}_W$.*

Estimators of the probability functions $p_k(x)$ are deduced from (5.5) as $\widehat{p}_{nk}(x) = \widehat{\pi}_{nk} \star f_U(x)$ and the conditional expectation r of Y has the consistent estimator

$$\widehat{r}_n(x) = \sum_{k \geq 0} k \widehat{p}_{nk}(x),$$

they have the same asymptotic behavior as $\widehat{\pi}_{nk}$ (Proposition 5.6). The function m is estimated by inversion of g.

5.2 Mixtures of exponential densities

Let X be an exponential variable with a random parameter on $\mathbb{R}_+$ and let G be the unknown distribution function of the parameter. A continuous mixture of exponential variables has the distribution function F given by (1.2), its density f is estimated from a n-sample $X_1, \ldots, X_n$ of independent and identically distributed variables, by projection on an orthonormal basis of infinitely differentiable functions. Introducing the estimator $\widehat{f}_n$ defined by (2.18), of the density f, and its derivatives in the inversion equation (1.3) of the Laplace transform provides a consistent estimator of the mixing distribution function G

$$\widehat{G}_n(\theta) = \lim_{x \to \infty} \sum_{k \leq \theta x} \frac{1}{k!} x^k \widehat{f}_n^{(k)}(x). \tag{5.6}$$

The estimator $\widehat{G}_n$ has the same rate of convergence as the estimator $\widehat{f}_n$ of the density, Proposition 2.2 and Proposition 2.3 apply to the estimator $\widehat{G}_n$.

In the exponential model (1.4), the density f_Y of the variable $Y = T(X)$ has an estimator $\widehat{f}_n$ defined by (2.18), its derivatives $\widehat{f_n^{(k)}}$, for $k \geq 0$, provide a consistent estimator of the function cL_H by (1.5). This estimator has the same convergence rate as the density estimator $\widehat{f}_n$. An estimator of the function H is deduced from (1.6) by plugging the estimator $\widehat{f}_n$ and its derivatives in the inversion formula (1.1), it has the convergence rate of $\widehat{f}_n$.

In the same way, the function H of model (1.7) is estimated by plugging the estimator $\widehat{f}_n$ of the sample with a continuous mixture density, and its derivatives, in the inversion formula (1.1).

Let X be a variable having a random exponential distribution F_θ conditionally on a positive random parameter θ, its density with respect to a known probability measure F_0 is

$$f_\theta(x) = e^{-\theta x - b(\theta)}, \ x \geq 0,$$

where $b(\theta) = \log \int_0^\infty e^{-\theta x} \, dF_0(x)$. Let g be mixing density of the random parameter θ on $\mathbb{R}_+$, the density of the variable X is

$$f(x) = \int_0^\infty f_\theta(x) g(\theta) \, d\theta.$$

Let $(\varphi_n)_{n \geq 0}$ be an orthonormal basis of function with respect to Lebesgue's measure on $\mathbb{R}_+$, such that $\varphi_0(x) \equiv 1$, the mixing density g has the expansion

$$g(\theta) = \sum_{k \geq 0} a_k(g) \varphi_k(\theta),$$

with $a_k(g) = \int g(\theta)\varphi_k(\theta) \, d\theta$ for $k \geq 0$. The density of X is written as

$$f(x) = \sum_{k \geq 0} a_k(g) \int f_\theta(x) \varphi_k(\theta) \, d\theta$$

where the integrals

$$P_k(x) = \int_0^\infty f_\theta(x) \varphi_k(\theta) \, d\theta$$

are known functions. They are not orthogonal and we have

$$b_j = \int_0^\infty f(x) P_j(x) \, dx = \sum_{k \geq 0} a_k(g) \int_0^\infty P_k(x) P_j(x) \, dx$$

so the coefficients $a_k(g)$ are not directly identifiable from the scalar products of the functions $P_j(x)$ and $P_k(x)$ which are known constants c_{jk}. The coefficients b_j are estimated from a n-sample $(X_i)_{i=1,\dots,n}$ as

$$\widehat{b}_{nj} = n^{-1} \sum_{i=1}^{n} P_j(X_i).$$

An approximation of the infinite series of f by $f_n = \sum_{k=0}^{k_n} a_k(g)P_k(x)$, where $k_n = o(n)$ tends to infinity as n tends to infinity, provides a system of k_n linear equations $A_n = C_n B_n$ defined by the vectors $A_n = (a_k)_{1\le k\le k_n}$ and $B_n = (b_j)_{1\le j\le k_n}$ of dimension k_n, and by the matrix $C_n = (c_{kj})_{1\le k,j\le k_n}$. If the matrix $C_n^T C_n$ is not singular, estimators $\widehat{a}_{nk}$ of the coefficients $a_k(g)$ are the solutions of a system of linear equation. From the properties of the estimators $\widehat{b}_{nj}$, $j = 0, \ldots, k_n$, they are a.s. consistent and have the convergence rate $n^{-\frac{1}{2}}$.

The mixing density of the random parameter θ is estimated by

$$\widehat{g}_n(\theta) = \sum_{k=0}^{k_n} \widehat{a}_{nk}\varphi_k(\theta),$$

which is a.s. consistent and has the convergence rate $k_n^{\frac{1}{2}} n^{-\frac{1}{2}}$. The estimator $\widehat{g}_n$ has the same form as the estimator defined in Section 2.4 for a density by projection on an orthonormal basis, their mean integrated square errors have the same order. Proposition 2.3 is still valid and the process $k_n^{-\frac{1}{2}} n^{\frac{1}{2}}(\widehat{g}_n - g)$ converges weakly to a Gaussian process with a finite covariance function.

5.3 Tests for mixtures of exponentials

The parameter of an exponential density $f_\theta(x) = \theta e^{-\theta x}$ has the maximum likelihood estimator $\widehat{\theta}_n = \bar{X}_n^{-1}$ where $\bar{X}_n$ is the empirical mean of a sample $(X_i)_{i\le n}$ with density f_θ. A test of the hypothesis H_0 of a true exponential density f_0 with unknown parameter value θ_0, against the alternative of a continuous mixture density

$$f(x) = \int_{\mathbb{R}_+} f_\theta(x)dG(\theta),$$

is performed by a comparison of the estimators $f_{\widehat{\theta}_n}$ of the exponential density and $f_{\widehat{G}_n} = \int_{\mathbb{R}_+} f_\theta(x)d\widehat{G}_n(\theta)$ for the mixture density, with the estimator (5.6) of the unknown mixing distribution G_0. Under the hypothesis H_0, G_0 reduces to the Dirac measure δ_{θ_0} and $f_{G_0} = f_0$.

The estimator $f_{\widehat{\theta}_n}$ has the convergence rate $n^{-\frac{1}{2}}$ and, under the conditions (2.19), the nonparametric estimator $f_{\widehat{G}_n}$ has the convergence rate $\alpha_n^{-1} = n^{-\frac{1}{2}} K_n^{\frac{1}{2}}$.

A Kolmogorov–Smirnov test for H_0 is defined by the statistic

$$T_n = \sup_{x\ge 0} \alpha_n |f_{\widehat{\theta}_n}(x) - f_{\widehat{G}_n}(x)|.$$

Proposition 5.7. *The statistic T_n converges weakly under H_0 to the supremum of a Gaussian process $T_0 = \sup_x |\int_{\mathbb{R}_+} f_\theta(x)\, dW_0(\theta)|$. Under fixed alternatives, the statistic T_n diverges.*

Proof. Under H_0, the statistic T_n is asymptotically equivalent to $\sup_x \alpha_n |f_{\widehat{G}_n}(x) - f_{G_0}(x)|$ due to the convergence rates $\alpha_n = o(n^{\frac{1}{2}})$ and by Proposition 2.3, the process $\alpha_n(\widehat{G}_n - G_0)$ converges weakly under H_0 to a Gaussian process W_0 on $\mathbb{R}_+$. The weak convergence of the process $\alpha_n\{f_{\widehat{G}_n} - f_{\widehat{\theta}_n}\}$ to $\int_{\mathbb{R}_+} f_\theta\, dW_0(\theta)$ follows. Under fixed alternatives, the difference of the estimators $f_{\widehat{\theta}_n} - f_{\widehat{G}_n}$ converges to a non zero limit $f_\theta - f_G$ and the statistic diverges. $\square$

We consider a sequence of local alternatives A_n defined by

$$A_n : \begin{cases} \theta_n = \theta_0 + n^{-\frac{1}{2}}\eta_n, \\ G_n(\theta) = G_0(\theta) + \alpha_n^{-1}\gamma_n(\theta),\ \theta \in \mathbb{R}_+, \end{cases}$$

where η_n converges to a limit η and the functions γ_n converge uniformly to a non zero limit γ_0.

Proposition 5.8. *The statistic T_n converges weakly under the local alternatives A_n to the variable $T_A = \sup_x |\int_{\mathbb{R}_+} f_\theta(x)\, d\{W_0(\theta) + \gamma_0(\theta)\}|$.*

A Cramer-von Mises statistic is a $L^2(\widehat{F}_{nX})$ distance between the estimators $f_{\widehat{\theta}_n}$ and $f_{\widehat{G}_n}$ of the density

$$T_{2n} = \alpha_n^2 n^{-1} \sum_{i=1}^n \{f_{\widehat{\theta}_n}(X_i) - f_{\widehat{G}_n}(X_i)\}^2 w_n(X_i),$$

where the sequence of weighting functions w_n converges uniformly to a function w.

Proposition 5.9. *Under H_0, the statistic T_{2n} converges weakly to*

$$T_{02} = \int_{\mathcal{I}_X} \Big\{\int_{\mathbb{R}_+} f_\theta(x)\, dW_0(\theta)\Big\}^2 w(x)\, dF_X(x).$$

Under fixed alternatives, it tends to infinity. Under the local alternatives A_n, T_{2n} converges weakly to the variable

$$T_{A2} = \int_{\mathcal{I}_X} \Big[\int_{\mathbb{R}_+} f_\theta(x)\, d\{W_0(\theta) + \gamma_0(\theta)\}\Big]^2 w(x)\, dF_X(x).$$

In the exponential model (1.4), the parameter has the maximum likelihood estimator $\widehat{\theta}_n = \bar{T}_n^{-1}$ where

$$\bar{T}_n = n^{-1} \sum_{i=1}^{n} T(X_i)$$

is the empirical mean of statistic T for a n-sample with density f_θ. A test of the hypothesis H_0 of a true exponential density (1.4) with parameter θ_0 is performed like previously for the exponential density (1.2).

A goodness of fit test of the hypothesis H_0 of a continuous exponential mixture model with conditional density (1.2) is performed as a comparison of the nonparametric estimator $\widehat{f}_n$ of the unknown density defined by (2.18) with the estimator $f_{\widehat{G}_n}$ of the exponential mixture density. Both series estimators have the same convergence rate α_n if they have the same size K_n. Proposition 2.3, the process $\alpha_n(f_{\widehat{G}_n} - \widehat{f}_n)$ converges weakly under H_0 to a Gaussian process W_0 on $\mathbb{R}_+$.

Local alternatives A_n converging to H_0 are defined as

$$A_n : \begin{cases} f_n(x) = f_0(x) + a_n^{-\frac{1}{2}} \eta_n(x), \\ G_n(\theta) = G_0(\theta) + \alpha_n^{-1} \gamma_n(\theta), \, \theta \in \mathbb{R}_+, \end{cases}$$

where the functions η_n, and respectively γ_n, converge uniformly to a limit η, and respectively γ such that the functions η and $\eta' = \int_{\mathbb{R}_+} f_\theta \gamma(\theta) \, d\theta$ differ.

Proposition 5.10. *Under H_0 the statistic*

$$T_n = \sup_{x \geq 0} \alpha_n |\widehat{f}_n(x) - f_{\widehat{G}_n}(x)|$$

converges weakly to the supremum of a Gaussian process W_0 on $\mathbb{R}_+$. Under fixed alternatives, the statistic T_n diverges. Under A_n, T_n converges weakly to $\sup_{x \geq 0} |W_0(x) + (\eta - \eta')(x)|$.

Proposition 5.11. *Under H_0 the statistic*

$$T_n = \alpha_n^2 \int_{\mathbb{R}_+} \{\widehat{f}_n(x) - f_{\widehat{G}_n}(x)\}^2 w_n(x) \, d\widehat{F}_{nX}(x)$$

converges weakly to $\int_{\mathbb{R}_+} W_0^2(x) w(x) \, dF_X(x)$. Under fixed alternatives, the statistic T_n diverges. Under the alternatives A_n, T_n converges weakly to $\int_{\mathbb{R}_+} \{W_0(x) + (\eta - \eta')(x)\}^2 w(x) \, dF_X(x)$.

5.4 Mixtures of bivariate exponential models

Let (X, Y) be a paired exponential variable with a random density

$$f_{\theta,\eta}(x,y) = \theta\eta^2 e^{-\eta x - \theta\eta y} \tag{5.7}$$

with respect to Lebesgue's measure on $\mathbb{R}_+^2$. The parameters are strictly positive and η is a random variable with an unknown distribution function G on $\mathbb{R}_+$. The survival function $\bar{F}_{\theta,\eta}(x,y) = P_{\theta,\eta}(X > x, Y > x)$ of the variable (X, Y) conditionally on η is the product of the marginal survival function $\bar{F}_{\theta,\eta}(x,y) = e^{-\eta x}e^{-\theta\eta y}$ and the unconditional survival function $\bar{F}_\theta = 1 - F_\theta$ of (X, Y) is the mixture function

$$\bar{F}_\theta(x,y) = \int_{\mathbb{R}_+} \bar{F}_{\theta,\eta}(x,y)\, dG(\eta) = L_G(x + \theta y), \tag{5.8}$$

where L_G is the Laplace transform of G. The marginal survival functions of X and Y are

$$\bar{F}_X(x) = L_G(x), \tag{5.9}$$

$$\bar{F}_{Y,\theta}(y) = L_G(\theta y) \tag{5.10}$$

and the distribution function G is solution of the inversion formula (1.1) of the Laplace transform $L_G(x)$ using the equality (5.9)

$$G(\eta) = \lim_{x\to\infty} \sum_{k \le \eta x} \frac{(-1)^k}{k!} \bar{F}_X^{(k)}(x).$$

Let $(X_i, Y_i)_{i=1,\dots,n}$ be a n-sample of the variable (X, Y), and let $\widehat{F}_{Xn}$ be the empirical distribution function of the sample $(X_i)_{i=1,\dots,n}$. The density of X is estimated by $\widehat{f}_{Xn}$ defined projections on a basis of infinitely differentiable functions as (2.18) and the mixing distribution G is estimated by plugging the derivatives of the estimator $\widehat{f}_{Xn}$ in the inversion formula

$$\widehat{G}_n(\eta) = \lim_{x\to\infty} \sum_{k=1}^{[\eta x]} \frac{1}{k!} \widehat{f}_{Xn}^{(k)}(x).$$

The estimator $\widehat{G}_n$ has the same convergence rate $n^{-\frac{1}{2}} K_n^{\frac{1}{2}}$ as the estimator $\widehat{f}_{Xn}$ (Proposition 2.3).

By (5.10), the density of the variable Y and its first derivative with respect to θ satisfy

$$f_\theta(y) = \theta \int_{\mathbb{R}_+} \eta e^{-\theta\eta y}\, dG(\eta) = -\theta L'_G(\theta y),$$

$$\begin{aligned} \dot{f}_\theta(y) &= \int_{\mathbb{R}_+} \eta e^{-\theta\eta y}\, dG(\eta) - \theta y \int_{\mathbb{R}_+} \eta^2 e^{-\theta\eta y}\, dG(\eta) \\ &= -L'_G(\theta y) - \theta y L''_G(\theta y), \end{aligned}$$

with the first derivatives of $L_G(t)$, $L'_G(t) = -\int_{\mathbb{R}_+} \eta e^{-\eta t}\, dG(\eta) = -f_X(t)$ and $L''_G(t) = \int_{\mathbb{R}_+} \eta^2 e^{-\eta t}\, dG(\eta) = f'_X(t)$. The maximum likelihood estimator of the parameter θ is solution of the equation

$$0 = \sum_{i=1}^n \frac{\dot{f}_\theta}{f_\theta}(Y_i),$$

$$\widehat{\theta}_n = \Big[n^{-1} \sum_{i=1}^n \frac{Y_i L''_{\widehat{G}_n}(\theta Y_i)}{L'_{\widehat{G}_n}(\theta Y_i)}\Big]^{-1} = -\Big[n^{-1} \sum_{i=1}^n \frac{Y_i \widehat{f}'_{nX}(\theta Y_i)}{\widehat{f}_{nX}(\theta Y_i)}\Big]^{-1}.$$

The maximum likelihood estimator is depend on the estimators of the density of X and its first derivative, they are defined by (2.18) and its first derivative, their convergence rate has the order $\alpha_n = n^{-\frac{1}{2}} K_n^{\frac{1}{2}}$ under the condition (2.19). It follows that $\widehat{\theta}_n^{-1}$ converges a.s. to

$$\int_{\mathbb{R}_+} y f'_X(\theta y) f_X^{-1}(\theta y) f_Y(y)\, dy = \theta \int_{\mathbb{R}_+} y f'_X(\theta y)\, dy = \theta_0^{-1}$$

where θ_0 is the parameter value under the probability P_0 of the sample.

Proposition 5.12. *Under the condition (2.19), the variable $n^{\frac{1}{2}} K_n^{-\frac{1}{2}}(\widehat{\theta}_n - \theta_0)$ converges weakly under P_0 to a centered Gaussian variable with finite variance.*

Proof. The variable $A_n = n^{-1} \sum_{i=1}^n \dot{f}_\theta(Y_i) f_\theta^{-1}(Y_i) = \int_{\mathbb{R}_+} \dot{f}_\theta f_\theta^{-1}\, d\widehat{F}_{nY}$ converge a.s. to $A = \int_{\mathbb{R}_+} \dot{f}_\theta f_\theta^{-1}\, dF_Y$ and its convergence rate α_n, then $\widehat{\theta}_n$ has the same nonparametric convergence rate. The variable $\alpha_n(\widehat{\theta}_n - \theta_0)$ is written as $\alpha_n(A_n - A)$ and its weak convergence is a consequence of Proposition 2.3. □

The density (5.7) of a two dimensional variable $X = (X_1, X_2)$ with respect to Lebesgue's measure has the form

$$f_{\theta,\eta}(x) = \exp\{-\eta T(x,\theta) - S(x,\theta) - b(\eta)\} \tag{5.11}$$

with a random parameter η of $\mathbb{R}_+$, bivariate real statistics $T(x,\theta)$ and $S(x,\theta)$, for $x = (x_1, x_2)$ in $\mathbb{R}_+^2$. It extends the univariate exponential model (1.4) of Example 2 with

$$S(x,\theta) = -\log T'_x(x,\theta)$$

and $b(\eta) = -\log \eta$, the distribution function of X is then

$$F_{\theta,\eta}(x) = 1 - \exp\{-\eta T(x,\theta)\}.$$

Let G be the distribution function of the parameter η, a continuous mixture of exponential variables with the density (5.11) has the distribution function

$$f_\theta(x) = \int_0^\infty f_{\theta,\eta}(x)\,dG(\eta),$$

it is proportional to the Laplace transform at $T(x,\theta)$ of the distribution function H having the density

$$\frac{dH(\eta)}{dG(\eta)} = c^{-1}\eta$$

with respect to G, where $c = E_G\eta$

$$cL_H \circ T(x,\theta) = e^{S(x,\theta)} f_\theta(x). \tag{5.12}$$

A kernel estimator $\widehat{f}_{nh}$ of the density f_{θ_0} of the variable X from a n-sample $(X_i)_{i=1,\dots,n}$ provides an estimator

$$\widehat{L}_{nh} \circ T(x,\theta) = c^{-1} e^{S(x,\theta)} \widehat{f}_{nh}(x)$$

of the function L_H at every value of the parameter θ. The order of differentiability of the estimator $\widehat{f}_{nh}$ is the order of differentiability of the kernel function. The derivatives of $\widehat{f}_{nh}$ define estimators of the derivatives of $\widehat{L}_{nh,\theta}$.

Assuming that $T(x,\theta)$ tends to infinity as x tends to infinity, the distribution function G is estimated by the inversion formula (1.1) of the Laplace transform

$$\widehat{H}_{nh}(\eta) = \lim_{x\to\infty} \sum_{k\le\eta T(x,\theta)} \frac{(-1)^k}{k!} T^k(x,\theta)\widehat{L}_{nh}^{(k)} \circ T(x,\theta).$$

The maximum likelihood estimator $\widehat{\theta}_{nh}$ of the parameter θ in the mixture of the exponential densities (5.11) is calculated from (5.12) as solution of the estimating equation

$$n^{-1}\sum_{i=1}^n \dot{S}_\theta(X_i,\theta) = n^{-1}\sum_{i=1}^n \dot{T}_\theta(X_i,\theta)\frac{\widehat{L}'_{nh}}{\widehat{L}_{nh}} \circ T(X_i,\theta),$$

it converges in probability to the parameter value θ_0 of the sample and its convergence rate to a normal variable is $n^{-\frac{1}{2}}$.

The continuous distribution function F of a two dimensional variable (X,Y) is defined by its marginals F_X and F_Y and a continuous dependence function C as

$$F = C(F_X(x), F_Y(y)).$$

The function $C = F(F_X^{-1}, F_Y^{-1})$ is a distribution function on $[0,1]^2$ with uniform marginal distributions. Models for a two dimensional variable (X,Y) defined by the marginal distribution functions F_X and F_Y and a dependence parameter θ belonging to a subset Θ of $\mathbb{R}^k$ determine the distribution function of (X,Y) in the form

$$F_\theta(x,y) = C_\theta(F_X(x), F_Y(y)), \tag{5.13}$$

with a dependence function C_θ on $[0,1]^2$. Their density is written as

$$f_\theta(x,y) = f_X(x) f_Y(y)\, c_\theta(F_X(x), F_Y(y)),$$

with the marginal densities f_X and f_Y, and the derivative c_θ of the function C_θ with respect to its components on $[0,1]^2$. The distribution function of the variable $(U,V) = (F_X^{-1}(X), F_Y^{-1}(Y))$ reduces to the function C_θ. Maximum likelihood estimators have been defined in various models for the dependence function C_θ of (5.13) (Clayton 1978, Oakes 1986, Glidden and Self 1999).

Considering the dependence parameter as a random variable extends the frailty models. Let G be the distribution function of θ, the distribution function of the uniform variable (U,V) on $[0,1]^2$ is

$$F_{U,V}(u,v) = \int_\Theta C_\theta(u,v)\, dG(\theta).$$

In Morgenstern's model, $C_\theta(u,v) = uv\{1 + \theta(1-u)(1-v)\}$ with $|\theta| < 1$, and the distribution function $F_{U,V}$ has the same form where the variable θ is replaced by its expectation. The models depending linearly on θ have the same property.

Let $\bar{u} = 1-u$ and $\bar{v} = 1-v$, the distribution function of a pair of independent uniform variables (U,V) satisfies $F_{U,V}(u,v) = 1 - \bar{u} - \bar{v} + \bar{u}\bar{v}$ on $[0,1]^2$. In the multiplicative model

$$C_\theta(u,v) = 1 - \bar{u} - \bar{v} + \theta\bar{u}\bar{v},$$

with a random parameter θ, the distribution function of the variable (U,V) is $F_{U,V}(u,v) = 1 - \bar{u} - \bar{v} + \mu\bar{u}\bar{v}$ where μ is the expectation of θ under the distribution function G, its density is $f_{U,V}(u,v) = \mu 1_{\{(u,v)\in[0,1]^2\}}$. In the random exponential model with the conditional distribution function

$$C_\theta(u,v) = 1 - \bar{u} - \bar{v} + (\bar{u}\bar{v})^\theta$$

on $[0,1]^2$, the distribution function of (U,V) is $F_{U,V}(u,v) = 1 - \bar{u} - \bar{v} + L_G(\log(\bar{u}\bar{v}))$ where L_G is the Laplace transform of G, and it density is

$f_{U,V}(u,v) = -\int_\Theta t^2(\bar{u}\bar{v})^t\,dG(t)$.

The estimation of G is more difficult than for a real parameter θ, the score function $\dot{l}_G$ for G in the model with density $\int_\Theta c_\theta(u,v)\,dG(\theta)$ is

$$\dot{l}_G a(u,v) = f_{U,V}^{-1}(u,v)\int_\Theta c_\theta(u,v)a(\theta)\,d\theta$$

for every function a of $L^2(\Theta)$ with integral zero, in the multiplicative exponential model it is

$$\dot{l}_G a(u,v) = -f_{U,V}^{-1}(u,v)\int_\Theta \theta^2(\bar{u}\bar{v})^{\theta-1}a(\theta)\,d\theta.$$

5.5 Tests in bivariate exponential mixture models

In the bivariate exponential model (5.7) with scalar parameters, the maximum likelihood estimators of the parameters are $\widehat{\theta}_n = \bar{Y}_n^{-1}\bar{X}_n$ and $\widehat{\eta}_n = \bar{X}_n^{-1}$. A test for the hypothesis H_0 of a degenerate distribution function $G_0 = \delta_{\eta_0}$ is performed with test statistics comparing the estimators $\bar{F}_{\widehat{\theta}_n,\widehat{\eta}_n}$ and $\bar{F}_{\widehat{\theta}_n,\widehat{G}_n}$ of the survival function of Y, where $\widehat{G}_n$ has the convergence rate $\alpha_n^{-1} = n^{-\frac{1}{2}}K_n^{\frac{1}{2}}$ like the estimator $\widehat{f}_{Xn}$, by Proposition 2.3. Let

$$T_n = \sup_{y\in\mathbb{R}_+} n^{\frac{1}{2}}K_n^{-\frac{1}{2}}|\bar{F}_{\widehat{\theta}_n,\widehat{G}_n}(y) - \bar{F}_{\widehat{\theta}_n,\widehat{\eta}_n}(y)|.$$

Proposition 5.13. *Under H_0, the statistic T_n converges weakly to the supremum of a Gaussian process W_0 on $\mathbb{R}_+^2$, under fixed alternatives, the statistic T_n diverges.*

Proof. Let $F_0 = F_{\theta_0,G_0}$ be the distribution function of (X,Y) under P_0, the process $W_n(y) = n^{\frac{1}{2}}K_n^{-\frac{1}{2}}(\bar{F}_{\widehat{\theta}_n,\widehat{G}_n} - \bar{F}_0)(y)$ is written as

$$W_n(y) = n^{\frac{1}{2}}K_n^{-\frac{1}{2}}\Big\{\int_{\mathbb{R}_+} e^{-\eta\widehat{\theta}_n y}\,d\widehat{G}_n(\eta) - \int_{\mathbb{R}_+} e^{-\eta\theta_0 y}\,dG_0(\eta)\Big\},$$

it converges weakly under P_0 to a Gaussian process W_0 and the process $n^{\frac{1}{2}}(\bar{F}_{\widehat{\theta}_n,\widehat{\eta}_n} - \bar{F}_0)$ converges weakly under P_0 to a centered Gaussian process, it follows that the statistic T_n is asymptotically equivalent to $\sup_{y\in\mathbb{R}_+}|W_n(y)|$ which converges weakly to $T_0 = \sup_{\mathbb{R}_+}|W_0|$. Under fixed alternatives, W_n diverges. □

Let A_n be a sequence of local alternatives defined by parameters θ_n and η_n, and by a distribution function G_n such that

$$A_n : \begin{cases} \theta_n = \theta_0 + n^{-\frac{1}{2}}\widetilde{\theta}_n, \\ \eta_n = \eta_0 + n^{-\frac{1}{2}}\widetilde{\eta}_n, \\ G_n(\eta) = G_0(\eta) + \alpha_n^{-1}\gamma_n(\eta),\ \eta \in \mathbb{R}_+, \end{cases}$$

where the parameters $\widetilde{\theta}_n$ and $\widetilde{\eta}_n$ converge to limits $\widetilde{\theta}$ and respectively $\widetilde{\eta}$, the functions γ_n converge uniformly to a limit γ. The statistic T_n is asymptotically equivalent to $\sup_{y\in\mathbb{R}_+} |W_{nA}(y)|$ where

$$\begin{aligned} W_{nA}(y) &= \alpha_n\Big\{\int_{\mathbb{R}_+} e^{-\eta\widehat{\theta}_n y}\, d\widehat{G}_n(\eta) - \int_{\mathbb{R}_+} e^{-\eta\theta_n y}\, dG_n(\eta)\Big\} \\ &= \alpha_n\Big\{\int_{\mathbb{R}_+} e^{-\eta\theta_n y}\, d(\widehat{G}_n - G_n)(\eta)\Big\} + o_p(1). \end{aligned}$$

Proposition 5.14. *Under local alternatives A_n, the statistic T_n converges weakly to the variable $T_A = \sup_{y\in\mathbb{R}_+} |W_0 - \int_{\mathbb{R}_+} \bar{F}_{\theta_0,\eta}(y)\, d\gamma(\eta)|$.*

A Cramer-von Mises statistic is a $L^2(\widehat{F}_{nY})$ distance between the estimators $\bar{F}_{\widehat{\theta}_n,\widehat{G}_n}$ and $\bar{F}_{\widehat{\theta}_n,\widehat{\eta}_n}$

$$T_{2n} = \sum_{i=1}^{n}\{\bar{F}_{\widehat{\theta}_n,\widehat{G}_n}(Y_i) - \bar{F}_{\widehat{\theta}_n,\widehat{\eta}_n}(Y_i)\}^2 w_n(Y_i),$$

where the sequence of weighting functions w_n converges uniformly to a function w.

Proposition 5.15. *Under H_0, the statistic T_{2n} converges weakly to*

$$T_{02} = \int_{\mathbb{R}_+} W_0^2(y) w(y)\, dF_Y(y),$$

under fixed alternatives, it tends to infinity. Under local alternatives A_n, T_{2n} converges weakly to the variable

$$T_A = \int_{\mathbb{R}_+} \Big\{W_0(y) - \int_{\mathbb{R}_+} \bar{F}_{\theta_0,\eta}(y)\, d\gamma(\eta)\Big\}^2 w(y)\, dF_Y(y).$$

In the bivariate exponential model (5.11), the parameter η has the maximum likelihood estimator $\widehat{\eta}_n = \bar{T}_n^{-1}$ where

$$\bar{T}_n = n^{-1}\sum_{i=1}^{n} T(X_i, \widehat{\theta}_n)$$

for a n-sample $(X_i)_{i\leq n}$ with density f_θ. The maximum likelihood estimator of the parameter θ is solution of the equation

$$\widehat{\eta}_n n^{-1}\sum_{i=1}^n \dot{T}_\theta(X_i,\widehat{\theta}_n) = n^{-1}\sum_{i=1}^n \dot{S}_\theta(X_i,\widehat{\theta}_n),$$

they are consistent and asymptotically Gaussian with the convergence rate $n^{-\frac{1}{2}}$. A test for the hypothesis H_0 of a true exponential density (5.11) on $\mathbb{R}_+^2$, with parameters θ_0, η_0, against a mixture of the exponential densities (5.11) with an unknown mixing distribution is performed like in Section 5.3 for the exponential density (1.2).

5.6 Large dimensional models

Let H be a Hilbert functional space in $L^2(\mathcal{I})$, for a sub-interval $\mathcal{I}$ of $\mathbb{R}$, and let A and X in H. We consider the regression model

$$Y = \int_{\mathcal{I}} A(t)X(t)\,dt + \varepsilon \tag{5.14}$$

with a centered error ε independent of X, with variance σ^2. The unknown function $A(t)$ has to be estimated from a n-sample $(Y_i, X_i)_{i=1\dots,n}$ of Y and $X(t)$. The variance of Y depends on the covariance of the process X as $v = \sigma^2 + \int_{\mathcal{I}^2} A(s)\operatorname{Cov}\{X(s), X(t)\}A(t)\,ds\,dt$.

Let $(\varphi_k)_{k\geq 0}$ be an orthonormal basis of functions of H such that the function $A(t)$ and $X(t)$ have convergent series

$$A(t) = \sum_{k\geq 0} a_k\varphi_k(t), \quad X(t) - EX(t) = \sum_{k\geq 0} x_k\varphi_k(t),$$

with $a_k = \int_{\mathcal{I}} A(t)\varphi_k(t)\,dt$ and $x_k = \int_{\mathcal{I}}\{X(t) - EX(t)\}\varphi_k(t)\,dt$. Let $X_n(t) - EX_n(t)$ and $A_n(t)$ be their approximations by the first k_n terms of the series, with $k_n = o(n)$ tending to infinity as n tends to infinity. The regression model has the approximation

$$Y_n - EY_n = \sum_{k=1}^{k_n} a_k x_k + \varepsilon.$$

Let A_n be the vector $(x_k)_{k=1,\dots,k_n}$, let $\mathbb{Y}_n$ be the vector $(Y_i)_{i=1,\dots,n}$ and let $\mathbb{X}_n$ be the $n\times k_n$ dimensional matrix with components x_{ik}, for $i = 1,\dots,n$ and $k = 1,\dots,k_n$, the approximation of (5.14) is still denoted

$$\mathbb{Y}_n - \mu 1_n = \mathbb{X}_n A_n + (\varepsilon_i)_{i=1,\dots,n}.$$

The coefficients of the significant components of the regressor are estimated by minimization of a penalized mean squared error of the sample $(Y_i, X_i)_{i=1\dots,n}$ under the model (5.14) and the constraint of the minimum norm for A_n

$$Q_n(A_n, \lambda_n) = \|\mathbb{Y}_n - \mu 1_n - \mathbb{X}_n A_n\|^2_{l^2_n} + \lambda_n^2 \|A_n\|^2_{l^2_{k_n}}, \qquad (5.15)$$

with bounded constants λ_n, hence $\lambda_n^2 k_n = o(n)$. The unknown vector A_n has the ridge estimator

$$\widehat{A}_n = \{\mathbb{X}_n^T \mathbb{X}_n + \lambda_n^2 I_{k_n}\}^{-1} \mathbb{X}_n^T (\mathbb{Y}_n - \widehat{\mu}_n 1_n), \qquad (5.16)$$

with the empirical mean $\widehat{\mu}_n = \bar{Y}_n$. The matrix $\{\mathbb{X}_n^T \mathbb{X}_n + \lambda_n^2 I_{k_n}\}^{-1}$ is a generalized inverse of $\mathbb{X}_n^T \mathbb{X}_n$ defined with the identity matrix I_{k_n} of dimension k_n. The variance of $\widehat{A}_n$ is

$$V_n = \sigma^2 \{\mathbb{X}_n^T \mathbb{X}_n + \lambda_n^2 I_{k_n}\}^{-1} \{1 + O(n^{-1})\} - \sigma^2 \lambda_n^2 \{\mathbb{X}_n^T \mathbb{X}_n + \lambda_n^2 I_{k_n}\}^{-2},$$

where $\mathbb{X}_n^T \mathbb{X}_n$ is a $O(n)$ and the second term is a $O(n^{-2}\lambda_n^2) = o(n^{-1})$, hence $V_n = O(n^{-1})$ uniformly in λ_n.

The components of the variable $U_n = \mathbb{X}_n^T (\mathbb{Y}_n - \widehat{\mu}_n 1_n)$ and of the matrix $H_n = \mathbb{X}_n^T \mathbb{X}_n + \lambda_n^2 I_{k_n}$ are sums of n independent variables such that $n^{-1} U_n$ converges a.s. to the matrix with components $\int_{\mathcal{I}^2} \mathrm{Cov}\{X(s), X(t)\} \varphi_k(s) \varphi_{k'}(t)\, ds\, dt$ and $n^{-1} H_n$ converges a.s. to the matrix with components $\int_{\mathcal{I}^2} \mathrm{Cov}\{X(s), X(t)\} \varphi_k(s) \varphi_{k'}(t)\, ds\, dt$, the estimator $\widehat{A}_n$ is therefore a.s. consistent.

The function $A(t)$ has the estimator $\widehat{A}_{n,k_n}(t) = \sum_{k=0}^{k_n} \widehat{a}_{kn} \varphi_k(t)$ and its mean integrated squared error is

$$\begin{aligned} \mathrm{MISE}_n(A) &= \int_{\mathcal{I}} E\{\widehat{A}_{n,k_n}(t) - A(t)\}^2\, dt \\ &= \sum_{k=0}^{k_n} \mathrm{Var}(\widehat{a}_{kn}) + \sum_{k \geq k_n + 1} a_k^2 \\ &= O(n^{-1} k_n) + o(1), \end{aligned}$$

where the components of $\widehat{A}_n$ satisfy $\widehat{a}_{kn} - a_k = O_p(n^{\frac{1}{2}})$, for $k = 0, \dots, k_n$, with $k_n = o(n)$ and uniformly in $\lambda_n = O(1)$.

Proposition 5.16. *Under the condition (2.19), $B_n = (k_n^{-1} n)^{\frac{1}{2}} (\widehat{A}_{n,k_n} - A)$ converges weakly to a Gaussian process with finite expectation and variance functions.*

Proof. The order of the MISE entail that the process $B_n = (k_n^{-1}n)^{\frac{1}{2}}$ is asymptotically equivalent to a normalized sum of n independent variables with convergent expectation and variance functions. By the central limit theorem, its finite dimensional distributions converge weakly to those of a Gaussian process. The process B_n converges weakly to a Gaussian process, as a continuous series estimator. □

A cross-validation estimator of k_n is defined by minimization of the integrated squared error for the estimator $\widehat{A}_{n,k_n}$ of the function A

$$\begin{aligned} I_n(\widehat{A}_{n,k_n}) &= \int_{\mathcal{I}} \{\widehat{A}_{n,k_n}(t) - A(t)\}^2 \, dt \\ &= \sum_{k=0}^{k_n} (\widehat{a}_{kn} - a_k)^2 + \sum_{k \geq k_n+1} a_k^2, \end{aligned}$$

which is asymptotically equivalent to the minimization of

$$\begin{aligned} I_n(k_n) - \int_{\mathcal{I}} A^2(t) \, dt &= \int_{\mathcal{I}} \widehat{A}_{n,k_n}^2(t) \, dt - 2 \int_{\mathcal{I}} \widehat{A}_{n,k_n}(t) A(t) \, dt \\ &= \sum_{k=0}^{k_n} \widehat{a}_{kn}^2 - 2 \sum_{k=0}^{k_n} a_k \widehat{a}_{kn}. \end{aligned}$$

The variable $I_n(k_n)$ is estimated by

$$CV_n(k_n) = n^{-1} \sum_{i=1}^{n} \int_{\mathcal{I}} \widehat{A}_{n,k_n,i}^2(t) \, dt - 2n^{-1} \sum_{i=1}^{n} \int_{\mathcal{I}} X_i(t) \widehat{A}_{n,k_n,i}(t) \, dt,$$

where $\widehat{A}_{n,k_n,i}$ is the estimator of A defined like $\widehat{A}_{n,k_n}$ without the observations (Y_i, X_i). The last integral is

$$\int_{\mathcal{I}} \{X_i(t) - EX_i(t)\} \widehat{A}_{n,k_n,i}(t) \, dt = \sum_{k=0}^{k_n} \widehat{a}_{kn,i} x_k$$

and its expectation converges to $\sum_{k=0}^{k_n} a_k x_k$. The minima of I_n and CV_n are therefore asymptotically equivalent in probability.

Let $\widehat{k}_n = \arg\min_{k \geq 0} CV_n(k)$, the process $\widehat{A}_n(t) = \sum_{k=0}^{\widehat{k}_n} \widehat{a}_{kn} \varphi_k(t)$ is asymptotically optimal i.e.

$$\frac{I_n(\widehat{A}_n)}{\inf_k I_n(k)} \to 1$$

in probability as n tends to infinity, like for the cross-validation of a density in Section 2.5.

The vector $\mathbb{Y}_n = \mu 1_n + \mathbb{X}_n A_n$ has the predictor

$$\widehat{\mathbb{Y}}_{n\lambda_n} = \widehat{\mu}_n 1_n + \mathbb{X}_n \widehat{A}_{n\lambda_n} = \widehat{\mu}_n 1_n + H_{n\lambda_n}(\mathbb{Y}_n - \widehat{\mu}_n 1_n),$$

with the $n \times n$ dimensional symmetric matrix $H_{n\lambda_n} = \mathbb{X}_n\{\mathbb{X}_n^T \mathbb{X}_n + \lambda_n^2 I_{k_n}\}^{-1}\mathbb{X}_n^T$ defined by (5.16). The penalization parameter λ_n is chosen by minimization of a generalized cross-validation criterion

$$GCV(\lambda_n) = \frac{n^{-1}\sum_{i=1}^n (\widehat{Y}_{n\lambda_n,i} - Y_i)^2}{n^{-1} \text{ tr } (I_n - H_{n\lambda_n})}.$$

The good performances of the estimator $\widehat{\lambda}_n$ obtained by minimization of a GCV criterion have been prove by simulations for spline estimators but its optimal properties are generally missing. The variance of $\widehat{Y}_{n\lambda_n,i} - Y_i$ is $v_{ni} = \sigma^2(n^{-1} + H_{nii,\lambda_n})$ where $H_{nii,\lambda_n} = O(n^{-1})$ like the variance V_n of $\widehat{A}_{n\lambda_n}$ and this rate does not depend on $\lambda_n = O(1)$, then $GCV(\lambda)$ converges a.s. to 1, uniformly with respect λ in a bounded interval.

5.7 Inverse problems

Let H and K be Hilbert spaces on a real interval I, let y be a function of K and let A be a linear bounded operator (1.11) from H to K

$$y(x) = Au(x) = \int_I A(x,s)u(s)\,ds, \tag{5.17}$$

where u is an unknown function of H and y belongs to K. A solution u of the equation is obtained by projections on orthonormal bases of the Hilbert spaces. Let $(\varphi_k)_{k\geq 0}$, and respectively $(\psi_k)_{k\geq 0}$, be an orthonormal basis of functions of H, and respectively K, the functions u and y have expansions $u = \sum_{k\geq 0} u_k \varphi_k$ and

$$y = \sum_{k\geq 0} z_k \psi_k = \sum_{k\geq 0} u_k A\varphi_k. \tag{5.18}$$

Let A^* be the adjoint of A such that A^*A is a strictly positive definite matrix, and let

$$a_k^2 = \|A\varphi_k\|_2^2 = \langle A^*A\varphi_k, \varphi_k\rangle,$$

a_k is then strictly positive for every k and by (5.18) we obtain $A\varphi_k = a_k\psi_k$. Let

$$A(x,s) = \sum_{k\geq 0} a_k \psi_k(x)\varphi_k(s),$$

$$Au(x) = \sum_{k\geq 0} u_k a_k \psi_k(x),$$

by the linear independence of the orthonormal functions ψ_k, it follows that

$$u_k = a_k^{-1} z_k$$

for every k such that a_k is not zero. If $a_k = 0$, $A\varphi_k \equiv 0$ and $z_k = 0$, we set $u_k = \frac{0}{0} := 0$ arbitrarily.

If the interval I is bounded, let $(x_i)_{i=1,\dots,n}$ be a regular grid on I such that $\delta_n = x_{i+1} - x_i$ tends to zero as n tends to infinity, the coefficients $z_k = \int_I y\psi_k$ and $a_k = \{\int_I (A\psi_k)^2\}^{\frac{1}{2}}$ are estimated from the knowledge of A and from observations of y at the points of the grid $(x_i)_{i=1,\dots,n}$, by approximation of the integrals with infinitesimal sums

$$\widehat{z}_{nk} = \delta_n \sum_{i=1}^{n} y(x_i)\psi_k(x_i),$$

$$\widehat{a}_{nk}^2 = \delta_n \sum_{i=1}^{n} \{A\varphi_k(x_i)\}^2.$$

It follows that

$$\widehat{u}_{nk} = \widehat{a}_{nk}^{-1} \widehat{z}_{nk},$$

for every k such that b_k is different from zero, in that case $\widehat{u}_{nk} = o(1)$. The estimators based on projections on m functions of the basis are

$$\widehat{y}_{nm}(x) = \sum_{k=0}^{m} \widehat{z}_{nk}\psi_k(x),$$

$$\widehat{A}_{nm}(x,s) = \sum_{k=0}^{m} \widehat{a}_{nk}\varphi_k(s)\psi_k(x),$$

$$\widehat{u}_{nm}(x) = \sum_{k=0}^{m} \widehat{u}_{nk}\varphi_k(x).$$

Lemma 5.1. *If the interval I is bounded, the estimators of the functions y, A and u are uniformly consistent and their biases are $o(\delta_n)$, as n and m tend to infinity.*

Proof. Let $a_i = \frac{1}{2}(x_{i+1} + x_i)$ and let h be an integrable function of $C^2(I)$, integrating its second order expansion $h(x) = h(a_i) + (x - a_i)h'(a_i) + \frac{1}{2}(x - a_i)^2 h''(a_i) + o(\delta_n^2)$ on $I_i =]x_i, x_{i+1}]$, we obtain

$$\begin{aligned}\int_{I_i} h(x)\,dx &= \delta_n h(a_i) + \delta_n h''(a_i)\Big\{\frac{1}{3}(x_{i+1}^2 + x_i x_{i+1} + x_i^2) - a_i^2\Big\} + o(\delta_n^3)\\ &= \delta_n h(a_i) + o(\delta_n),\end{aligned}$$

it follows that

$$\widehat{z}_{nk} - z_k = \delta_n \sum_{i=1}^n y(x_i)\psi_k(x_i) - \int_I y(x)\psi_k(x)\,dx = o(\delta_n),$$

$$\widehat{a}^2_{nk} - a^2_k = \delta_n \sum_{i=1}^n \{A\varphi_k(x_i)\}^2 - \int_I \{A\varphi_k\}^2 = o(\delta_n)$$

and $\widehat{a}_{nk} - a_k = o(\delta_n^{\frac{1}{2}})$, the consistency follows. □

By the argument of the proof of Lemma 5.1, the integrated squared errors of the estimators based on a projection on m functions of the basis are

$$\begin{aligned}
\|\widehat{y}_{nm} - y\|_I^2 &= \sum_{k>m} z_k^2 + \sum_{k=0}^m (\widehat{z}_{nk} - z_k)^2 \\
&= \sum_{k>m} z_k^2 + o(m\delta_n^2) \\
\|\widehat{A}_{nm} - A\|_{I^2}^2 &= \sum_{k>m} a_k^2 + o(m\delta_n) \\
\|\widehat{u}_{nm} - u\|_I^2 &= \sum_{k>m} u_k^2 + o(m\delta_n).
\end{aligned}$$

Choosing $m_n = o(n)$ we have $m_n\delta_n = o(1)$ and the integrated squared errors of the estimators converge to zero, as n tends to infinity.

The solution of the equation $y = Au$ with an expression of A as $\sum_{k\geq 0} a_k\psi_k\varphi_k$ is restrictive, more generally with convergent expansion

$$y(x) = \sum_{j\geq 0} z_j\psi_j(x) = \sum_{k\geq 0} u_k A\varphi_k(x)$$

and for the function $B_k(x) = A\varphi_k(x)$

$$B_k(x) = \sum_{j\geq 0} b_{jk}\psi_j(x),$$

we have

$$y(x) = \sum_{j\geq 0}\sum_{k\geq 0} u_k b_{jk}\psi_j(x) = \sum_{j\geq 0} z_j\psi_j(x).$$

Considering approximations of the functions y by $y_n = \sum_{j=0}^{j_n} z_j\psi_j$, u by $u_n = \sum_{k=0}^{k_n} u_k\varphi_k$ and B_k by $B_{nk} = \sum_{j=0}^{j_n} b_{jk}\psi_j$, for $k = 0, \dots, k_n$, such that j_n and k_n are tend to infinity, we denote Z_n the (j_n+1) dimensional vector $(z_k)_{j=0,\dots,j_n}$, U_n the (k_n+1) dimensional vector $(u_k)_{k=0,\dots,k_n}$, and B_n the

$(j_n+1) \times (k_n+1)$ dimensional matrix $(b_{jk})_{j \le j_n, k \le k_n}$. The approximation of the linear equation is equivalent to $Z_n = B_n U_n$ and its solution under the constraint of a minimum norm for U_n is

$$\widehat{U}_n = \{B_n^T B_n + \lambda_n^2 I_{n+1}\}^{-1} B_n^T Z_n, \tag{5.19}$$

defined like in (5.16) with bounded penalization constants λ_n. The equation (5.17) has the solution

$$\widehat{u}_n(s) = \sum_{k=0}^{k_n} \widehat{u}_{nk} \varphi_k(s) \tag{5.20}$$

with the components $\widehat{u}_{nk}$ of the vector $\widehat{U}_n$.

Proposition 5.17. *The integrated squared error of the function* $\widehat{u}_n$ *is*

$$\mathrm{ISE}_n(u) = \sum_{k=0}^{k_n} (\widehat{u}_{kn} - u_k)^2 + \sum_{k \ge k_n+1} u_k^2 = o(1).$$

Proof. Let $\widehat{Z}_n = B_n \widehat{U}_n$ and let $H_n = B_n^T B_n + \lambda_n^2 I_{k_n+1}$, we have

$$\begin{aligned} \widehat{Z}_n - Z_n &= B_n \widehat{U}_n - Z_n = B_n(\widehat{U}_n - U_n), \\ \widehat{U}_n - U_n &= H_n^{-1} B_n^T (\widehat{Z}_n - Z_n). \end{aligned}$$

The matrices $B_n^T B_n$ and H_n have the dimensions $k_n \times k_n$ and their terms are $O(j_n)$.

For a function y in K with a finite norm $\|y\|_K = (\int_I y^2(x)\,dx)^{\frac{1}{2}}$, the integral $\int_I y^2(x)\,dx$ approximated by $\sum_{j \le j_n} z_j^2$ is bounded. The norm of the approximations are

$$\begin{aligned} \sum_{j \le j_n} (\widehat{z}_{nj} - z_j)^2 &= (\widehat{Z}_n - Z_n)^T (\widehat{Z}_n - Z_n) \\ &= (\widehat{U}_n - U_n)^T B_n^T B_n (\widehat{U}_n - U_n), \end{aligned}$$

as $\sum_{j \le j_n} (\widehat{z}_{nj} - z_j)^2$ is bounded, it follows that $\sum_{k \le k_n} (\widehat{u}_{kn} - u_k)^2 = O(j_n^{-1})$ and the integrated squared error for $\widehat{U}_n$ is

$$\mathrm{ISE}_n(u) = \int_{\mathcal{I}} \{\widehat{u}_n(s) - u(s)\}^2\,ds = \sum_{k=0}^{k_n} (\widehat{u}_{kn} - u_k)^2 + \sum_{k \ge k_n+1} u_k^2$$

where $\sum_{k \ge k_n+1} u_k^2 = o(1)$. □

The model (5.17) extends to a stochastic model with error

$$Y(t) = AU(t) + \varepsilon(t) \tag{5.21}$$

with random processes U and Y and a centered noise process ε independent of U. The variance of the process $Y(t)$ is

$$v(t) = \sigma^2(t) + \int_{\mathcal{I}\times\mathcal{I}} A(t,s)A(t,s') \operatorname{Cov}\{U(s), U(s')\}\, ds\, ds'.$$

For a sample of the process Y observed on a regular grid $(t_i)_{i=1,\dots,p_n}$ on I such that δ_n tends to zero as n tends to infinity, the discrete observations of (5.21) provide the regression model

$$Y_i = Y(t_i) = AU(t_i) + \varepsilon_i := A_i U + \varepsilon_i$$

where ε_i is a centered random variable with variance σ_i^2, for $i = 1, \dots, p_n$, they are multivariate observations of the regression model (5.14). Let $(\varphi_k)_{k\geq 0}$ be an orthonormal basis of functions of H such that the series

$$A_{ni}(s) = A_n(t_i, s) = \sum_{k=0}^{k_n} a_{ik}\varphi_k(s), \quad U_n(s) = \sum_{k=0}^{k_n} u_k\varphi_k(s)$$

converge to $A_i(s)$ and respectively $U(s)$, as k_n tend to infinity. The variables Y_i are approximated by $Y_{ni} = \sum_{k=0}^{k_n} a_{ik}u_k + \varepsilon_i$.

Let $\mathbb{Y}_n$ be the $k_n \times p_n$ dimensional matrix $(Y_{ij})_{i\leq p_n, j\leq k_n}$, let $\mathbb{U}_n$ be the (k_n+1) dimensional vector with components $(u_k)_{k\leq k_n}$, and let $\mathbb{A}_n$ be the $p_n \times (k_n+1)$ dimensional matrix $(a_{ik})_{i\leq p_n, k\leq k_n}$, $\mathbb{U}_n$ is estimated by minimization of the penalized quadratic risk under a constraint of a minimum norm $\mathbb{U}_n$

$$Q_n(\mathbb{U}_n, \lambda_n) = \|\mathbb{Y}_n - \mathbb{A}_n\mathbb{U}_n\|_{l^2_{p_n}}^2 + \lambda_n^2\|U_n\|_2^2,$$

with bounded penalization constants λ_n. Using (5.16), $\mathbb{U}_n$ has the estimator

$$\widehat{\mathbb{U}}_n = \{\mathbb{A}_n^T\mathbb{A}_n + \lambda_n^2 I_{k_n}\}^{-1}\mathbb{A}_n^T\mathbb{Y}_n,$$

where I_{k_n} is the identity matrix of dimension k_n+1. Let V_n the variance of $\mathbb{Y}_n$ and let $H_n = \mathbb{A}_n^T\mathbb{A}_n + \lambda_n^2 I_{k_n}$. The variance of the vector $\widehat{U}_n$ is

$$\Sigma_{nU} = H_n^{-1}\mathbb{A}_n^T V_n \mathbb{A}_n H_n^{-1}$$

where the elements of the matrix $\mathbb{A}_n^T\mathbb{A}_n$ and H_n are $O(p_n)$. For a process ε with independent increments on a bounded interval, the variances σ_i^2 of the variables $\varepsilon(t_i)$ is bounded by their maximum value σ^2 and $\Sigma_{nU} = O(p_n^{-1})$.

The process $U(s)$ is estimated by

$$\widehat{U}_n(s) = \sum_{k=0}^{k_n} \widehat{U}_{nk}\varphi_k(s)$$

using the kth component of $\widehat{\mathbb{U}}_n$, it is a.s. consistent on a compact interval I. The MISE error of $\widehat{U}_n$ is

$$\begin{aligned}
\mathrm{MISE}_n(U) &= \int_I E\|\widehat{U}_n(s) - U(s)\|_{p_n}^2 \, ds \\
&= \sum_{k=0}^{k_n} \mathrm{Var}(\widehat{U}_{nk}) + \sum_{k \geq k_n+1} u_k^2 \\
&= O(p_n^{-1}k_n) + o(1).
\end{aligned}$$

By Proposition 5.16 and if $k_n = o(p_n)$, the estimator $\widehat{U}_n$ is L^2 consistent. The optimal k_n is obtained as the estimation and approximation errors have the same order and we assume

$$k_n = o(p_n), \qquad \sum_{k \geq k_n+1} u_k^2 = O(p_n^{-1}k_n). \tag{5.22}$$

Proposition 5.18. *Under the condition (5.22) and if the variances σ_i^2 are uniformly bounded, the process $k_n^{-\frac{1}{2}} p_n^{\frac{1}{2}}(\widehat{U}_n - U)$ converges weakly on I to a Gaussian process.*

A cross-validation estimator $\widehat{k}_n$ of k_n is defined by minimization of the integrated squared error $I_n(k)$ of the estimator $\widehat{A}_{n,k_n}$, $\widehat{k}_n$ is asymptotically optimal uniformly in λ_n, like in Section 5.6. The minimization of a cross-validation criterion for the choice of the parameter provides an optimal estimator for k_n.

In the stochastic model with error (5.21), the expectation of $Y(t)$ is $y(t) = Au(t)$ in K, follows the model (5.17) with $u(s) = E\{U(s)\}$ in H and a solution is the series estimator (5.20) of the function u, with (5.19). Extending the problem to the estimation of the process U in the stochastic model, an expansion of Y on an orthonormal basis $(\psi_k)_{k\geq 0}$ of $L^2(\mathbb{R}_+)$ is approximated by $Y_T(t) = \sum_{k=0}^{K_T} y_k\psi_k(t)$, from the observation of Y on $[0,T]$, with $K_T = o(T)$ as T tends to infinity. The coefficients $y_k = \int_{\mathbb{R}_+} Y(t)\psi_k(t)\,dt$ are unbiasedly estimated by

$$\widehat{y}_{Tk} = \int_0^T Y(t)\psi_k(t)\,dt = \int_0^T Au(t)\psi_k(t)\,dt + \int_0^T \varepsilon(t)\psi_k(t)\,dt$$

such that $y_{Tk} = E\widehat{y}_{Tk}$ converges to y_k, as T tends to infinity, and

$$\widehat{y}_{Tk} - y_{Tk} = \int_0^T \varepsilon(t)\psi_k(t)\,dt,$$

it is denoted e_{Tk}. The process Y is approximated on the interval $[0, T]$ by $\widehat{Y}_T = \sum_{k=0}^{K_T} \widehat{y}_{Tk}\psi_k$ with the mean integrated squared error

$$\text{MISE}_T(Y) = \int_{\mathbb{R}_+} E\{Y(t) - \widehat{Y}_T(t)\}^2\,dt = \sum_{k=0}^{K_T} E\{\widehat{y}_{Tk} - y_k)^2 + \sum_{k>K_T} y_k^2.$$

We assume that the sum of the variances $\sigma_k^2 = \int_{\mathbb{R}_+} \varepsilon^2(t)\psi_k^2(t)\,dt$ is bounded, the variance σ_{Tk}^2 of $\widehat{y}_{Tk}$ is asymptotically equivalent to σ_k^2 as T tends to infinity, and the process $\widehat{Y}_T$ has a bounded error

$$\text{MISE}_T(Y) = \sum_{k=0}^{K_T} \sigma_{Tk}^2 + o(1).$$

Let Z_T be the $(J_T + 1)$ dimensional vector $(y_j)_{j \leq J_T}$. The operator A has an expansion on an orthonormal basis $(\varphi_k\psi_j)_{j,k\geq 0}$ of $H \times K$, it is approximated by $A_T(t,s) = \sum_{k=0}^{K_T}\sum_{j=0}^{J_T} a_{jk}\varphi_k(t)\psi_j(x)$, and the process $U(t)$ is approximated by $U_T(s) = \sum_{k=0}^{K_T} u_k\varphi_k(s)$. The coefficients of Y are denoted $y_j = B_{Tj}^T V_T$ where B_{Tj} is the jth column of the $(J_T + 1) \times (K_T + 1)$ dimensional matrix $B_T = (a_{jk})_{j\leq J_T, k\leq K_T}$ and V_T is the $(K_T + 1)$ dimensional vector $(u_k)_{k\leq K_T}$ such that

$$Z_T = B_T V_T + e_{Tk}.$$

A solution that minimizes the penalized quadratic risk Q_T has the form $\widehat{u}_T(t) = \sum_{k=0}^{K_T} \widehat{u}_{Tk}\varphi_k(t)$ with the components $\widehat{u}_{Tk}$ of the vector

$$\widehat{V}_T = \{B_T^T B_T + \lambda_T^2 I_{K_T+1}\}^{-1} B_T^T Z_T, \tag{5.23}$$

like in (5.19) with bounded penalization constants λ_T. Arguing like for Proposition 5.16, the mean integrated squared error of the estimator $\widehat{u}_T$ is

$$\text{MISE}_T(u) = \sum_{k=0}^{K_T} \text{Var}(\widehat{u}_{Tk}) + \sum_{k>K_T} u_k^2 = O(J_T^{-1}K_T) + o(1)$$

and it converges to zero as $K_T = o(J_T)$, K_T and J_T tend to infinity as T tends to infinity. The MISE is optimum if the estimation and approximation errors have the same order, i.e.

$$K_T = o(J_T), \qquad \sum_{k>K_T} u_k^2 = O(J_T^{-1}K_T). \tag{5.24}$$

Proposition 5.19. *Under the condition (5.24), the process $J_T^{\frac{1}{2}} K_T^{-\frac{1}{2}}(\widehat{u}_T - u_T)$ converges weakly to a Gaussian process.*

The choice of the basis of orthonormal functions depends on the Hilbert space and on the operator, we can choose them as the eigenfunctions of the operator. Let A be continuous, bounded, symmetric and strictly positive definite kernel operator on $S \times S$, it has an $L^2(S)$-complete orthogonal system of eigenfunctions $(\phi_k)_{k\geq 1}$ and strictly positive and bounded eigenvalues $(\lambda_k)_{k\geq 1}$ such that

$$A(s,t) = \sum_{k\geq 1} \lambda_k \phi_k(s)\phi_k(t).$$

Let $\gamma_k = \|\phi_k\|_2$ and let y belong to $L^2(S)$, then $\sum_{k\geq 1} \gamma_k^2 \lambda_k^2 \langle u, \varphi_k\rangle^2$ is finite and

$$A\phi_k(s) = \int A(s,t)\phi_k(t)\,dt = \lambda_k\phi_k(s)$$

has the norm $\lambda_k\gamma_k$ and the norm of A is

$$\|A\|_2 = \Big\{\sum_{k\geq 1} \lambda_k^2\gamma_k^2\Big\}^{\frac{1}{2}}.$$

Let u in $L^2(S)$ have an expansion $u = \sum_{k\geq 1} u_k\phi_k$, with a finite norm $\|u\|_2 = \{\sum_{k\geq 1} u_k^2\gamma_k^2\}^{\frac{1}{2}}$, the function

$$Au(s) = \int A(s,t)u(t)\,dt = \sum_{k\geq 1} u_k\lambda_k\phi_k(s)$$

has the norm $\|Au\|_2 = \{\sum_{k\geq 1} u_k^2\lambda_k^2\gamma_k^2\}^{\frac{1}{2}}$ which is the norm of y. The function y has also a convergent expansion $y = \sum_{k\geq 1} y_k\phi_k$ and the solution u is defined in the basis $(\gamma_k)_{k\geq 1}$ by its projections

$$u_k = \lambda_k^{-1} y_k.$$

For a bounded operator with infinitely many eigenvalues, the sequence $(\lambda_k\gamma_k)_{k\geq 1}$ converges to zero as k tends to infinity, and the convergence of the solution u depends on their convergence rates. Model (5.21) with $E\{\varepsilon(s)\varepsilon(t)\} = \sigma^2 1_{\{s=t\}}$ entails

$$y_k = \int_S y\phi_k = \lambda_k u_k + \varepsilon_k, \tag{5.25}$$

with $\varepsilon_k = \int_S \varepsilon\phi_k$ such that $E\varepsilon_k = 0$ and $E(\varepsilon_k^2) = \sigma^2\gamma_k^2$ for every $k \geq 1$.

For $k = 1, \ldots, m$, let

$$|u_k| \leq C_1 k^{-\alpha}, \quad |\gamma_k| \leq C_2 k^{-\beta}, \quad |\lambda_k| \geq C_3 k^{-\rho}, \tag{5.26}$$

with α, β and $\rho > 0$ such that $\alpha + \beta > \frac{1}{2}$, $\alpha + \rho > \beta$ and $\rho > 2\beta$.

Proposition 5.20. *Under the conditions (5.26) in model (5.21), the mean integrated squared error for $\widehat{u}_m = \sum_{k=1}^m \lambda_k^{-1} y_k\phi_k$, as σ tends to zero, is*

$$E\|\widehat{u}_n - u\|_2^2 = O\Big(\sigma^{\frac{4(\alpha+\beta)}{1+2(\alpha+\rho-\beta)}}\Big).$$

Proof. By (5.25), the expectation of $\widehat{u}_m$ is $u_m = \sum_{k=1}^m u_k\phi_k$ and $\lambda_k^{-1}y_k - u_k = \lambda_k^{-1}\varepsilon_k$. Under the conditions, the mean integrated squared error of $\widehat{u}_m$ is

$$\begin{aligned} E\|\widehat{u}_m - u\|_2^2 &= \sum_{k=m+1}^{\infty} u_k^2\gamma_k^2 + \sum_{k=1}^{m} E(\varepsilon_k^2)\lambda_k^{-2}\gamma_k^2 \\ &= \sum_{k=m+1}^{\infty} u_k^2\gamma_k^2 + \sigma^2\sum_{k=1}^{m}\lambda_k^{-2}\gamma_k^4 \\ &= O(m^{-2(\alpha+\beta)}) + O(\sigma^2 m^{1+2(\rho-2\beta)}). \end{aligned}$$

The error is minimum as $m = O(\sigma^{-\frac{2}{1+2(\alpha+\rho-\beta)}})$ when σ tends to zero, and it is a $O(\sigma^{\frac{4(\alpha+\beta)}{1+2(\alpha+\rho-\beta)}})$. □

In model (1.11) without error and under the same conditions for u_k and γ_k, the mean integrated squared error for $\widehat{u}_m$ is a $O(m^{-2(\alpha+\beta)})$.

Proposition 5.20 applies to the estimation using an orthonormal basis of functions where the error develops as

$$\begin{aligned} E\|\widehat{u}_m - u\|_2^2 &= \sum_{k=m+1}^{\infty} u_k^2 + \sigma^2\sum_{k=1}^{m}\lambda_k^{-2} \\ &= O(m^{-2\alpha}) + O(\sigma^2 m^{1+2\rho}) = O(\sigma^{\frac{4\alpha}{1+2(\alpha+\rho)}}), \end{aligned}$$

for $m = O(\sigma^{-\frac{2}{1+2(\alpha+\rho)}})$, with α and $\rho > 0$, as σ tends to zero. It reduces to a $O(m^{-2\alpha})$ in the model without error.

Let R be an operator on $S \times S$ with an orthogonal system of eigenfunctions $(\phi_k)_{k\geq 1}$ in $L^2(S)$ and with eigenvalues $\lambda_k > 0$, it is written $R(s,t) = \sum_{k\geq 1}\lambda_k\phi_k(s)\phi_k(t)$. Extending the notations of Wabha (1973), let

$$H_R = \Big\{g : g \in L^2(S), \sum_{k\geq 1}\lambda_k^{-1}\gamma_k^{-2}g_k^2 < \infty, g_k = \langle g, \phi_k\rangle\Big\},$$

H_R is a Hilbert space with scalar product $\langle f, g\rangle_R = \sum_{k\geq 1}\lambda_k^{-1}\gamma_k^{-2}f_k g_k$ and the norm of a function $u = \sum_{k\geq 1}u_k\phi_k$ of H_R is

$$\|u\|_R = \sum_{k\geq 1}\lambda_k^{-1}\gamma_k^{-2}u_k^2.$$

For every s in S, the function $R_s(\cdot) = R(s,\cdot)$ belongs to H_R, it has the norm

$$\|R_s\|_R = \Big\{\sum_{k\geq 1}\lambda_k\phi_k^2(s)\Big\}^{\frac{1}{2}}$$

and it satisfies $\langle R_s, u\rangle_R = u(s)$ for every function u of H. By linearity and by the Riesz theorem, there exists ξ_s in H_R such that

$$u(s) = \langle \xi_s, u\rangle_R,$$

and for every t in S

$$\langle R_s, \xi_t\rangle_R = \langle \xi_s, \xi_t\rangle_R.$$

The linear operator A of (1.11) has a representation in H_R as

$$Au(t) = \langle A_t, u\rangle = \langle \eta_t, u\rangle_R \tag{5.27}$$

with η_t and u in H_R and

$$\eta_t(s) = \langle \eta_t, R_s\rangle_R = \int_S A(t,x)R(s,x)\,dx.$$

Let V be the space generated by the functions $(\eta_t)_{t\in S}$, a solution u of the equation (1.11) is the sum of its projection $P_V u$ on V and $P_{V^c}u$ on the orthogonal to V in H, then $Au = AP_V u$ by (5.27), and u may be defined in V. An approximation has been defined by discretization on a regular grid $(t_i)_{i\leq n}$ with path δ (Wabha 1973), the space V is approximated by V_n generated by the functions $(\eta_{t_i})_{i=1,\ldots,n}$, let $\widetilde{\eta}_n$ be the vector with these functions as components and let $Q_n = (Q(t_i,t_j))_{i,j=1,\ldots,n}$ where $Q = ARA^*$ and $A^*u(t) = \int_S A(s,t)u(s)\,ds$. An approximate solution of (1.11) is defined by

$$u_n = \widetilde{\eta}_n(Q_n + cI)^{-1}\widetilde{y}_n^T$$

where $\widetilde{y}_n$ is the vector with components $(y_{t_i})_{i=1,\ldots,n}$ and c is a constant which may be estimated by cross-validation. The choice of R is not related to K, Wabha (1973) assumes that the norm $\|u\|_R$ of the solution u is small. The approximation error has the order δ^n.

5.8 Applications

An important application of the equation (1.11) is the restoration of a signal. Let $X(x)$ denote an image variable in a bounded rectangle R of the plane and let $Z(x)$ be the observed image modified by a blurring function K and a symmetric noise function U. The problem is to recover X from grouped observation on the rectangle $R_{ij} = [s_{i-1}, s_i[\times[t_{j-1}, t_j[$

$$Z_{ij} = \int_{R_{ij}} X(r_1,r_2)K_{1i}(r_1)K_{2j}(r_2)\,dr_1\,dr_2 + U(s_i,t_i), \tag{5.28}$$

with an integral on $R_{ij} = [s_{i-1}, s_i[\times[t_{j-1}, t_j[$, and for $i = 1, \ldots, I_n$ and $j = 1, \ldots, J_n$. In the discretized problem, the kernel K may be defined as $K(s_i, t_j, r_1, r_2) = (r_1^2 - s_{i-1}^2)^{\frac{1}{2}} + (r_2^2 - t_{j-1}^2)^{\frac{1}{2}}$, for $r = (r_1, r_2)$ in R_{ij}, $i = 1, \ldots, I_n$ and $j = 1, \ldots, J_n$.

In a real interval, the function X is solution of the equation

$$Z(s_i) = \int_{[s_{i-1}, s_i[} X(r) K_i(r)\, dr + U(s_i) \qquad (5.29)$$

where $K_i(r) = K(s_i, r) = (r^2 - s_{i-1}^2)^{\frac{1}{2}}$ for r in $[s_{i-1}, s_i[$ (Sullivan 1986), and U is a Gaussian process with variance σ_i^2, for $i = 1, \ldots, I_n$. Let $a = s_{I_n}$ and let $(\varphi_k)_{k\geq 0}$ be an orthonormal basis of $H = L^2([0, a])$ such that the functions X and $K_{ij}(r) := K(s_i, t_j, r)$ have convergent expansions

$$X(r) = \sum_{k\geq 0} x_k \varphi_k(r),$$

$$K_i(r) = \sum_{k\geq 0} b_{ik} \varphi_k(r),$$

with $x_k = \int_{[0,a]} X(r)\varphi_k(r)\, dr$ and $b_{ik} = \int_{[s_{i-1}, s_i[} K_i(r)\varphi_k(r)\, dr$, such that $Z_i = \sum_{k\geq 0} x_k b_{ik}$. The expansions of X and K_i are approximated as

$$X_n(r) = \sum_{k=0}^{k_n} x_k \varphi_k(r), \quad K_{ni}(r) = \sum_{k=0}^{k_n} b_{ik} \varphi_k(r)$$

and the vector of coefficients $X_n = (x_k)_{k\leq k_n}$ is estimated by the solution of the equation $Z_i = B_{ni}^T X_n$, with $B_{ni} = (b_{ik})_{k\leq k_n}$.

Let Z_n denote the vector $(Z_i)_{i\leq I_n}$ and let B_n be the $(k_n + 1) \times I_n$ dimensional matrix $(b_{ki})_{k\leq k_n, i\leq I_n}$, X_n may be estimated by the minimum $\widehat{X}_n$ of the quadratic risk

$$Q_n(X_n, \lambda_n) = \|Z_n - B_n^T X_n\|_2^2 + \lambda_n^2 \|X_n\|_2^2,$$

under the constraint of a minimum norm $\|X_n\|_2^2$, with bounded penalization constants λ_n. Using (5.16), we obtain

$$\widehat{X}_n = \{B_n B_n^T + \lambda_n^2 Id_n\}^{-1} B_n Z_n, \qquad (5.30)$$

with the identity matrix Id_n of dimension $k_n + 1$, then the process X is estimated by $\widehat{X}_n(r) = \sum_{k=0}^{k_n} \widehat{x}_{nk}\varphi_k(r)$, with the components of $\widehat{X}_n$, their variances are $O(I_n)$.

The mean integrated squared error of the estimator $\widehat{X}_n$ has the same behaviour as in Section (5.6).

Proposition 5.21. *The mean integrated squared error of* $\widehat{X}_n$ *is*

$$\text{MISE}_n(X) = O(I_n^{-1} k_n) + o(1).$$

As $k_n = o(I_n)$, Proposition 5.16 extends to the weak convergence of the process $(k_n^{-1} I_n)^{\frac{1}{2}} (\widehat{X}_n - X)$ to a Gaussian process on I.

For X and Z in a bounded rectangle R of the plane, let $(\varphi_k)_{k \geq 0}$ be an orthonormal basis of $H = L^2(R)$ such that the functions $K_{ij}(r) = K_{1i}(r_1) K_{2j}(r_2)$ and X have convergent expansions in H

$$K_{ij}(r) = \sum_{k_1, k_2 \geq 0} b_{1ik_1} b_{2jk_2} \varphi_{k_1}(r_1) \varphi_{k_2}(r_2),$$

$$X(r) = \sum_{k_1, k_2 \geq 0} x_{k_1 k_2} \varphi_{k_1}(r_1) \varphi_{k_2}(r_2),$$

with the coefficients

$$b_{1ik_1} b_{2jk_2} = \int_{R_{ij}} K_{ij}(r) \varphi_{k_1}(r_1) \varphi_{k_2}(r_2)\, dr_1\, dr_2,$$

$$x_{k_1 k_2} = \int_{R_{ij}} X(r) \varphi_{k_1}(r_1) \varphi_{k_2}(r_2)\, dr_1\, dr_2.$$

The functions X and K_{ij} are approximated by sums over $k_1 \leq k_{n1}$ and $k_2 \leq k_{n2}$, where k_{n1} and k_{n2} are $o(I_n)$, let X_n denote the $(k_{n1}+1) \times (k_{n2}+1)$ dimensional matrix $(x_{k_1 k_2})_{k_1 \leq k_{n1}, k_2 \leq k_{n2}}$. Let $B_{1n} = (b_{1ik_1})_{k_1 \leq k_{n1}, i \leq I_n}$ and let $B_{2n} = (b_{2ik_2})_{k_2 \leq k_{n2}, i \leq J_n}$, the observation matrix $Z_n = (Z_{ij})_{i \leq I_n, j \leq J_n}$ is solution of the equation

$$Z_n = B_{1n}^T X_n B_{2n}$$

under a constraint of minimum dimensions, it minimizes the risk

$$Q_n(X_n, \lambda_n) = \|Z_n - B_{1n}^T X_n B_{2n}\|_2^2 + \lambda_n^2 \|X_n\|_2^2$$

with bounded constants λ_n. Using the generalized inverses

$$H_{jn} = B_{jn} B_{jn}^T + \lambda_n^2 I_{jn}$$

with the identity matrices I_{jn} of dimension $k_{nj} + 1$, for $j = 1, 2$, we obtain

$$\widehat{X}_n = H_{1n}^{-1} B_{1n} Z_n B_{2n}^T H_{2n}^{-1T}. \tag{5.31}$$

The process X is estimated by

$$\widehat{X}_n(r) = \sum_{k_1=0}^{k_{n1}} \sum_{k_2=0}^{k_{n2}} \widehat{x}_{nk_1k_2} \varphi_{k_1}(r_1) \varphi_{k_2}(r_2),$$

with the components of $\widehat{X}_n$.

Let $K_T = o(J_T)$, K_T and J_T tend to infinity as T tends to infinity, Proposition 5.21 extends to the plane.

Proposition 5.22. *The estimator $\widehat{X}_n$ has mean integrated squared error*

$$\mathrm{MISE}_n(X) = O(I_n^{-2} k_{1n} k_{2n}) + o(1).$$

Proof. Let $\widehat{Z}_n = B_{1n}^T \widehat{X}_n B_{2n} = B_{1n}^T H_{1n}^{-1} B_{1n} Z_n B_{2n}^T H_{2n}^{-1T} B_{2n}$, we have

$$\widehat{Z}_n - Z_n = B_{1n}^T (\widehat{X}_n - X_n) B_{2n},$$

its solution under the constraint on the norm $\widehat{X}_n - X_n$ is

$$\widehat{X}_n - X_n = H_{1n}^{-1} B_{1n} (\widehat{Z}_n - Z_n) B_{2n}^T H_{2n}^{-1}.$$

The matrices $B_{jn} B_{jn}^T$ and H_{jn} have the dimensions $k_{jn} \times k_{jn}$, $H_{1n} = O(I_n)$ and $H_{2n} = O(J_n)$.

For a process Z with a finite norm in a Hilbert space, the integral $\int Z^2(x,y)\,dx\,dy$ is bounded. The norm of $\widehat{Z}_n - Z_n$ is

$$\begin{aligned}\|(\widehat{Z}_n - Z_n)^T (\widehat{Z}_n - Z_n)\| &= \sum_{i \le I_n} \sum_{j \le j_n} (\widehat{Z}_{nij} - Z_{ij})^2 \\ &= \|B_{2n}^T (\widehat{X}_n - X_n)^T B_{1n} B_{1n}^T (\widehat{X}_n - X_n) B_{2n}\|\end{aligned}$$

and it is bounded, it follows that

$$(\widehat{X}_n - X_n)^T (\widehat{X}_n - X_n) = \sum_{k_1 \le k_{n1}} \sum_{k_2 \le k_{n2}} (\widehat{x}_{nk_1k_2} - x_{k_1k_2j})^2 = O(k_{n1} k_{n2} I_n^{-1} J_n^{-1}).$$

The mean integrated squared error for $\widehat{X}_n$ is

$$\mathrm{MISE}_n(X) = \sum_{k_1=0}^{k_{n1}} \sum_{k_2=0}^{k_{n2}} (\widehat{x}_{nk_1k_2} - x_{k_1k_2})^2 + \sum_{k_1 > k_{n1}} \sum_{k_2 > k_{n2}} u_{k_1k_2}^2$$

where $\sum_{k_1 > k_{n1}} \sum_{k_2 > k_{n2}} u_{k_1k_2}^2 = o(1)$. □

The MISE error of $\widehat{X}_n$ is optimum if the estimation and approximation errors have the same order, i.e.

$$K_{n1} k_{n2} = o(I_n J_n), \qquad \sum_{k_1 > k_{n1}} \sum_{k_2 > k_{n2}} u_{k_1k_2}^2 = O(k_{n1} k_{n2} I_n^{-1} J_n^{-1}). \quad (5.32)$$

Proposition 5.23. *Under the condition (5.32), the process* $(I_n J_n)^{\frac{1}{2}} (k_{n1} k_{n2})^{-\frac{1}{2}} (\widehat{X}_n - X)$ *converges weakly to a Gaussian process.*

We consider the convolution model

$$Z(x) = \int_S X(x-u) T(u)\,du + U(x) \quad (5.33)$$

defined on a d dimensional bounded set S as the sum of an blurred process X and a symmetric noise process $U(x)$. Deconvolution methods based on the inverse of Fourier transforms have been used in models with parametric or nonparametric functions T (Hall and Titterington 1986, Hall and Qiu 2007, Benhaddou, Pensky and Picard 2006,). The model defined by (5.33) with $X(x,u) = X(x-u)$ is a linear inverse problem with error like in equation (5.21).

Let X belong to a functional Hilbert space H provided with an orthonormal basis of functions $(\varphi_k)_{k\geq 0}$, and let Z belong to a functional Hilbert space K provided with an orthonormal basis of functions $(\psi_k)_{k\geq 0}$ so that $X = \sum_{k\geq 0} x_k \varphi_k$ and

$$EZ(x) = \sum_{k\geq 0} z_k \psi_k(x) = \sum_{k\geq 0} x_k \int_S \varphi_k(x-u)T(u)\,du. \tag{5.34}$$

The convolution $\varphi_k \star T(x)$ has an expansion on the basis $(\psi_k)_{k\geq 0}$ of K as $\varphi_k \star T(x) = \sum_{j\geq 0} v_{jk}\psi_j(x)$. Using approximations on $(\varphi_k)_{k\leq k_n}$ and $(\psi_j)_{j\leq j_n}$, the equation (5.33) is equivalent to

$$Z_n = V_n X_n$$

with the vectors $Z_n = (z_j)_{j\leq j_n}$ and $X_n = (x_k)_{k\leq k_n}$, and with the matrix $V_n = (v_{jk})_{j\leq j_n, k\leq k_n}$. It is solved like previously the equation (5.29).

5.9 Fredholm equations

Fredholm equations have the general form

$$u(s) - \lambda \int_a^b K(s,t)u(t)\,dt = f(s) \tag{5.35}$$

where K and f are known functions of a Hilbert space H, K is a continuous function on $H \times H$ and λ is a constant. In the homogeneous equation, the function f is zero and (5.35) defines an implicit equation.

Proposition 5.24. *Solutions of the homogeneous equation (5.35) with a symmetric kernel K and distinct constants are orthogonal.*

Proof. Let u, and respectively u', be solution of (5.35), with the constant λ, and respectively λ', then $(\lambda - \lambda')\int u(s)u'(s)\,ds$ is written as

$$\lambda\lambda' \int K(s,t)u(t)u'(s)\,dt\,ds - \lambda\lambda' \int K(s,t)u(s)u'(t)\,dt\,ds = 0.$$

□

For a symmetric and strictly positive kernel K, there exists an orthogonal system of eigenfunctions $(\phi_k)_{k\geq 1}$ in H and strictly positive eigenvalues $(\lambda_k)_{k\geq 1}$ such that

$$K(s,t) = \sum_{k\geq 1} \gamma_k \phi_k(s)\phi_k(t).$$

If the series $u(t) = \sum_{k\geq 1} u_k\phi_k(t)$ and $f(s) = \sum_{k\geq 1} y_k\phi_k(s)$ in H, the equation (5.35) is equivalent to

$$u_k(1 - \lambda\gamma_k) = y_k$$

and its solution is straightforward under boundedness conditions

$$\|K\|_\infty < M, \quad \lambda M|I| < 1. \tag{5.36}$$

If K is not symmetric, the search of a solution relies on the reciprocal kernel of K, it is a function k such that

$$K(s,t) + k(s,t) = \int_I k(s,\tau)K(\tau,t)\,d\tau = \int_I K(s,\tau)k(\tau,t)\,d\tau. \tag{5.37}$$

Integrating a solution u of (5.35) with respect to k yields

$$\begin{aligned} v(s) &:= \int_I u(t)k(s,t)\,dt \\ &= \lambda \int_I \int_I k(s,t)K(t,\tau)u(\tau)\,d\tau\,dt + \int_I f(t)k(s,t)\,dt \\ &= \lambda \int_I \int_I K(s,t)k(t,\tau)u(\tau)\,d\tau\,dt + \int_I f(t)k(s,t)\,dt \end{aligned}$$

and the function v is a solution of the equation

$$g(s) = v(s) - \lambda \int_I K(s,t)v(t)\,dt +$$

where $g(s) = \int_I f(t)k(s,t)\,dt$. This equation and the definition (5.37) imply

$$\begin{aligned} v(s) &= \lambda \int_I \{K(s,t) + k(s,t)\}u(t)\,dt + g(s), \\ &= u(s) - f(s) + \lambda v(s) + g(s), \end{aligned}$$

equivalently

$$f(s) - \int_I f(t)k(s,t)\,dt = u(s) + (1-\lambda)\int_I u(t)k(s,t)\,dt. \tag{5.38}$$

Proposition 5.25. *With $\lambda = 1$, the function f is the unique solution of the Fredholm equation $u(s) = f(s) - \int_I f(t)k(s,t)\,dt$.*

Fredholm's solution of (5.35) is defined under boundedness conditions (5.36). Let $K_1(s,t) = \lambda K(s,t)$ and for $n > 1$

$$K_n(s,t) = \lambda \int_I K_{n-1}(s,\tau)K(\tau,t)\,d\tau,$$

the function

$$k(s,t) = -\{K_1(s,t) + \cdots + K_n(s,t) + \cdots\}$$

is a reciprocal kernel of $K(s,t)$ and the solution of (5.35) is

$$\varphi(s) = f(s) - \int_I k(s,t) f(t)\, dt.$$

Following Fredholm, Heywood and Fréchet (1912) proved the existence of solutions of (5.35) defined, according to the values of λ, by the convergent series

$$D(\lambda) = 1 - \lambda \int_I K(s,s)\, ds + \frac{\lambda^2}{2} \int_{I^2} K\left(\begin{smallmatrix} s_1, & s_2 \\ s_1, & s_2 \end{smallmatrix}\right) ds_1\, ds_2 + \cdots$$
$$+ \frac{(-\lambda)^n}{2} \int_{I^n} K\left(\begin{smallmatrix} s_1, & s_2, & \dots & s_n \\ s_1, & s_2, & \dots & s_n \end{smallmatrix}\right) ds\, ds_1 \dots ds_n + \cdots,$$

$$D(s,t,\lambda) = 1 - \lambda \int_{I^2} K\left(\begin{smallmatrix} s, & s_1 \\ t, & s_1 \end{smallmatrix}\right) ds\, ds_1 + \frac{\lambda^2}{2} \int_{I^3} K\left(\begin{smallmatrix} s, & s_1, s_2 & \\ t, & s_1, & s_2 \end{smallmatrix}\right) ds\, ds_1 + \cdots$$
$$+ \frac{(-\lambda)^n}{2} \int_{I^{n+1}} K\left(\begin{smallmatrix} s, & s_1, & \dots & s_n \\ t, & s_1, & \dots & s_n \end{smallmatrix}\right) ds\, ds_1 \dots ds_n + \cdots$$

where $K\left(\begin{smallmatrix} s_1, & \dots & s_n \\ t_1, & \dots & t_n \end{smallmatrix}\right)$ is the determinant of the $n \times n$ dimensional matrix $(K(s_i,t_j) =)_{i,j}$. If λ is not a zero of $D(\lambda)$, let

$$K(s,t,\lambda) = \frac{D(s,t,\lambda)}{D(\lambda)}$$

be the resolvent of the kernel K.

Proposition 5.26. *If λ is not a zero of $D(\lambda)$, then $k(s,t) = -\lambda K(s,t,\lambda)$ is the reciprocal function of $\lambda K(s,t)$*

A solution of the equation (5.35) is solution of (5.38). If λ is a zero of $D(\lambda)$, it is a pole of the function $K(s,t,\lambda)$. Furthermore, if $D(\lambda)$ has no zero, the homogeneous equation (5.35) has the unique solution zero and if $D(\lambda)$ has zeros, there exists at least a solution of the homogeneous equation (5.35).

Let $f(x) = Au(x)$ in H, a solution of the Fredholm equation (5.35) in H is obtained by projections of u, f and A on an orthonormal basis $(\varphi_k)_{k\geq 0}$ of H. Let $u = \sum_{k\geq 0} u_k \varphi_k(s)$

$$a_k = \int_I f(x) \varphi_k(x)\, dx,$$
$$b_k(s) = \int_I K(s,t) \varphi_k(t)\, dt,$$

then $f(s) = \sum_{k\geq 0} a_k \varphi_k(s)$ and

$$K(s,t) = \sum_{k\geq 0} b_k(s) \varphi_k(t).$$

Integrating the equation (5.35) with respect to K implies

$$\int_I K(s,t)f(t)\,dt = \int_I K(s,t)u(t)\,dt - \lambda \int_{I^2} K(s,\tau)K(\tau,t)u(t)\,d\tau\,dt$$

$$\sum_{k\geq 0} a_k b_k(s) = \sum_{k\geq 0} u_k b_k(s) - \lambda \sum_{k\geq 0} u_k \int_I K(s,\tau) b_k(\tau)\,d\tau$$

$$= \sum_{k\geq 0} b_k(s)\Big\{u_k - \lambda \sum_{j\geq 0} u_j B_{jk}\Big\},$$

where $B_{jk} = \int_I \varphi_j(\tau) b_k(\tau)\,d\tau$. Approximating the series expansions of f, K and u by their projections on m terms, with m sufficiently large, this equation is equivalent to the system of linear equations

$$a_k = u_k - \lambda \sum_{j\geq 0} u_j B_{jk},\ k = 0, \dots, m,$$

and the m linear equations may be solved for every λ which is not a root of the discriminant of this system of equation.

Another approach for solving the Fredholm equation (1.12) is its discretization on a partition of the interval I into n subintervals $I_i =]x_i, x_{i+1}]$ of the same length yields

$$\begin{aligned} f(x_i) &= u(x_i) - \lambda \int_I K(x_i, y) u(y)\,du \\ &:= u(x_i) - \lambda \int_I k_i(y) u(y)\,dy, \end{aligned} \tag{5.39}$$

for $i = 1, \dots, n$. Heywood and Fréchet (1912) proved the existence of functions $\alpha_1, \dots, \alpha_n$ defined according to the partition such that the kernel K has the approximation

$$K_n(x, y) = \sum_{i=1}^n \alpha_i(x) k_i(y),$$

with an approximation order $|K(x,y) - K_n(x,y)| = O(\delta)$ under boundedness conditions. Let $b_{ki} = \int_I k_i(y)\varphi_k(y)\,dy$, the equations (5.39) are approximated as

$$f(x_i) = u(x_i) - \lambda \delta_n \sum_{k=1}^n k_i(x_k) u(x_k) = o(\delta_n).$$

and these linear equations may be solved for every λ which is not a root of their discriminant.

Using approximations of convergent expansions of the functions u, f and K on an orthonormal basis $(\varphi_k)_{k\geq 0}$ of H by

$$u_n(x) = \sum_{k=0}^{k_n} u_k \varphi_k(x), \quad f(x) = \sum_{k=0}^{k_n} y_k \varphi_k(x)$$

and for the function $B_k(x) = \int_{\mathcal{I}} K(x,t)\varphi_k(t)\,dt$

$$B_k(x) = \sum_{j=0}^{k_n} b_{kj} \psi_j(x),$$

the $k_n + 1$ dimensional vectors $U_n = (u_k)_{k\leq k_n}$ and $Y_n = (y_k)_{k\leq k_n}$, and the matrix $B_n = (b_{kj})_{k,j\leq k_n}$ satisfy the equation

$$(I_{k_n+1} - \lambda B_n^T)U_n = Y_n,$$

with the identity matrix I_{k_n+1}.

A solution is obtained by minimization of the penalized mean squared error

$$Q_n(U_n, \lambda) = \|(I_{k_n+1} - \lambda B_n^T)U_n - Y_n\|_2^2 + \mu_n^2 \|U_n\|_2^2,$$

where μ_n is a penalization parameter, as

$$\widehat{U}_n = (B_n + \lambda B_n B_n^T + \mu^2 I_{k_n+1})^{-1} B_n Y_n,$$

and the function u is estimated by $\widehat{u}_n(x) = \sum_{k=0}^{k_n} \widehat{u}_{nk} \varphi_k(x)$ with the components $\widehat{u}_{nk}$ of the vector $\widehat{U}_n$.

Chapter 6

Hazard functions under censoring and truncation

6.1 Introduction

Let T be a random time variable on $\mathbb{R}_+$ with distribution function F and survival function $\bar{F} = 1 - F$, the cumulative hazard function of T is $\Lambda = -\log \bar{F}$ so that $\bar{F} = e^{-\Lambda}$ and $\Lambda(t)$ tends to infinity as t tends to infinity. Let $(T_i)_{i=1,\dots,n}$ be a sample of independent and identically distributed observations of T, the counting process $N_n(t) = \sum_{i=1}^n 1_{\{T_i \le t\}}$ has the predictable compensator

$$\widetilde{N}_n(t) = \int_0^t Y_n(s)\, d\Lambda(s)$$

where $Y_n(t) = \sum_{i=1}^n 1_{\{T_i \ge t\}}$ and

$$M_n = N_n - \widetilde{N}_n \tag{6.1}$$

is a local martingale with respect to the filtration generated by the processes N_n and Y_n. The process

$$\widehat{\Lambda}_n(t) = \int_0^t Y_n^{-1}(s) 1_{\{Y_n(s)>0\}}\, dN_n(s) \tag{6.2}$$

is a consistent estimator of the cumulative hazard function, it is unbiased on every interval where $Y_n > 0$ and the process $n^{\frac{1}{2}}(\widehat{\Lambda}_n - \Lambda)$ converges weakly to a centered Gaussian process with variance function $v(t) = \int_0^t \bar{F}^{-1} 1_{\{\bar{F}>0\}}\, d\Lambda$ and covariance function $v(s,t) = v(s \wedge t)$, at (s,t), due to the martingale property of M_n (Rebolledo 1980).

If F has a density f, the hazard function $\lambda(t) = \bar{F}^{-1} f$ is the derivative of Λ. By an expansion of λ on Laguerre's basis (2.3) as $\lambda = \sum_{k=0}^{\infty} c_k L_k$, with $c_k = \int_0^{\infty} L_k(t)\, d\Lambda(t)$, the function λ is estimated by

$$\widehat{\lambda}_n(t) = \sum_{k=0}^{K_n} \widehat{c}_{nk} L_k(t) \tag{6.3}$$

where $K_n = o(n)$ tends to the infinity as n tends to the infinity and with the estimators of the coefficients

$$\widehat{c}_{nk} = \int_0^\infty L_k(t)\, d\widehat{\Lambda}_n(t),\ k = 0, \ldots, K_n. \tag{6.4}$$

The estimators $\widehat{c}_{nk}$ are asymptotically unbiased and consistent and the variables $n^{\frac{1}{2}}(\widehat{c}_{nk} - c_k)$ converge weakly to centered Gaussian process with variances $\int_0^\infty L_k^2 \bar{F}^{-1} 1_{\{\bar{F}>0\}}\, d\Lambda$.

The mean squared error $\mathrm{MISE}_n(\lambda)$ of the estimator (6.3) is the sum of the approximation error $\|\lambda - \lambda_n\|_2^2 = o(1)$ of the function λ by $\lambda_n = \sum_{k=0}^{K_n} c_k L_k$, and the estimation error

$$E\|\widehat{\lambda}_n - \lambda_n\|_2^2 = E\int_0^\infty \{\widehat{\lambda}_n(t) - \lambda_n(t)\}\, dt = 0(K_n n^{-1}).$$

As n tends to infinity, the size K_n defining the approximating function λ_n is optimum as both errors have the same order so K_n is chosen in order that

$$K_n = o(n), \quad \|\lambda_n - \lambda\|_2 = O(n^{-\frac{1}{2}} K_n^{\frac{1}{2}}). \tag{6.5}$$

Proposition 6.1. *Under the condition (6.5), the process $(K_n^{-1} n)^{\frac{1}{2}}(\widehat{\lambda}_n - \lambda)$ converges weakly to a Gaussian process with finite mean and variance functions.*

By the same arguments as in Proposition 2.2, the rate of convergence $(K_n n^{-1})^{\frac{1}{2}}$ can be compared to the optimal rate of the kernel estimator for the hazard function (Pons 1986).

Proposition 6.2. *The series estimator of an intensity belonging to the space W_{2s}, defined by (2.20), has the optimal convergence rate $n^{-\frac{s}{2s+1}}$.*

The proofs are the same as for the series density estimator.

A weighted estimator for an intensity λ of the space W_{2s} is defined like (2.23) as

$$\widehat{\lambda}_{n\nu_n}(x) = \sum_{k=1}^n \frac{\mu_k}{\nu_n + \mu_k} \widehat{c}_{nk} \psi_k(x)$$

with the parameters $\mu_k = k^{-2s}$ and $\nu_n = O(n^{-\frac{2s}{2s+1}})$, and with orthonormal functions ψ_k of $L^2(\mathbb{R}_+)$ provided with the Lebesgue measure, satisfying the condition (2.22). Proposition 2.4 applies to the estimator $\widehat{\lambda}_{n\nu_n}$.

Under right-censoring, the variables T_i are censored by variables C_i and the observations are $X_i = T_i \wedge C_i$ and the indicators $\delta_i = 1_{\{T_i \le C_i\}}$. The observed counting processes are

$$N_n(t) = \sum_{i=1}^n \delta_i 1_{\{X_i \le t\}},$$
$$Y_n(t) = \sum_{i=1}^n 1_{\{X_i \ge t\}},$$

N_n has the predictable compensator $\widetilde{N}_n(t) = \int_0^t Y_n(s)\, d\Lambda(s)$ and $M_n = N_n - \widetilde{N}_n$ is a local martingale with respect to the filtration generated by the processes N_n and Y_n. The cumulative hazard function is still consistently estimated by the process $\widehat{\Lambda}_n$ defined in (6.2) and the hazard function $\lambda = \bar{F}^{-1} f$ has the series estimator $\widehat{\lambda}_n$ given by (6.3), with the same convergence properties.

With independent censoring variables having a common distribution function G, the processes $n^{-1}N_n(t)$, and respectively $n^{-1}Y_n(t)$ converge a.s. to the function $\int_0^t \bar{G}\, dF$, and respectively $\bar{G}(t)\bar{F}(t)$. The estimator $\widehat{\Lambda}_n$ is $n^{-\frac{1}{2}}$-consistent under right-censoring and $n^{\frac{1}{2}}(\widehat{\Lambda}_n - \Lambda)$ converges weakly to a centered Gaussian process on every interval strictly included in the support of the variable X. The derivative λ of Λ is consistently estimated by $\widehat{\lambda}_n$ defined like in (6.3) and its asymptotic properties are similar to the properties of the estimator without censoring.

Cox's model (Cox 1972) with proportional hazard functions has been extended to describe the hazard function of n right-censored individuals depending on a vector of left-continuous covariates in the form

$$\lambda_i(t, \beta \mid Z_i) = \lambda_0(t) e^{\beta^T Z_i(t)},\ i = 1, \ldots, n. \tag{6.6}$$

Let $N_i(t) = \delta_i 1_{\{T_i \le t\}}$ and $\bar{N}_n(t) = \sum_{i=1}^n N_i(t)$ be the counting processes of the uncensored time variables in the population and let $S_n^{(0)}(t, \beta) = \sum_{i=1}^n 1_{\{X_i \ge t\}} e^{\beta^T Z_i(t)}$. The cumulative hazard function $\Lambda_0(t) = \int_0^t \lambda_0(s)\, ds$ is estimated by the process

$$\widehat{\Lambda}_n(t) = \int_0^t S_n^{(0)-1}(s, \widehat{\beta}_n) 1_{\{S_n^{(0)}(s, \widehat{\beta}_n) > 0\}}\, d\bar{N}_n(s) \tag{6.7}$$

where the estimator $\widehat{\beta}_n$ of the parameter β maximizes the logarithm of the partial likelihood ratio

$$l_n(\beta) = n^{-1} \sum_{i=1}^n \delta_i \Big\{ (\beta - \beta_0)^T Z_i(T_i) - \log \frac{S_n^{(0)}(T_i, \beta}{S_n^{(0)}(T_i, \beta_0)} \Big\}$$
$$= n^{-1} \sum_{i=1}^n \int_0^\tau \Big\{ (\beta - \beta_0)^T Z_i(s) - \log \frac{S_n^{(0)}(s, \beta}{S_n^{(0)}(s, \beta_0)} \Big\} \, dN_i(s),$$

it is solution of the score equation

$$\sum_{i=1}^n \int_0^\tau Z_i \, dN_i = \sum_{i=1}^n \int_0^\tau \frac{S_n^{(1)}(s, \beta)}{S_n^{(0)}(s, \beta)} \, dN_i(s)$$

where $S_n^{(1)}(t, \beta)$ is the derivative of $S_n^{(0)}(t, \beta)$ with respect to β. The asymptotic properties of the estimators $\widehat{\beta}_n$ and $\widehat{\Lambda}_n$ in the model (6.6) have been studied by several authors, in particular by Andersen and Gill (1986) for covariate processes. The hazard function λ_0 is estimated by projection of the estimator of the cumulative hazard on an orthonormal basis of functions like in (6.3) and (6.4), it has he same asymptotic properties.

In a proportional hazards model (6.6), the cumulative hazard function Λ_0 is estimated by the process $\widehat{\Lambda}_n$ given by (6.7), which depends on the estimator of the regression parameter $\widehat{\beta}_n$. The baseline hazard function λ_0 is estimated like a hazard function $\lambda(t)$ in the form (6.3), with the estimated coefficients (6.4) defined with the estimator (6.7) of Λ_0

$$\widehat{c}_{nk} = \int_0^\infty L_k(t) \, d\widehat{\Lambda}_n(t), \; k = 0, \ldots, K_n.$$

The asymptotic properties of the process $\widehat{\Lambda}_n$ imply the weak convergence of the estimators $\widehat{c}_{nk}$, with the rate $n^{-\frac{1}{2}}$ and the estimator $\widehat{\lambda}_n$ of the baseline hazard function λ_0 has the same rate of convergence as the estimator defined in (6.3), it converges weakly to a Gaussian process.

The survival function $\bar{F}_0(t) = \exp\{-\Lambda_0(t)\}$ is consistently estimated by

$$\widehat{\bar{F}}_n(t) = \prod_{T_i \le t} \Big\{ 1 - S_n^{(0)-1}(T_i, \widehat{\beta}_n) 1_{\{S_n^{(0)}(T_i, \widehat{\beta}_n) > 0\}} \Big\}^{\delta_i}$$

and the process $n^{\frac{1}{2}}(\widehat{\bar{F}}_n - \bar{F}_0)$ converges weakly on every compact interval where $\bar{F}_0$ is strictly positive to a centered Gaussian process with a finite variance function.

In a proportional hazards model (6.6), the conditional cumulative hazard function of the ith individual under the probability P_0 of the observations and at the parameter value β_0 is

$$\Lambda_0(t \mid Z_i) = \int_0^t e^{\beta_0^T Z_i(s)}\, d\Lambda_0(s),$$

the conditional survival function under P_0 of the observed times T_i with a covariate vector Z_i is the exponential survival function

$$\bar{F}_0(t \mid Z_i) = \exp\{-\Lambda_0(t \mid Z_i)\}.$$

Their estimators are deduced from the estimators (6.7) for the baseline cumulative hazard function Λ_0 and $\widehat{\beta}_n$ for β_0

$$\widehat{\Lambda}_n(t \mid Z_i) = \int_0^t e^{\widehat{\beta}_n^T Z_i(s)}\, 1_{\{S_n^{(0)}(s,\widehat{\beta}_n)>0\}} S_n^{(0)-1}(s, \widehat{\beta}_n)\, d\bar{N}_n(s),$$

$$\widehat{\bar{F}}_n(t \mid Z_i) = \prod_{T_i \le t} \Big\{1 - \Delta\widehat{\Lambda}_n(T_i \mid Z_i)\Big\}^{\delta_i}$$

where $\Delta H(t)$ is the jump of a function H at t. Its conditional density

$$f_0(t \mid Z_i) = \lambda_0(t) e^{\beta_0^T Z_i(t)} \bar{F}_0(t \mid Z_i)$$

is estimated using the series estimator (6.3) of the baseline hazard function λ_0 by

$$\widehat{f}_n(t) = \widehat{\lambda}_n(t) e^{\widehat{\beta}_n^T Z_i(t)} \widehat{\bar{F}}_n(t \mid Z_i).$$

6.2 Nonparametric conditional hazard functions

Cox's model has been extended to a model with a parametric expression of the covariates

$$\lambda_i(t \mid Z_i) = \lambda_0(t) e^{r_\theta(Z_i(t))},\ i = 1, \ldots, n, \tag{6.8}$$

with a function $r_\theta(z)$ of $C^2(\Theta)$ for every z in the range of the covariates. The estimator of the parameter θ maximizes the partial likelihood

$$l_n(\theta) = n^{-1} \sum_{i=1}^n \delta_i \{r_\theta(Z_i(T_i)) - \log S_n(T_i, \theta)\},$$

where $S_n(t,\theta) = \sum_{i=1}^n 1_{\{X_i \ge t\}} e^{r_\theta(Z_i(t))}$, it is $n^{-\frac{1}{2}}$-consistent and asymptotically Gaussian. The cumulative hazard function Λ_0 is still estimated by the process (6.7) and its asymptotic properties are the same as in the proportional hazards model, according to the limits of the process $n^{-1}S_n$ and its first two derivatives. The model (6.8) is generalized to a nonparametric conditional hazard function on a vector of p explanatory variables Z_i with distribution function F_Z on a subset $\mathbb{Z}$ of $\mathbb{R}^d$

$$\lambda_i(t \mid Z_i) = \lambda_0(t) e^{r(Z_i)},\ i = 1, \ldots, n, \tag{6.9}$$

with a nonparametric function r of $C^2(\mathbb{Z})$.

For real valued processes Z_i, the function r is approximated by a function $r_n = \sum_{j=0}^{K_n} b_j \psi_j$, on an orthonormal basis of functions $(\psi_k)_{k\geq 0}$ in $L^2(\mathbb{R}, F_Z)$. The functions $(\psi_k)_{k\geq 0}$ depend on the distribution function F_Z and they have the empirical estimators $\widehat{\psi}_{nk}$ defined by (3.2) which satisfy the convergence of Lemma 3.1. The hazard functions λ_i has an approximation

$$\lambda_{K_n,i}(t \mid Z_i) = \lambda_0(t) e^{B_{K_n}^T \widehat{\Psi}_{nK_n}(Z_i)}, \tag{6.10}$$

where $B_{K_n} = (b_0, \ldots, b_{K_n})^T$ is the vector of the parameters of $\lambda_{K_n,i}$ and $\widehat{\Psi}_{nK_n} = (\psi_0, \widehat{\psi}_{n1}, \ldots, \widehat{\psi}_{nK_n})^T$ is the vector of the $K_n + 1$ first estimated functions of the basis, with $K_n = o(n^{\frac{1}{2}})$ and tending to infinity as n tends to infinity. The equation (6.10) has the same form as the proportional hazards models with covariates $\widehat{\Psi}_{nK_n}(Z_i)$, we can estimate B_n and the nonparametric baseline hazard function λ_0 like in the model (6.6).

The estimators $\widehat{b}_{n,k}$ of the coefficients b_k and $\widehat{\Lambda}_n$ of the cumulative hazard function Λ_0 are $n^{\frac{1}{2}}$-consistent and asymptotically Gaussian, their asymptotic distributions are like in the proportional hazards model. The nonparametric regression function r is approximated by $r_n = \sum_{k=0}^{K_n} b_k \psi_k$ and it is estimated by

$$\widehat{r}_n = \sum_{k=0}^{K_n} \widehat{b}_{n,k} \widehat{\psi}_{nk},$$

its mean integrated squared error is the sum

$$\text{MISE}_n(r) = E\|\widehat{r}_n - r_n\|_2^2 + \|r - r_n\|_2^2$$

of the squared mean integrated estimation error of the function r_n

$$E\|\widehat{r}_n - r_n\|_2^2 = 0(K_n^2 n^{-1}),$$

by Lemma 3.3.

For a regression function r in the space W_{4s}, $s > 1$, the squared error of approximation of r by r_n is $\|m - m_K\|_2^2 \leq CK^{-4s}$ and the squared error of approximation of r by r_n is

$$\|r_n - r\|_2^2 = \sum_{k=K_n+1}^{\infty} b_k^2 = O(K_n^{-4s}),$$

it converges to zero as K_n tends to infinity.

Choosing K_n such that the estimation and approximation errors have the same order and $K_n = o(n^{\frac{1}{2}})$

$$E\|\widehat{r}_n - r_n\|_2 = O(K_n^{-4s}), \quad \|r_n - r\|_2 = O(n^{-\frac{1}{2}} K_n), \tag{6.11}$$

like in (3.7) for a nonparametric regression function, the estimator $\widehat{r}_n$ is L^2-consistent and its convergence rate is $n^{-\frac{s}{2s+1}}$ as $K_n = O(n^{\frac{1}{2(2s+1)}})$. Propositions 3.3 and 3.4 are still true for $\widehat{r}_n$.

Proposition 6.3. *Under the condition (6.11), the process $K_n^{-1} n^{\frac{1}{2}} (\widehat{r}_n - r)$ converges weakly to a Gaussian process with finite expectation and variance functions.*

The estimators extends a nonparametric function r on a metric space $L^2(\mathcal{I}_Z, F_Z)$ on a subset $\mathcal{I}_Z$ of R^d. The observation matrix $\mathbb{Z}_n = (Z_{ij})_{i \le n, j \le d}$ is mapped to the $n \times d$ dimensional matrix $\widetilde{\mathbb{Z}}_n$ of a n independent and identically distributed d dimensional vectors of independent covariates, by (3.17). The regression function $r(Z)$ is mapped from $\mathcal{I}_Z$ onto a subset $\mathcal{I}_{\widetilde{Z}}$ of $\mathbb{R}^d$ by (3.18) as

$$\mu(\widetilde{Z}) = r(Z)$$

with a regression function μ of $L^2(\mathcal{I}_{\widetilde{Z}}, F_{\widetilde{Z}})$. It is expanded on the orthonormal basis defined as a transform of the Laguerre orthonormal basis $\psi_k = L_k \circ H_{\widetilde{Z}}$, by (3.19), on $\mathbb{R}^d$ and the functions ψ_k have the empirical estimators $\widehat{\psi}_{nk}(z) = L_k \circ \widehat{H}_{n(\widetilde{Z}}(z)$. The function μ has an expansion $\mu(\widetilde{z}) = \sum_{k=0}^{\infty} b_k \psi_k(\widetilde{z})$ which is approximated by the sum of the first K_n terms of the series and the conditional hazard function λ_i has again the approximation (6.10). The vector of coefficients $B_{K_n} = (b_k)_{k \le K_n}$ is estimated by maximization of the partial likelihood like in the proportional hazards model and the estimators of $n^{\frac{1}{2}}$ consistent and asymptotically Gaussian. The function μ is estimated by

$$\widehat{\mu}_n = \sum_{k=0}^{K_n} \widehat{b}_{n,k} \widehat{\psi}_{nk},$$

and the mean integrated squared error of $\widehat{\mu}_n$ has the same rate as in the univariate case, Proposition 6.3 applies.

In a proportional hazard model with a nonparametric additive regression function on a subset $\mathcal{I}_Z$ of $\mathbb{R}^d$, the function μ is the sum

$$\mu(\widetilde{Z}) = \sum_{j=1}^{d} \mu_j(\widetilde{Z}_j) \tag{6.12}$$

of regression functions on the independent components of $\widetilde{Z}$. The functions μ_j are estimated by the marginals of the multivariate estimator $\widehat{\mu}_n$ of the function μ like in (3.24). The squared mean integrated estimation error of the additive estimator $\widehat{\mu}_n$ of (6.12) is

$$E\|\widehat{\mu}_n - \mu_n\|_2^2 = E\left\|\sum_{j=1}^{d}(\widehat{\mu}_{jn} - \mu_{jn})\right\|_2^2 = 0\left(n^{-1}\sum_{j=1}^{d} K_{jn}^2\right)$$

and the approximation error of μ is

$$\sum_{j=1}^{d}\int(\mu_j - \mu_{jn})^2\, dF_{\widetilde{Z}_j} = \sum_{j=1}^{d}\sum_{k=K_{jn}+1}^{\infty} b_{jk}^2 = o(1).$$

Let $K_{jn} = o(n)$, for $j = 1, \ldots, d$, satisfy

$$\sum_{j=1}^{d}\|\mu_j - \mu_{jn}\|_2 = O\left(n^{-\frac{1}{2}}\left(\sum_{j=1}^{d} K_{jn}^2\right)^{\frac{1}{2}}\right), \tag{6.13}$$

then the estimator $\widehat{\mu}_n$ is L^2-consistent and satisfies Proposition 3.11.

Proposition 6.4. *Under (6.13), the process $n^{\frac{1}{2}}(\sum_{j=1}^{d} K_{jn}^2)^{-\frac{1}{2}}(\widehat{r}_n - r)$ converges weakly to a Gaussian process.*

The optimal convergence rate of $\widehat{r}_n$ in W_{4s} is again $n^{-\frac{s}{2s+1}}$ as $(\sum_{j=1}^{d} K_{jn}^2)^{-\frac{1}{2}} = O(n^{\frac{1}{2(2s+1)}})$.

A functional ANOVA model for hazard functions extends the model of section 3.8 for nonparametric regressions in the form (6.9)

$$r(Z_i(t)) = \sum_{k=1}^{p} r_k(Z_{ik}(t)) + \sum_{k_1 \neq k_2 = 1}^{p} r_{k_1k_2}(Z_{ik_1}(t), Z_{ik_2}(t)) + \cdots, \tag{6.14}$$

for a p dimensional vector of covariate processes, with an expansion of the function r as the sum of regression functions on each components of the regressor and their interactions. For the identifiability of the functions r_k, $r_{k_1k_2}, \ldots$, the regression function $r(x)$ is reparametrized as a model with the constraints (3.32) and (3.33) for the components of the transformed covariates $\widetilde{Z}_j$, $j = 1, \ldots, d$, and the estimation follows by the same methods as previously.

6.3 Estimation in frailty models

In frailty models with proportional hazards, the conditional hazard function of the time variable T_{ij} for the jth individual of the ith group depends on the vector of observed covariates Z_{ij} of $\mathbb{R}^p$ and on a random variable u_i of group effect, shared by all individuals of the group and depending on covariates omitted in the model. The hazard function of the variable T_{ij} follows a multiplicative model

$$\lambda_{ij}(t,\beta \mid u_i, Z_{ij}) = u_i \lambda_0(t) e^{\beta^T Z_{ij}(t)},\ i = 1,\ldots,I,\ j = 1,\ldots,n_i. \qquad (6.15)$$

with a parameter β in an open subset B of $\mathbb{R}^p$. The components of the covariates Z_i in $\mathbb{R}^p$ are supposed to be linearly independent. The total sample size is $n = \sum_{i=1}^I n_i$ and the ratios $n^{-1}n_i$ converges to strictly positive limits p_i as n tends to infinity, for $i = 1,\ldots,I$. The conditional survival function

$$\bar{F}_{ij}(t,\beta \mid Z_{ij}) = \int \exp\{-\Lambda_{ij}(t,\beta \mid u_i, Z_{ij})\}\, dF(u_i) \qquad (6.16)$$

is the Laplace transform of the frailty variable at $\exp\{\Lambda_{ij}(t,\beta \mid Z_{ij})\}$ where $\Lambda_{ij}(t,\beta \mid Z_{ij}) = \int_0^t e^{\beta^T Z_{ij}(s)} \lambda_0(s)\, ds$ is the conditional cumulative hazard function without frailty.

In a model with an exponential frailty density $f_\theta(u) = \theta e^{-\theta x}$, the survival function of T_{ij} conditionally on Z_{ij} is

$$\bar{F}_{ij}(t,\beta,\theta \mid Z_{ij}) = \frac{\theta}{\theta + \exp\{\Lambda_{ij}(t,\beta \mid Z_{ij})\}}.$$

The observations under independent right censoring are the independent time events $X_{ij} = T_{ij} \wedge C_{ij}$ and the censoring indicators $\delta_{ij} = 1_{\{X_{ij} \le C_{ij}\}}$, where T_{ij} have the conditional distribution functions F_{ij} and C_{ij} are the censoring variables. Let $Y_{ij}(t) = 1_{\{X_{ij} \le t\}}$, let $N_i = \sum_{j=1}^{n_i} \delta_{ij}$ and let

$$\Lambda_{ni,\beta} = \sum_{j=1}^{n_i} \int_0^T Y_{ij} e^{\beta^T Z_{ij}}\, d\Lambda_0 := \int_0^T S^{(0)}_{i,n_i}(\beta)\, d\Lambda_0.$$

The conditional survival functions are

$$\bar{F}_{ij}(t,\beta \mid u_i, Z_{ij}) = \exp\Big\{ - u_i \int_0^t \lambda_0(s) e^{\beta^T Z_{ij}(s)}\, ds \Big\} = \bar{F}^{u_i}(t,\beta \mid Z_{ij}).$$

The same model is used for the dependence of time variables to a latent variable U, let T have a conditional survival function $\bar{F}(t \mid U = u) = \bar{F}^u(t)$, this is the simpler form of the frailty models. In models for two competing risks or correlated time variables, pairs of random variables $T = (T_1, T_2)$

independent conditionally on a latent variable U follow a frailty model with a survival function

$$\bar{F}(t \mid U = u) = \{\bar{F}_{T_1}(t_1)\bar{F}_{T_2}(t_2)\}^u, \ t = (t_1, t_2).$$

Several parametric distributions have been proposed for the variable u_i, the most commonly used is the Gamma-frailty model where the frailty variable u has a Gamma distribution $\Gamma(\alpha, \gamma)$ with density

$$f_{\alpha,\gamma}(u) = \frac{\gamma^\alpha}{\Gamma(\alpha)} u^{\alpha-1} e^{-\gamma u}, \ u > 0,$$

$\alpha > 0$ and $\gamma > 0$, its mean is $\mu = \gamma^{-1}\alpha$ and its variance is $\sigma^2 = \gamma^{-2}\alpha$. The conditional survival function of the variables T_{ij} is

$$\bar{F}_{ij}(t, \beta, \alpha, \gamma \mid Z_{ij}) = \frac{\gamma^\alpha}{\{\gamma + \exp\{\Lambda_{ij}(t, \beta \mid Z_{ij})\}^\alpha}.$$

The properties of the estimators in a Gamma-frailty model without covariates have mainly been studied by Nielsen et al. (1992) and Murphy (1995) using differentiability properties of the likelihood process. Hougaard (1984,1986, 1995) introduced frailty variables with stable, inverse Gaussian and other distributions, Oakes (1989) considered normal and Poisson frailty variables.

In the model with hazard function (6.15) observed on a time interval $[0, \tau]$ and with independent and identically distributed frailty variables u_i, $i = 1, \ldots, i$ following a $\Gamma(\alpha, \gamma)$ distribution, the likelihood of the observations is

$$\begin{aligned}
L_n(\alpha, \beta, \gamma, \Lambda_0) &= \int_0^\infty \prod_{i=1}^I \prod_{j=1}^{n_i} \{u\lambda_{ij}(T_{ij}, \beta \mid Z_{ij})\}^{\delta_{ij}} \bar{F}^u(t, \beta \mid Z_{ij})\, f_{\alpha,\gamma}(u)\, du \\
&= \prod_{i=1}^I \prod_{j=1}^{n_i} \lambda_{ij}^{\delta_{ij}}(T_{ij}, \beta \mid Z_{ij}) \int_0^\infty u^{N_i} e^{-u\Lambda_{ni,\beta}}\, f_{\alpha,\gamma}(u)\, du \\
&= \frac{\gamma^\alpha}{\Gamma(\alpha)} \prod_{i=1}^I \int_0^\infty u^{\alpha+N_i-1} e^{-u(\gamma+\Lambda_{ni,\beta})}\, du \prod_{j=1}^{n_i} \lambda_{ij}^{\delta_{ij}}(T_{ij}, \beta \mid Z_{ij}) \\
&= \frac{\gamma^\alpha}{\Gamma(\alpha)} \prod_{i=1}^I \frac{\Gamma(\alpha + N_i)}{(\gamma + \Lambda_{ni,\beta})^{\alpha+N_i}} \prod_{j=1}^{n_i} \lambda_{ij}^{\delta_{ij}}(T_{ij}, \beta \mid Z_{ij}).
\end{aligned}$$

Most authors consider an EM algorithm for the estimation of the parameters α, γ and β, and eventually the parameter of a parametric baseline hazard function. In the model with the frailty distribution $F_{\alpha,\gamma}$, the maximum likelihood estimators of the parameters are solutions of the score equations

$$\frac{\widehat{\alpha}_n}{\widehat{\gamma}_n} = n^{-1} \sum_{i=1}^{I} \frac{\widehat{\alpha}_n + N_i}{\widehat{\gamma}_n + \Lambda_{ni,\widehat{\beta}_n}}, \tag{6.17}$$

$$\sum_{i=1}^{I} \sum_{j=1}^{n_i} \int_0^{\tau} Z_{ij} \, dN_{ij} = \sum_{i=1}^{I} \frac{\widehat{\alpha}_n + N_i}{\widehat{\gamma}_n + \Lambda_{ni,\widehat{\beta}_n}} \dot{\Lambda}_{ni,\widehat{\beta}_n} \tag{6.18}$$

where $\Lambda_{ni,\beta} = \int_0^{\tau} S_{i,n_i}^{(0)}(s,\beta)\, d\Lambda_0(s)$. Let

$$S_{i,n_i}^{(k)}(t,\beta) = \sum_{j=1}^{n_i} Y_{ij}(t) Z_{ij}^{\otimes k}(t) e^{\beta^T Z_{ij}(t)},$$

for $k = 1, 2$ then (6.18) is equivalent to

$$\sum_{i=1}^{I} \sum_{j=1}^{n_i} \int_0^{\tau} Z_{ij} \, dN_{ij} = \sum_{i=1}^{I} \frac{\widehat{\alpha}_n + N_i}{\widehat{\gamma}_n + \Lambda_{ni,\widehat{\beta}_n}} \int_0^{\tau} S_{i,n_i}^{(1)}(s,\widehat{\beta}_n)\, d\Lambda_0(s).$$

Let $J_n(s)$ be the indicator of $\sum_{l=1}^{I} \frac{\widehat{\gamma}_n + \Lambda_{nl,\widehat{\beta}_n}}{\widehat{\alpha}_n + N_l} S_{l,n_l}^{(1)}(s,\widehat{\beta}_n)$ strictly positive, the estimator of the cumulative baseline hazard function is deduced as

$$\widehat{\Lambda}_n(t,\widehat{\beta}_n) = \int_0^{t} J_n(s) \frac{\sum_{i=1}^{I} \sum_{j=1}^{n_i} Z_{ij} \, dN_{ij}(s)}{\sum_{l=1}^{I} \frac{\widehat{\gamma}_n + \Lambda_{nl,\widehat{\beta}_n}}{\widehat{\alpha}_n + N_l} S_{l,n_l}^{(1)}(s,\widehat{\beta}_n)}.$$

In a frailty model (6.15) with the constraint $\alpha = \gamma$, the expectation of the hazard function conditionally on the observed covariate is the same as in the proportional hazards model with a two-stage sampling and its variance depends on the variance α^{-1} of the frailty variable. It follows that $EN_i = E\Lambda_{ni,\beta_0} = E \int_0^{\tau} S_{i,n_i}^{(0)}(s,\beta_0)\, d\Lambda_0(s)$, where β_0 is the true parameter value. The score equation for the estimation of β_0 becomes

$$\sum_{i=1}^{I} \sum_{j=1}^{n_i} \int_0^{\tau} Z_{ij} \, dN_{ij} = \sum_{i=1}^{I} \frac{\alpha + N_i}{\alpha + \Lambda_{ni,\beta}} \int_0^{\tau} S_{i,n_i}^{(1)}(s,\beta)\, d\Lambda_0(s).$$

Introducing N_i as a consistent estimator of Λ_{ni,β_0} in the score equation at β_0, it reduces to the equation

$$\sum_{i=1}^{I} \sum_{j=1}^{n_i} \int_0^{\tau} Z_{ij} \, dN_{ij} = \Big\{ \sum_{i=1}^{I} \int_0^{\tau} S_{i,n_i}^{(1)}(s,\beta_0)\, d\Lambda_0(s) \Big\} \{1 + o_p(1)\}$$

which provides the estimator

$$\widehat{\Lambda}_n(t, \widehat{\beta}_n) = \int_0^t 1_{\{\sum_{l=1}^I S_{l,n_l}^{(1)}(s,\widehat{\beta}_n)>0\}} \frac{\sum_{i=1}^I \sum_{j=1}^{n_i} Z_{ij}\, dN_{ij}(s)}{\sum_{l=1}^I S_{l,n_l}^{(1)}(s, \widehat{\beta}_n)}$$

for the cumulative baseline hazard function, where $\widehat{\beta}_n$ is an estimator of β_0. This estimator converges uniformly in probability to the function $\Lambda_0(t)$ but it differs from the usual estimator of the cumulative baseline hazard function $\Lambda(t, \beta)$ in the proportional hazards model (6.6) with a two-stage sampling. The process $n^{\frac{1}{2}}\{\widehat{\Lambda}_n(t, \widehat{\beta}_n) - \Lambda_0(t)\}$ converges weakly to a centered Gaussian process with asymptotic variance $v(t) = \sum_{i=1}^I p_i \int_0^t 1_{\{s_i^{(1)}(\beta_0)>0\}} s_i^{(1)-1}(\beta_0)\, d\Lambda_0$.

The logarithm of the likelihood ratio $\log L_n(\alpha, \beta, \lambda_0) - \log L_n(\alpha_0, \beta_0, \lambda_0)$ depends on the parameter β in the form

$$\begin{aligned} l_n(\alpha, \beta) - l_n(\alpha_0, \beta_0) &= \sum_{i,j} \int_0^\tau (\beta - \beta_0)^T Z_{ij}\, dN_{ij} \\ &\quad - \sum_i \{(\alpha + N_i) \log(\alpha + \Lambda_{ni,\beta}) - \log \Gamma(\alpha + N_i)\}. \end{aligned}$$

As the sub-sample sizes n_i tend to infinity, the processes $n_i^{-1} S_{i,n_i}^{(k)}(t, \beta)$ converge uniformly in probability to functions $s_i^{(k)}(t, \beta)$ and $n^{-1} \sum_i S_{i,n_i}^{(k)}(t, \beta)$ is asymptotically equivalent to $n^{-1} \sum_i n_i s_i^{(k)}(t, \beta)$ which converges to a function $s^{(k)}(t, \beta)$.

By Stirling's formula, as n_i tends to infinity

$$\Gamma(\alpha + N_i) \sim \{2\pi(\alpha + N_i)\}^{\frac{1}{2}} \Big\{\frac{(\alpha + N_i)}{e}\Big\}^{\alpha + N_i}$$

and the sum of the last two terms in the expression of $l_n(\alpha, \beta) - l_n(\alpha_0, \beta_0)$ is asymptotically equivalent to

$$\sum_i (\alpha + N_i) \log \frac{\alpha + N_i}{\alpha + \Lambda_{ni,\beta}} - \sum_i (\alpha_0 + N_i) \log \frac{\alpha_0 + N_i}{\alpha_0 + \Lambda_{ni,\beta_0}} = O(n).$$

The variable $n^{-1}\{l_n(\alpha, \beta) - l_n(\alpha_0, \beta_0)\}$ converges in probability to

$$\begin{aligned} l(\beta) - l(\beta_0) &= \int_0^\tau (\beta - \beta_0) s^{(1)}(\beta_0)\, d\Lambda_0 \\ &\quad + \sum_{i=1}^I p_i \int_0^\tau s_i^{(0)}(\beta_0)\, d\Lambda_0 \log \frac{\int_0^\tau s_i^{(0)}(\beta_0)\, d\Lambda_0}{\int_0^\tau s_i^{(0)}(\beta)\, d\Lambda_0}, \end{aligned}$$

its first derivative is

$$l'(\beta) = \int_0^\tau s_i^{(1)}(\beta_0)\,d\Lambda_0 - \sum_{i=1}^I p_i \int_0^\tau s_i^{(0)}(\beta_0)\,d\Lambda_0 \frac{\int_0^\tau s_i^{(1)}(\beta)\,d\Lambda_0}{\int_0^\tau s_i^{(0)}(\beta)\,d\Lambda_0},$$

and it is zero at β_0. The second derivative of the log-likelihood with respect to β is asymptotically equivalent to

$$-\sum_{i=1}^I p_i \frac{\{\int s_{i,\beta}^{(0)}\,d\Lambda_0\}\{\int s_{i,\beta}^{(2)}\,d\Lambda_0\} - \int s_{i,\beta}^{(1)}\,d\Lambda_0\}^2}{\int s_{i,\beta}^{(0)}\,d\Lambda_0},$$

the ratios are positive definite matrices in the parameter set, for regressors Z_i with linearly independent components. The log-likelihood is therefore maximum at β_0, and the maximum likelihood estimator $\widehat{\beta}_n$ of β_0 is consistent. The usual estimator (6.7) of Λ_0 is deduced, depending on $\widehat{\beta}_n$. The function l differs from the limit of the log-partial likelihood ratio in the proportional hazards model however it is simpler to study the properties of the estimator of the parameters α and β through the asymptotic properties of the process l_n.

The process

$$M_{ij} = N_{ij} - \int_0^\cdot Y_{ij}(s)e^{\beta_0^T Z_{ij}(s)}\,d\Lambda_0(s) \tag{6.19}$$

is a local martingale with respect to the filtration $\mathbb{F}$ generated by the processes (N_{ij}, Y_{ij}, Z_{ij}), since $E_{\alpha_0} u_i = 1$. We denote $M_i = \sum_{j=1}^{n_i} M_{ij}$ so $M_i(\tau) = N_i - \Lambda_{ni,\beta_0}$.

The first two derivatives of the log-likelihood l_n with respect to β are the processes $n^{\frac{1}{2}}U_n$ and $nI_n(\alpha,\beta)$ defined by

$$\begin{aligned}U_n(\alpha,\beta) &= n^{-\frac{1}{2}} \sum_{i=1}^I \sum_{j=1}^{n_i} \int_0^\tau Z_{ij}\,dM_{ij} + n^{-\frac{1}{2}} \sum_{i=1}^I \int_0^\tau S_{i,n_i}^{(1)}(\beta_0)\,d\Lambda_0 \\ &\quad -n^{-\frac{1}{2}} \sum_{i=1}^I (\alpha + N_i) \frac{\int_0^\tau S_{i,n_i}^{(1)}(\beta)\,d\Lambda_0}{\alpha + \int_0^\tau S_{i,n_i}^{(0)}(\beta)\,d\Lambda_0}, \\ I_n(\alpha,\beta) &= n^{-1} \sum_{i=1}^I (\alpha + N_i) \frac{\dot\Lambda_{ni,\beta}^2 - (\alpha + \Lambda_{ni,\beta})\ddot\Lambda_{ni,\beta}}{(\alpha + \Lambda_{ni,\beta})^2}.\end{aligned}$$

Let $W_{n,ij}$ be predictable process with respect to the filtration $\mathbb{F}$, then the process $n^{-\frac{1}{2}} \sum_{i=1}^I \sum_{j=1}^{n_i} \int_0^t W_{n,ij}\,dM_{ij}$ is a centered local martingale with respect to $\mathbb{F}$ and its value at τ converges weakly to a centered Gaussian variable with a finite variance, if $n^{-1} \sum_{i=1}^I \sum_{j=1}^{n_i} \int_0^t E\{W_{n,ij}^2 Y_{ij} e^{\beta_0^T Z_{ij}}\}\,d\Lambda_0$

converges to a finite limit (Rebolledo 1980). In particular, the variables $n_i^{-\frac{1}{2}} \sum_{j=1}^{n_i} \int_0^\tau Z_{ij}\, dM_{ij}$ converge weakly to a centered Gaussian variable with variance $\int_0^\tau s_i^{(2)}(\beta_0)\, d\Lambda_0$, for $i = 1, \ldots, I$. The variable $n^{-\frac{1}{2}} \sum_{i=1}^{I} M_i(\tau)$ converges weakly to a centered Gaussian variable with finite variance $v_0 = \sum_{i=1}^{I} p_i v_{0i}$ where $v_{0i} = \int_0^\tau s_i^{(0)}(\beta_0)\, d\Lambda_0$, it follows that the variables

$$U_{2ni} = n^{\frac{1}{2}}\{(\alpha + N_i)(\alpha + \Lambda_{ni,\beta_0})^{-1} - 1\}$$

converge weakly to a centered Gaussian variable with finite variance $(p_i v_{0i})^{-1}$. At β_0 the variable $U_n(\alpha) = U_n(\alpha, \beta_0)$ is written as

$$U_n(\alpha, \beta_0) = n^{-\frac{1}{2}} \sum_{i=1}^{I} \sum_{j=1}^{n_i} \int_0^\tau Z_{ij}\, dM_{ij} - n^{-1} \sum_{i=1}^{I} \dot{\Lambda}_{ni,\beta_0} U_{2ni}$$

and it converges weakly to a centered Gaussian process with a finite variance by Rebolledo's theorem (1980). By the convergence in probability of the variables $n_i^{-1}\Lambda_{i,\beta_0}$ and their derivatives, the process $I_n(\alpha, \beta)$ converges in probability to the matrix

$$I_0(\beta) = \sum_i p_i \frac{(\int_0^\tau s_{i\beta}^{(1)}\, d\Lambda_0)^{\otimes 2} - (\int_0^\tau s_{i\beta}^{(0)}\, d\Lambda_0)(\int_0^\tau s_{i\beta}^{(2)}\, d\Lambda_0)}{(\int_0^\tau s_{i\beta_0}^{(0)}\, d\Lambda_0)^{-1}(\int_0^\tau s_{i\beta}^{(0)}\, d\Lambda_0)^2}.$$

Proposition 6.5. *The variable $n^{\frac{1}{2}}(\widehat{\beta}_n - \beta_0)$ converges weakly to a centered Gaussian variable with a finite variance. The estimator $\widehat{\Lambda}_n$ of the cumulative hazard function is consistent and $n^{\frac{1}{2}}(\widehat{\Lambda}_n - \Lambda_0)$ converges weakly to a centered Gaussian process.*

Proof. The consistency of $\widehat{\beta}_n$ implies that $\widehat{\Lambda}_n$ defined by (6.7) is consistent and it enables an expansion of the variable $U_n(\alpha, \widehat{\beta}_n)$

$$n^{\frac{1}{2}}(\widehat{\beta}_n - \beta_0) = I_n^{-1}(\alpha, \beta_0) U_n(\alpha, \beta_0) + o_p(1)$$

where $I_n(\alpha, \beta_0)$ converges in probability to $I_0(\beta_0)$ and the asymptotic distribution of $U_n(\alpha, \widehat{\beta}_n)$ does not depend on α. □

From the expansion of the variable $U_n(\widehat{\alpha}_n, \widehat{\beta}_n)$, the asymptotic variance of $n^{\frac{1}{2}}(\widehat{\beta}_n - \beta_0)$ is the limit $I_0^{-1}\{\operatorname{Var} U_0\} I_0^{-1}$ of

$$I_n^{-1}(\widehat{\alpha}_n, \widehat{\beta}_n)\{\operatorname{Var} U_n(\widehat{\alpha}_n, \widehat{\beta}_n)\} I_n^{-1}(\widehat{\alpha}_n, \widehat{\beta}_n).$$

The maximum likelihood estimators of the parameters α and γ are solutions of the score equations (6.17) and (6.18).

Proposition 6.6. *The maximum likelihood estimators of the parameters α and γ in bounded sets satisfy $\widehat{\alpha}_n - \widehat{\gamma}_n = o_p(1)$ as n tends to infinity.*

Proof. The variables N_i and $\widehat{\Lambda}_{ni\beta_0}$ are $O(n_i)$ and $n_i^{-1}(N_i - \widehat{\Lambda}_{ni}\beta_0)$ is a $o_p(1)$. Under the condition that the parameters α and γ belong to bounded sets, the score equation (6.18) is asymptotically equivalent to

$$\sum_{i=1}^{I} p_i \int s_{i,\beta_0}^{(1)}\, d\Lambda_0 = \sum_{i=1}^{I} p_i \frac{\int s_{i,\beta_0}^{(0)}\, d\Lambda_0}{\int s_{i,\widehat{\beta}_n}^{(0)}\, d\Lambda_0} \int s_{i,\widehat{\beta}_n}^{(1)}\, d\Lambda_0 + o_p(1)$$

this expression is free of the parameters α and γ and the estimator $\widehat{\beta}_n$ of β_0 is consistent, for all α and γ. Then, for every $i = 1, \ldots, i$, we get

$$n_i^{-1}(N_i - \widehat{\Lambda}_{ni,\widehat{\beta}_n}) = o_p(1)$$

and the right-hand term of equation (6.17) is also a $o_p(1)$. □

Proposition 6.6 justifies the assumption $\gamma = \alpha$ for large sub-sample sizes.

For the estimation of the parameter α, the derivative of the partial log-likelihood at $\widehat{\beta}_n$ reduces to

$$\begin{aligned}\widetilde{l}_n(\alpha) &= \log \alpha + 1 - \frac{\Gamma'(\alpha)}{\Gamma(\alpha)} + \sum_i \frac{\Gamma'(\alpha + N_i)}{\Gamma(\alpha + N_i)} \\ &\quad - \sum_i \Big\{ \frac{\alpha + N_i}{\alpha + \widehat{\Lambda}_{ni}} + \log(\alpha + \widehat{\Lambda}_{ni}) \Big\},\end{aligned}$$

where $\widehat{\Lambda}_{ni} = \Lambda_{ni,\widehat{\beta}_n}$. By Stirling's formula, the derivative of $\log \Gamma(\alpha + N_i)$ with respect to α is approximated by $\log(\alpha + N_i)$ as n tends to infinity, then

$$\begin{aligned}\widetilde{l}_n(\alpha) &= \log \alpha + 1 - \frac{\Gamma'(\alpha)}{\Gamma(\alpha)} + \sum_i \log \frac{\alpha + N_i}{\alpha + \widehat{\Lambda}_{ni}} - \sum_i \frac{\alpha + N_i}{\alpha + \widehat{\Lambda}_{ni}} + o_p(1) \\ &= \log \alpha + 1 - \frac{\Gamma'(\alpha)}{\Gamma(\alpha)} + o_p(1)\end{aligned}$$

and a maximum likelihood estimator of the parameter α is solution of the equation $\widetilde{l}_n(\alpha) = 0$. Assuming that α is unbounded and applying again Stirling's formula to $\log \Gamma(\alpha)$ would provide the approximation

$$\widetilde{l}_n(\alpha) = \log \alpha + 1 - \alpha^{-1} + o_p(1) = 0$$

so the assumption of an infinite parameter α is not relevant.

In frailty models with a nonparametric regression function r of explanatory variables, the conditional hazard function for the jth individual of the ith group has the form

$$\lambda_{ij}(t, \beta \mid u_i, Z_{ij}) = u_i \lambda_0(t) e^{r(Z_{ij})},\ i = 1, \ldots, I,\ j = 1, \ldots, n_i. \tag{6.20}$$

The observations are independent time variables $X_{ij} = T_{ij} \wedge C_{ij}$ conditionally on the variables u_i, under an independent right censoring by the variables C_{ij}. The variables T_{ij} have the conditional survival functions

$$\bar{F}_{ij}(t, r \mid u_i, Z_{ij}) = \exp\{-u_i \int_0^t \lambda_0(s) e^{r(Z_{ij})}\, ds\} = \bar{F}^{u_i}(t, r \mid Z_{ij}).$$

The notations $\delta_{ij} = 1_{\{X_{ij} \leq C_{ij}\}}$, $Y_{ij}(t) = 1_{\{X_{ij} \leq t\}}$, $N_i = \sum_{j=1}^{n_i} \delta_{ij}$ are unchanged and $\Lambda_{ni,r} = \sum_{j=1}^{n_i} \int_0^\tau Y_{ij} e^{r(Z_{ij})} \lambda_0$. The log-likelihood of the observations in a Gamma-frailty model with density f_α is

$$\begin{aligned} L_n(\alpha, r) &= \prod_i \prod_j \lambda_{ij}^{\delta_{ij}}(T_{ij}, r \mid Z_{ij}) \int_0^\infty u^{N_i} e^{-u\Lambda_{ni,r}} f_\alpha(u)\, du \\ &= \frac{\alpha^\alpha}{\Gamma(\alpha)} \prod_i \int_0^\infty u^{\alpha+N_i-1} e^{-u(\alpha+\Lambda_{ni,r})}\, du \prod_j \lambda_{ij}^{\delta_{ij}}(T_{ij}, r \mid Z_{ij}) \\ &= \frac{\alpha^\alpha}{\Gamma(\alpha)} \prod_i \frac{\Gamma(\alpha + N_i)}{(\alpha + \Lambda_{ni,r})^{\alpha+N_i}} \prod_j \lambda_{ij}^{\delta_{ij}}(T_{ij}, r \mid Z_{ij}). \end{aligned}$$

For real valued covariates, the function r is approximated by a function $r_n = \sum_{k=0}^{K_n} b_k \psi_k$ with an orthonormal basis of functions $(\psi_j)_{j\geq 0}$ in $L^2(F_Z)$, which yields approximations of the hazard functions (6.10) by

$$\lambda_{n,ij}(t, B_n \mid u_i, Z_{ij}) = u_i \lambda_0(t) e^{B_n^T \Psi_n(Z_i)},\ i = 1, \ldots, I,\ j = 1, \ldots, n_i.$$

In the expression of the log-likelihood, the sum of the cumulative hazards $\Lambda_{ni,r}$ is replaced by Λ_{ni,B_n} with an approximation of the conditional hazard functions like in (6.10) where the regressors are the estimated functions of the basis $\widehat{\Psi}_n(\widetilde{Z}_i)$, at the transformed covariates values by orthogonalization. The estimation of the function r is performed like with the parametric log-likelihood $L_n(\alpha, B_n, \lambda_0)$, the asymptotic behaviour of the previous estimators is still valid. An estimator of the function r follows and it has the same convergence rate as the estimator of the nonparametric regression function of model (6.9).

6.4 Time-varying coefficients in Cox model

In a proportional hazards model with a time-varying coefficient, we assume that conditionally on a p dimensional covariate process Z, a time variable T^0 has a hazard function

$$\lambda(t \mid Z) = \lambda(t) e^{\beta(t)^T Z(t)}. \tag{6.21}$$

with a time dependent coefficient $\beta(t)$. Under an independent and non informative right censorship, T^0 is censored by a variable C, the observed variables are $T = T^0 \wedge C$, on a finite time interval $[0, \tau]$, and the censoring indicator $\delta = 1_{\{T^0 \leq C\}}$. Under the probability P_0 of a n-sample $(T_i, \delta_i, Z_i)_{1 \leq i \leq n}$, the functions of the model are denoted λ_0 and β_0.

By projection on a vector of p orthonormal functions $(\phi_k)_{k \geq 0}$ with components $(\phi_{jk})_{j=1,\dots,p,k_j \geq 0}$ in $L^2(\mathbb{R}_+)$ with the Lebesgue measure, the components of a vector function β of $L^2(\mathbb{R}_+)$ have expansions

$$\beta_j(t) = \sum_{k=0}^{\infty} b_{jk}\phi_{jk}(t), \; j = 1, \dots, p.$$

Replacing in (6.21) the functions β_j by approximations

$$\beta_{jK_{nj}}(t) = \sum_{k_j=0}^{K_{nj}} b_{jk_j}\varphi_{jk_j}$$

where $K_{nj} = m_{jn}$ tends to infinity as n tends to infinity, and denoting $W_{ij}(t)$ the K_{nj} dimensional vector function with components

$$W_{ijk_j}(t) = \varphi_{jk_j}(t) Z_{ij}(t)$$

for $i = 1, \dots, n$, $j = 1, \dots, p$ and $k_j = 0, \dots, K_{nj}$, the hazard function is approximated by

$$\lambda_m(t \mid Z_i) = \lambda(t) \exp\Big\{\sum_{j=1}^{p} \sum_{k_j=0}^{K_{nj}} b_{jk_j} W_{ijk_j}\Big\}.$$

This expression is similar to (6.6) for the proportional hazards model with a K dimensional parameter $B_K = (b_{jk_j})_{j=1,\dots,p,k_j=0,\dots,K_{nj}}$, where $K = \sum_{j=1}^{p} K_{nj}$. With the same notations and

$$S_{nj}^{(k)}(t, \beta) = \sum_{i=1}^{n} 1_{\{T_i \geq t\}} W_{ij}^{\otimes k}(t) \exp\Big\{\sum_{j=1}^{p} \sum_{k_j=0}^{K_{nj}} b_{jk_j} W_{ijk_j}(t)\Big\}, \tag{6.22}$$

for $k = 0, 1$, the estimators of the cumulative hazard and the parameter β are defined like in Cox's model as

$$\widehat{\Lambda}_n(t) = \int_0^t S_n^{(0)-1}(s, \widehat{\beta}_n) 1_{\{S_n(s, \widehat{\beta}_n) > 0\}} \, d\bar{N}_n(s)$$

and $\widehat{\beta}_n$ maximizes the partial likelihood

$$l_n(\beta) = n^{-1} \sum_{i=1}^{n} \delta_i \Big\{\sum_{j=1}^{p} \sum_{k_j=0}^{K_{nj}} b_{jk_j} W_{ijk_j}(T_i) - \log S_n^{(0)}(T_i, \beta)\Big\}.$$

The asymptotic properties of the processes $\widehat{\Lambda}_n$ and $\widehat{\beta}_{jk_j}$ are the same as in the proportional hazards model (6.6), they are consistent and their convergence rate is $n^{\frac{1}{2}}$. The mean integrated square error of the process $\widehat{\beta}_n$ is the sum

$$\mathrm{MISE}_n(\beta) = E\|\widehat{\beta}_n - \beta_{0K}\|_2^2 + \|\beta_0 - \beta_{0K}\|_2^2$$

of the squared mean integrated estimation error of the vector function β_K,

$$E\|\widehat{\beta}_n - \beta_K\|_2^2 = \sum_{k=1}^{K} \|\operatorname{Var} \widehat{b}_{nk}\| = 0(n^{-1}K)$$

for β in $L^2(\mathbb{R})_+$, and the squared error of approximation of β by β_K

$$\|\beta_{0K} - \beta_0\|_2^2 = \sum_{j=1}^{p} \sum_{k_j=K_{nj}+1}^{\infty} b_{jk_j}^2$$

which converges to zero as K tends to infinity. Choosing the sizes K_{nj} such that

$$K_{nj} = o(n), \quad \|\beta_{0jK_{nj}} - \beta_{0j}\|_2 = O(n^{-\frac{1}{2}} K_{nj}^{\frac{1}{2}}), \tag{6.23}$$

the estimator $\widehat{\beta}_n$ is L^2-consistent. Propositions 2.2 and 2.3 are modified according to the dimension of the covariate.

Proposition 6.7. *Under the conditions (6.23) and with a uniformly bounded basis of functions $(\phi_k)_k$, the processes $\widehat{\beta}_{nj}$ are such that $(K_{nj}^{-1}n)^{\frac{1}{2}}(\widehat{\beta}_{nj} - \beta_{0j})$ converge weakly to Gaussian processes with finite means and variances.*

The asymptotic behavior of the process $\widehat{\Lambda}_n - \Lambda_0$ relies on an expansion of $S_n^{(0)}(\widehat{\beta}_n)$ for $\widehat{\beta}_n$ close to β_0. Let $\alpha_j = \lim_{n\to\infty} K_{nj}K^{-1} > 0$. As the variable $\|n^{-1}S_n^{(0)} - s^{(0)}\|_{[0,\tau]\times B\times \mathcal{I}_X}$ converges in probability to zero, the process

$$\begin{aligned} G_n &= (nK)^{-\frac{1}{2}}\{S_n^{(0)}(\widehat{\beta}_n) - S_n^{(0)}(\beta_0)\} \\ &= \sum_{j=1}^{p} \alpha_j^{\frac{1}{2}} (nK_{nj}^{-1})^{\frac{1}{2}} (\widehat{\beta}_{nj} - \beta_{0j}) n^{-1} S_{nj}^{(1)}(\beta_0)\} + o_p(1) \end{aligned}$$

converges weakly to a Gaussian process G, by Proposition 6.7.

Proposition 6.8. *On every finite interval $[0,\tau]$ where $P_0(T > \tau)$ is strictly positive, the process $(nK^{-1})^{\frac{1}{2}}(\widehat{\Lambda}_n - \Lambda_0)$ converges weakly to the centered Gaussian process*

$$-\int_0^t \frac{G(s)}{s^{(0)2}(s,\beta_0)}\, d\Lambda_0(s).$$

Proof. On $[0, \tau]$, the martingale $\bar{M}_n(t) = \bar{N}_n(t) - \int_0^t S_n^{(0)}(s, \beta_0)\, d\Lambda_0(s)$ is such that $n^{-\frac{1}{2}}\bar{M}_n$ converges weakly to a centered Gaussian process with variance $\int_0^t s^{(0)}(s, \beta_0)\, d\Lambda_0(s)$. The asymptotic distribution of $\widehat{\Lambda}_n$ is deduced from the first order expansion

$$(nK^{-1})^{\frac{1}{2}}(\widehat{\Lambda}_n - \Lambda_0)(t) = (nK)^{-\frac{1}{2}} \int_0^t nS_n^{(0)-1}(\widehat{\beta}_n) 1_{\{S_n(\widehat{\beta}_n)>0\}}\, d\bar{M}_n - \int_0^t \frac{G_n}{n^{-1}S_n^{(0)}(\widehat{\beta}_n)}\, d\Lambda_0 + o_p(1),$$

and the first term of the right member is a $o_p(1)$ by the weak convergence of $\bar{M}_n$. □

6.5 Stochastically varying coefficients in Cox model

The model (6.21) extends to a function β depending on a random variable X with distribution function F_X on a subset $\mathcal{I}_X$ of $\mathbb{R}^p$. On a probability space $(\Omega, \mathcal{F}, P_0)$, we consider the right-censored time variables $T = T^0 \wedge C$, the censoring indicator $\delta = 1_{\{T^0 \leq C\}}$, X is a real random variable and Z is a p dimensional vector of left-continuous processes. Let $(T_i, \delta_i, X_i, Z_i)_{1 \leq i \leq n}$ be a sample of n independent and identically distributed observations of (T, δ, X, Z), we assume that conditionally on (X_i, Z_i), T_i^0 has a hazard function defined by

$$\lambda_i(t \mid X_i, Z_i) = \lambda(t) e^{\beta(X_i)^T Z_i(t)}. \tag{6.24}$$

with a function β of $L^2(F_X)$. Under P_0, the parameter values are (β_0, λ_0). The censoring variable and C is independent of T^0 conditionally on (X, Z). We suppose that the time observations T_i belong to a finite time interval $[0, \tau]$ where $P_0(T > \tau)$ is strictly positive and that the X_i's have a density f_X. The problem is to estimate the function β_0 on a compact subset J_X of the support of the distribution function F_X of X, so that $\beta_0(x)$ belongs to a compact set B for every x in J_X. A kernel estimator $\widehat{\beta}_n(t)$ of $\beta_0(t)$ has been defined by maximization of a local log-likelihood (Pons 2000).

By projection on a vector of p orthonormal functions $(\phi_k)_{k \geq 0}$ in $L^2(F_X)$, the components β_j of a vector function β of $L^2(F_X)$ have expansions $\beta_j(x) = \sum_{k_j=0}^{\infty} b_{jk_j}\varphi_{jk}$, then the functions β_j are approximated by

$$\beta_{jK_{nj}}(x) = \sum_{k_j=0}^{K_{nj}} b_{jk_j}\varphi_{jk_j}(x)$$

for $j = 1, \ldots, p$, where K_{nj} tends to infinity as n tends to infinity. The unknown functions φ_{jk_j} are estimated by $\widehat{\varphi}_{njk_j}$ defined from (3.2) with the empirical cumulative hazard function of the n-sample $(X_i)_{i=1,\ldots,n}$, they satisfy Lemma 3.1. Denoting $B_K = (b_{jk_j})_{k_j=0,\ldots,K_{nj},j=1,\ldots,p}$ the parameter set, for $K = \sum_{j=1}^p K_{nj}$

$$W_{ijk_j}(t) = \widehat{\varphi}_{njk_j}(X_i) Z_{ij}(t), \tag{6.25}$$

$$\bar{S}_n^{(k)}(t, B_K) = \sum_{i=1}^n 1_{\{T_i \geq t\}} W_{ijk_j}^{\otimes k}(t) \exp\Big\{\sum_{j=1}^p \sum_{k_j=0}^{K_{nj}} b_{jk_j} W_{ijk_j}(t)\Big\}, \tag{6.26}$$

for $k = 0, 1, 2$, $i = 1, \ldots, n$, $j = 1, \ldots, p$ and $k_j = 0, \ldots, K_{nj}$, the coefficient b_{jk_j} are estimated by $\widehat{b}_{njk_j}$ which maximizes the estimated partial likelihood

$$l_n(B_K) = \sum_{i=1}^n \delta_i \Big\{\sum_{j=1}^p \sum_{k_j=0}^{K_{nj}} b_{jk_j} W_{ijk_j}(T_i) - \log \bar{S}_n^{(0)}(T_i, B_K)\Big\}$$

and the function β is estimated by $\widehat{\beta}_n$ with components

$$\widehat{\beta}_{nj} = \sum_{k_j=0}^{K_{nj}} \widehat{b}_{njk_j} \widehat{\varphi}_{njk_j}(x)$$

for $j = 1, \ldots, p$. An estimator of the cumulative baseline hazard function follows as

$$\widehat{\Lambda}_n(t) = \int_0^\tau 1_{\{\bar{S}_{nK}^{(k)}(s, \widehat{B}_{nK}) > 0\}} \frac{d\bar{N}_n(s)}{\bar{S}_n^{(0)}(s, \widehat{B}_{nK})}$$

with the notation $\bar{N}_n(t) = \sum_{i=1}^n 1_{\{T_i \leq t\}}$. The asymptotic properties of the process $\widehat{\Lambda}_n$ and of the components of $\widehat{B}_{nK}$ are the same as in the proportional hazards model (6.6). As n tends to infinity, $n^{-1} l_n$ is asymptotically equivalent to

$$l(B_K) = E \int_0^\tau \Big\{\sum_{j=1}^p \sum_{k_j=0}^{K_{nj}} b_{jk_j} W_{ijk_j}(t) - \log \bar{S}_n^{(0)}(t, B_K)\Big\}$$
$$\times e^{\sum_{j=1}^p \sum_{k_j=0}^{K_{nj}} b_{0jk_j} W_{ijk_j}(t)} \, d\Lambda_0(t)$$

such that at the parameter value B_{0K} under P_0

$$l(B_{0K}) = \int_0^\tau \{\bar{s}^{(1)}(s, B_{0K}) - \bar{s}^{(0)}(t, B_{0K}) \log \bar{s}^{(0)}(t, B_{0K})\} \, d\Lambda_0(t),$$

where the functions $\bar{s}^{(k)}(t, B_{0K}) = E\bar{S}_n^{(k)}(t, B_K)$ converge to the functions $\bar{s}^{(k)}(t, \beta_0)$, for β_0 in $L^2(F_X)$. Let $U_{njk_j}(B_K)$ and $-I_{njk_j}(B_K)$ be normalized first two derivatives of l_n with respect to the components of B_m

$$U_{njk_j}(B_K) = \sum_{i=1}^n \int_0^\tau \Big\{W_{ijk_j}(t) - \frac{\bar{S}_{njk_j}^{(1)}(t, B_K)}{\bar{S}_n^{(0)}(t, B_K)}\Big\}\, dN_i(t)$$

$$U_{njk_j}(B_{0K}) = \sum_{i=1}^n \int_0^\tau \Big\{W_{ijk_j}(t) - \frac{\bar{S}_{njk_j}^{(1)}(t, B_{0K})}{\bar{S}_n^{(0)}(t, B_{0K})}\Big\}\, dM_i(t)$$

where $M_i(t) = N_i(t) - \int_0^\tau \exp\{\sum_{j=1}^p \sum_{k_j=0}^{K_{nj}} b_{0jk_j} W_{ijk_j}(t)\}\, \Lambda_0(t)$ is defined by (6.25) under P_0, and

$$I_{njk_jlk_l}(B_K) = n^{-1} \sum_{i=1}^n \Big\{\frac{\bar{S}_{njk_j}^{(1)}(B_K)\bar{S}_{nlk_l}^{(1)}(B_K)}{\bar{S}_n^{(0)2}(B_K)} - \frac{\bar{S}_{njk_jlk_l}^{(2)}(B_K)}{\bar{S}_n^{(0)}(B_K)}\Big\}\, dN_i,$$

$$I_{njk_jlk_l}(B_{0K}) = n^{-1} \sum_{i=1}^n \Big\{\frac{\bar{S}_{njk_j}^{(1)}(B_{0K})\bar{S}_{nlk_l}^{(1)}(B_K)}{\bar{S}_n^{(0)}(B_{0K})} - \bar{S}_{njk_jlk_l}^{(2)}(B_{0K})\Big\}\, d\Lambda_0 + o_p(1).$$

The process U_n is a m dimensional vector and $I_n(B_{0K})$ is is a strictly positive definite $m \times m$ dimensional matrix under P_0. As n tends to infinity, the components of $n^{-1}U_n(B_{0K})$ converges in probability to zero, it follows that the function l is maximum at B_{0K}. The inequality

$$l(B_{0K}) \le \sup_{B_m} |l_n(B_K) - l(B_K)| + l_n(B_{0K}) = l_n(B_{0K}) + o_p(1),$$

the uniform convergence of l_n to l and Lemma 3.1 imply that $\widehat{\beta}_n - \beta_{0K}$ converges to zero in probability as n tends to infinity, where the function β_{0K} has the components $\beta_{0jK_{nj}}(x) = \sum_{k_j=0}^{K_{nj}} b_{jk_j}\varphi_{jk_j}(x)$, for $j = 1, \dots, p$. The MISE of the estimator $\widehat{\beta}_n$ is

$$\text{MISE}_n(\beta) = \sum_{j=1}^p \{E\|\widehat{\beta}_{njK_{nj}} - \beta_{jK_{nj}}\|_2^2 + \|\beta_j - \beta_{jK_{nj}}\|_2^2\},$$

where $\sum_{j=1}^p \|\beta_j - \beta_{jK_{nj}}\|_2^2 = o(1)$ as the sizes K_{nj} tend to infinity.

Lemma 6.1. *Let K_{nj} tend to infinity as n tends to infinity, then*

$$\sum_{j=1}^p E\|\widehat{\beta}_{nj} - \beta_{0jK_{nj}}\|_2^2 = O\Big(n^{-1}\sum_{j=1}^p K_{nj}^2\Big),$$

Proof. The variable $n^{-\frac{1}{2}}U_n(B_{0K})$ is the value at τ of a sum of martingales integrals of predictable processes with respect to the local martingales M_i. For β_0 in $L^2(F_X)$, $\sum_{j=1}^p \sum_{k_j=0}^{K_{nj}} b_{0jk_j} W_{ijk_j}(t)$ converges to $\beta_0^T(X_i)Z_{ij}(t)$ and the variance of the components of $n^{-\frac{1}{2}}U_n(B_{0K})$ converge to finite limits

$$I_{0jk_jlk_l} = \int_0^\tau \Big\{ \bar{s}^{(2)}_{jk_jlk_l}(\beta_0) - \frac{\bar{s}^{(1)}_{0jk_j}(\beta_0)\bar{s}^{(1)}_{0lk_l}(\beta_0)}{\bar{s}^{(0)}(\beta_0)} \Big\} \, d\Lambda_0$$

which is the limit of $I_{njk_jlk_l}(B_{0K})$ as n tends to infinity. Then the weak convergence of the components of $n^{-\frac{1}{2}}U_n(B_{0K})$ to centered Gaussian variables with variances the diagonal components of I_0 is a consequence of Rebolledo's theorem (1980).

By a first order asymptotic expansion of $U_n(\widehat{B}_{nK})$, there exists B_{nK} between $\widehat{B}_{nK}$ and B_{0K} such that

$$n^{-\frac{1}{2}}U_n(B_{0K}) = n^{\frac{1}{2}}(\widehat{B}_{nK} - B_{0K})^T I_n(B_{nK}),$$

and the weak convergence of the components of the variable $U_n(B_{0K})$ implies that the components of the vector $n^{\frac{1}{2}}(\widehat{B}_{nK} - B_{0K})$ converge weakly to centered Gaussian variables with a finite variances.

Like in the proof of Lemma 3.3, the square estimation error of the function β_{0K_n} develops as the sum of the p errors

$$\begin{aligned}
E\|\widehat{\beta}_{nj} - \beta_{0jK_{nj}}\|_2^2 &= E\int_{J_X} \Big\{ \sum_{k_j=0}^{K_{nj}} (\widehat{b}_{njk_j}\widehat{\varphi}_{njk_j} - b_{jk_j}\varphi_{jk_j}) \Big\}^2 dF_X \\
&= E\sum_{k_j=0}^{K_{nj}} \Big[(\widehat{b}_{njk_j} - b_{jk_j})^2 \\
&\quad + \int_{J_X} \widehat{b}^2_{njk_j}(\widehat{\varphi}_{njk_j} - \varphi_{jk_j})^2 \, dF_X \\
&\quad + \sum_{k_l=0}^{K_l} \Big\{ 2(\widehat{b}_{njk_j} - b_{jk_j})(\widehat{b}_{nlk_l} - b_{lk_l}) \int_{J_X} \widehat{\varphi}_{nlk_l}\varphi_{jk_j} \, dF_X \\
&\quad + \widehat{b}_{njk_j}\widehat{b}_{nlk_l} \int_{J_X} (\widehat{\varphi}_{njk_j} - \varphi_{jk_j})(\widehat{\varphi}_{nlk_l} - \varphi_{lk_l}) \, dF_X. \Big\} \Big]
\end{aligned}$$

For every $k_j \leq K_{nj}$, $E(\widehat{b}_{njk_j} - b_{jk_j})^2$ is a $O(n^{-1})$ and

$$E\int_{J_X} (\widehat{\varphi}_{njk_j} - \varphi_{jk_j})(\widehat{\varphi}_{nlk_l} - \varphi_{lk_l}) \, dF_X = O(n^{-1}).$$

Then $\int_{J_X} (\widehat{\varphi}_{njk_j} - \varphi_{jk_j})^2 \, dF_X$ is a $O_p(n^{-1})$ and $\int_{J_X} \widehat{\varphi}_{nlk_l}\varphi_{jk_j} \, dF_X$ is a $O_p(n^{-\frac{1}{2}})$ by the orthogonality of the functions φ_{jk_j}. It follows that $E\|\widehat{\beta}_{nj} - \beta_{0jK_{nj}}\|_2^2 = O(K_{nj}^2 n^{-1})$. □

As n tends to infinity, the number K_{nj} of functions of the basis of the approximating functions is optimal as it satisfies the conditions

$$K_{nj} = o(n^{\frac{1}{2}}), \quad \|\beta_0 - \beta_{0K}\|_2 = O\Big(n^{-\frac{1}{2}}\Big(\sum_{j=1}^{p} K_{nj}^2\Big)^{\frac{1}{2}}\Big), \tag{6.27}$$

then the MISE is a $O(n^{-1}\sum_{j=1}^{p} K_{nj}^2)$ as n tends to infinity. The optimal convergence rate of the process $\widehat{\beta}_n$ for a time dependent function of β in W_{4s}, defined by (3.8), is $n^{-\frac{s}{2s+1}}$ as the sizes K_{jn} are $O(n^{\frac{1}{2(2s+1)}})$.

Proposition 6.9. *Under P_0 and the condition (6.27), the process $K^{-1}n^{\frac{1}{2}}(\widehat{\beta}_n - \beta_0)$ converges weakly to a Gaussian process with a finite mean and variance functions on J_X.*

The proof is similar to the proof of Proposition 3.4 and the convergence rate of $\widehat{\beta}_n$ is compared to the optimal convergence rate of a kernel estimator for β_0 under the conditions of Proposition 3.3.

The asymptotic behavior of the process $\widehat{\Lambda}_n - \Lambda_0$ is studied according to the convergence rate of the estimator $\widehat{\beta}_n$. Let $\alpha_j = \lim_{n\to\infty} K_{nj}K_n^{-1} > 0$, the variables $\|n^{-1}S_n^{(k)} - s^{(k)}\|_{[0,\tau]\times B\times \mathcal{I}_X}$ converge in probability to zero, for $k = 0, 1$, and the process

$$\begin{aligned} G_n &= n^{-\frac{1}{2}}K_n^{-1}\{S_n^{(0)}(\widehat{\beta}_n) - S_n^{(0)}(\beta_0)\} \\ &= \sum_{j=1}^{p} \alpha_j n^{\frac{1}{2}} K_{nj}^{-1}(\widehat{\beta}_{nj} - \beta_{0j})n^{-1}S_{nj}^{(1)}(\beta_0)\} + o_p(1) \end{aligned}$$

converges weakly to a Gaussian process G, by Proposition 6.9 and a first order asymptotic expansion of $S_n^{(0)}(\widehat{\beta}_n)$.

Proposition 6.10. *On every finite interval $[0,\tau]$ where $P_0(T > \tau)$ is strictly positive, the process $n^{\frac{1}{2}}K^{-1}(\widehat{\Lambda}_n - \Lambda_0)$ converges weakly to the Gaussian process $-\int_0^t \frac{G(s)}{s^{(0)2}(s,\beta_0)}\, d\Lambda_0(s)$.*

The proof is the same as for Proposition 6.8 in models with time-varying coefficients.

6.6 Bivariate hazard functions

Let (S,T) be a pair of failure times with a distribution function F and a density f. The partial derivatives of the function $\log \bar{F}$ determine the partial hazard functions

$$\lambda_{S|T\geq y}(x) = \frac{\partial}{\partial x}\log \bar{F}(x,y)$$
$$= \lim_{h\to 0} h^{-1}P(x \leq S < x+h \mid S \geq x, T \geq y),$$
$$\lambda_{T|S\geq x}(y) = \frac{\partial}{\partial y}\log \bar{F}(x,y)$$
$$= \lim_{h\to 0} h^{-1}P(y \leq T < y+h \mid S \geq x, T \geq y),$$
$$\lambda_{S,T}(x,y) = \frac{\partial^2}{\partial x \partial y}\log \bar{F}(x,y) = \frac{f(x,y)}{\bar{F}(x,y)}.$$

When S and T are independent, the functions $\lambda_{S|T\geq y}(x)$ and respectively $\lambda_{T|S\geq x}(y)$ reduce to the first and second marginal hazard functions $\lambda_S(x)$ and respectively $\lambda_T(y)$, and the function $\lambda_{S,T}(x,y)$ is their product. The marginal survival functions are $\bar{F}(x,0) = \bar{F}_X(x)$ and $\bar{F}(0,y) = \bar{F}_Y(y)$, the conditional hazard function $\lambda_{S|T\geq 0}(x)$ is the marginal hazard function $\lambda_S(x)$ for S, in the same way the conditional hazard function $\lambda_{T|S\geq 0}(y)$ is the marginal hazard function $\lambda_T(x)$ for T. A bivariate survival function is the unique solution of the equation

$$\bar{F}_{S,T}(s,t) = \int_0^s \int_0^s \bar{F}_{S,T}(x,y)\Lambda_{S,T}(dx,dy) + \bar{F}_S(s) + \bar{F}_T(t) - 1$$

where $\Lambda_{S,T}(s,t) = \int_0^s \int_0^s \lambda_{S,T}(x,y)\,dx\,dy$ is the joint cumulative hazard function.

Let $(X_i,Y_i)_{i=1,\dots,n}$ be a sample of n independent and identically distributed censored variables with a survival function $\bar{F}_{X,Y}\bar{G}$ such that $X_i = S_i \wedge C_{1i}$ and $Y_i = T_i \wedge C_{2i}$ with a bivariate censoring variable $C = (C_1, C_2)$, and let $\delta_{1i} = 1_{\{S_i \leq C_{1i}\}}$ and $\delta_{2i} = 1_{\{T_i \leq C_{2i}\}}$ be the indicators of censorship. The two dimensional observations generate the processes

$$\bar{N}_{1n}(x,y) = \sum_{i=1}^n \delta_{1i}1_{\{X_i\leq x, Y_i\geq y\}},$$
$$\bar{N}_{2n}(x,y) = \sum_{i=1}^n \delta_{2i}1_{\{X_i\geq x, Y_i\leq y\}},$$
$$\bar{N}_{n}(x,y) = \sum_{i=1}^n \delta_{1i}\delta_{2i}1_{\{X_i\leq x, Y_i\leq y\}},$$
$$Y_n(x,y) = \sum_{i=1}^n 1_{\{X_i\geq x, Y_i\geq y\}}.$$

The conditional cumulative hazard functions

$$\Lambda_{S|T\geq y}(x) = \int_0^x \lambda_{S|T\geq y}(s)\,ds, \quad \Lambda_{T|S\geq x}(y) = \int_0^y \lambda_{T|S\geq x}(t)\,dt$$

and the bivariate hazard function $\Lambda_{S,T}$ are estimated from the censored observations as

$$\begin{aligned}
\widehat{\Lambda}_{n,S|T\geq y}(x) &= \int_0^x Y_n^{-1}(x,y)1_{\{Y_n(x,y)>0\}}\,\bar{N}_{1n}(dx,y), \\
\widehat{\Lambda}_{n,T|S\geq x}(y) &= \int_0^y Y_n^{-1}(x,y)1_{\{Y_n(x,y)>0\}}\,\bar{N}_{2n}(x,dy), \\
\widehat{\Lambda}_{n,ST}(x,y) &= \int_0^x \int_0^y Y_n^{-1}(s,t)1_{\{Y_n(s,t)>0\}}\,\bar{N}_n(ds,dt).
\end{aligned}$$

Series estimators of their derivatives $\lambda_{S|T\geq y}(x)$ and respectively $\lambda_{T|S\geq x}(y)$ are deduced by projections of $\widehat{\Lambda}_{n,S|T\geq y}(x)$ and respectively $\widehat{\Lambda}_{n,T|S\geq x}(y)$ on an orthonormal basis of functions on $\mathbb{R}_+$, the estimators are asymptotically normal with the same convergence rate as $\widehat{\lambda}_n$ defined by (6.3). An estimator of $\lambda_{S,T}(x,y)$ is obtained by projections of $\widehat{\Lambda}_{n,ST}(x,y)$ on the product orthonormal basis on $\mathbb{R}_+^2$. Let

$$\lambda_{S,T}(x,y) = \sum_{j,k=0}^{\infty} c_{jk} L_k(x) L_j(y)$$

with $c_{jk} = \int_0^\infty \int_0^\infty L_k(x)L_j(y)\,\Lambda_{S,T}(dx,dy)$, the function λ is estimated by

$$\widehat{\lambda}_{n,ST}(x,y) = \sum_{j=0}^{J_n}\sum_{k=0}^{K_n} \widehat{c}_{n,jk} L_k(x) L_j(y), \tag{6.28}$$

where J_n and $K_n = o(n)$ tend to the infinity as n tends to the infinity, with the estimators of the coefficients

$$\widehat{c}_{n,jk} = \int_{\mathbb{R}_+^2} L_k(x)L_j(y)\,\widehat{\Lambda}_{n,ST}(dx,dy). \tag{6.29}$$

The estimators $\widehat{c}_{n,jk}$ are asymptotically unbiased and consistent and the variables $n^{\frac{1}{2}}(\widehat{c}_{n,jk} - c_{jk})$ converge weakly to centered Gaussian process with variances $\int_{\mathbb{R}_+^2} L_k^2(x)L_j^2(y)\bar{F}(x,y)^{-1}1_{\{\bar{F}(x,y)>0\}}\,\Lambda_{S,T}(dx,dy)$.

The mean squared error $\text{MISE}_n(\lambda_{S,T})$ of the estimator (6.28) is the sum of the approximation error $\|\lambda_{S,T} - \lambda_{n,ST}\|_2^2 = o(1)$ of the function $\lambda_{S,T}$ by

$$\lambda_{n,ST}(x,y) = \sum_{j=0}^{J_n}\sum_{k=0}^{K_n} c_{jk} L_k(x) L_j(y)$$

which converges to λ, and the estimation error

$$E\|\widehat{\lambda}_{n,ST} - \lambda_{n,ST}\|_2^2 = E\int_{\mathbb{R}_+^2} \{\widehat{\lambda}_{n,ST}(x,y) - \lambda_{n,ST}(x,y)\}^2\,dx\,dy$$
$$= 0(J_n K_n n^{-1}).$$

Proposition 6.11. *The process $(J_n^{-1}K_n^{-1}n)^{\frac{1}{2}}(\widehat{\lambda}_n - \lambda)$ converges weakly to a centered Gaussian process with finite variance and covariance functions.*

By the same arguments as in Proposition 2.2, the convergence rate of $\widehat{\lambda}_n$ is the optimal rate $n^{-\frac{s}{2s+1}}$ for a hazard function in W_{2s}, defined by (2.20), as the sizes J_n and K_n are such that $J_n K_n = O(n^{\frac{1}{2s+1}})$.

A non-stationary hazard function has been defined for two consecutive time variables S and T^0, conditionally on a p dimensional vector Z of left-continuous process of covariates with right-hand limits. We assume that the time variable S is uncensored and T^0 may be right-censored at a random time C independent of (S, T^0) conditionally on Z and non informative for the distribution of T_0. The observed variables are S, $T = T^0 \wedge C$ belonging to a finite time interval $[0,\tau]$, and $\delta = 1_{\{T^0 \leq C\}}$. The variable (S, T^0) belongs to the triangle $I_\tau = \{(s,x) \in [0,\tau] \times [0,\tau]; s + x \leq \tau\}$ and we denote $X = T - S > 0$.

The conditional hazard function of $X^0 = T^0 - S$ given (S, Z) has the form

$$\lambda_{X|S,Z}(x \mid S, Z) = \lambda_{X|S}(x; S)\, e^{\beta^T Z(S+x)}, \tag{6.30}$$

with a parameter β in a subset B of $\mathbb{R}^p$ which contains the true parameter value β_0 under P_0 and an unknown baseline hazard function $\lambda_{X|S}$ with true value $\lambda_{0,X,S}$. The observations are a n-sample $(S_i, T_i, \delta_i, Z_i)_{1 \leq i \leq n}$. Up to additive terms constant in $(\beta, \lambda_{X|S})$, the log-likelihood of $(S_i, T_i, \delta_i, Z_i)$ under (6.30) is

$$l_{(i)} = \delta_i\{\log \lambda_{X|S}(X_i; S_i) + \beta^T Z_i(T_i)\}$$
$$- \int_0^\tau Y_i(x) e^{\beta^T Z_i(S_i+x)} \lambda_{X|S}(x; S_i)\,dx$$

where the process $Y_i(x) = 1_{\{X_i \geq x\}}$ counts the unobserved events. Denoting

$$N_i(s,x) = \delta_i 1_{\{S_i \leq s, X_i \leq x\}},$$
$$N_i^{(1)}(s) = 1_{\{S_i \geq s\}},$$

the log-likelihood of the sample is expressed as

$$l_n(\beta, \lambda_{X|S}) = \sum_{i=1}^{n} \int_0^T \{\log \lambda_{X|S}(x; s) + \beta^T Z_i(s+x)\} N_i(ds, dx) \\ - \sum_{i=1}^{n} \int_0^T Y_i(x) e^{\beta^T Z_i(s+x)} \lambda_{X|S}(x; s)\, dx\, dN_i^{(1)}(s).$$

Let β_0 and $\lambda_{0,X|S}$ be the parameter values under the probability measure P_0 of the observations. As n tends to infinity, the process $n^{-1} l_n$ converges a.s. to

$$l(\beta, \lambda_{X|S}) = \int_0^T E\Big[Y(x)\Big\{\log \lambda_{X|S}(x; s) + \beta^T Z(s+x)\}e^{\beta_0^T Z(s+x)} \\ -e^{\beta^T Z(s+x)} \frac{\lambda_{X|S}(x; s)}{\lambda_{0,X|S}(x; s)} \Big| S = s\Big\}\Big] \lambda_{0,X|S}(x; s) f_S(s)\, dx\, ds.$$

Let $\lambda_{\theta,X|S}(x; s) = \lambda_{0,X|S}(x; s) + h(x; s)\theta$, with a real θ, the first derivatives of the function l with respect to θ and β are zero at zero and β and l is maximum at β_0 and $\lambda_{0,X|S}$. The second derivative $I(\beta, \lambda_{\theta,X|S})$ of the function $-l(\beta, \lambda_{\theta,X|S})$ with respect to θ and β is strictly positive definite at β_0 and $\lambda_{0,X|S}$, this entails the concavity of the function l at the true parameter values. A kernel estimator of the baseline hazard function $\lambda_{0,X|S}$ was defined in Pons and Visser (2000), here we define a series estimator.

The conditional hazard function $\lambda_{X|S}$ is expanded on the product $(\varphi_{jk})_{jk} = (\varphi_j \varphi_k)_{jk}$ of the functions of an orthonormal basis $(\varphi_j)_j$, such as Laguerre's polynomials for the hazard function of Weibull distributions, $\lambda_{X|S} = \sum_{j,k=0}^{\infty} c_{jk} \varphi_{jk}$, it is approximated by the sum λ_{J_n,K_n} of its projections on the functions $(\varphi_{jk})_{j \le J_n, k \le K_n}$ such that J_n and K_n tend to infinity as n tends to infinity, J_n and K_n are $o(n)$. Let A_n be the matrix of the coefficients of this approximation and let ϕ_{J_n}, and respectively ϕ_{K_n}, be the vector of the functions $(\varphi_j)_{j \le J_n}$, and respectively $(\varphi_k)_{k \le K_n}$. The log-likelihood of the sample is approximated by

$$l_n(\beta, A_n) = \sum_{i=1}^{n} \int_0^T \log\{\phi_{J_n}^T(x) A_n \phi_{K_n}(s) + \beta^T Z_i(s+x) N_i(ds, dx) \\ - \sum_{i=1}^{n} \int_0^T Y_i(x) e^{\beta^T Z_i(s+x)} \phi_{J_n}^T(x) A_n \phi_{K_n}(s)\, dN_i^{(1)}(s)\, dx$$

and $n^{-1} l_n(\beta, A_n)$ converges a.s. to the function $l(\lambda_{X|S}, \beta)$. The estimators $\widehat{\beta}_n$ and $\widehat{a}_{n,jk}$ of the parameters maximize the process l_n they are a.s.

consistent. It follows that the function $\lambda_{X|S}$ is a.s. consistently estimated by

$$\widehat{\lambda}_{n,X|S}(x;s) = \sum_{j=0}^{J_n}\sum_{k=0}^{K_n} \widehat{a}_{n,jk}\varphi_{jk}(s,x)$$

and the second derivative $I_n(\widehat{\beta}_n, \widehat{A}_n)$ of the function $-n^{-1}l_n$ with respect to β and A_n, at the estimates, converges a.s. to $I_0 = I(\beta_0, \lambda_{0,X|S})$.

Proposition 6.12. *The variable $n^{\frac{1}{2}}(\widehat{\beta}_n - \beta_0)$ converges weakly to a centered Gaussian variable with variance the first components of the matrix I_0^{-1}, the variables $n^{\frac{1}{2}}(\widehat{a}_{n,jk} - a_{jk})$ converges weakly to centered Gaussian variables with variances on the diagonal of the matrix I_0^{-1}.*

Proof. Let $U_{n,\beta}$ be the first derivative of the process $l_n(\beta, \lambda_{0,X|S})$ with respect to β. The asymptotic normality of $n^{\frac{1}{2}}(\widehat{\beta}_n - \beta_0)$ relies on a classical one-step Taylor expansion of the score function $U_{n,\beta}$ for large n

$$n^{\frac{1}{2}}U_{n,\widehat{\beta}_n} = \{I_0 + o_p(1)\}\, n^{\frac{1}{2}}(\widehat{\beta}_n - \beta_0).$$

The proof is similar for each components of the matrix A_n. □

The mean integrated squared error of the density estimator is the sum

$$\text{MISE}_n(\lambda_{X|S}) = E\|\widehat{\lambda}_{n,X|S} - \lambda_{J_n,K_n}\|_2^2 + \|\lambda_{X|S} - \lambda_{J_n,K_n}\|_2^2,$$

by Proposition 6.12, the squared mean integrated estimation error is such that

$$E\|\widehat{\lambda}_{n,X|S} - \lambda_{J_n,K_n}\|_2^2 = O(n^{-1}J_nK_n)$$

and the squared error of approximation of $\lambda_{0,X|S}$ is

$$\|\lambda_{X|S} - \lambda_{J_n,K_n}\|_2^2 = \sum_{j>J_n}\sum_{k>K_n} a_k^2 = o(1).$$

As n tends to infinity, the sizes J_n and K_n defining the approximating function $\lambda_{X|S}$ is optimum for the L^2 norm if both errors have the same order so J_n and K_n are chosen in order that

$$J_nK_n = o(n), \quad \|f_n - f\|_2 = O(n^{-\frac{1}{2}}(J_nK_n)^{\frac{1}{2}}). \tag{6.31}$$

Under the condition (6.31), the estimator $\widehat{\lambda}_{n,X|S}$ is L^2-consistent. The mean squared error of $\widehat{\lambda}_n$ is

$$\text{MSE}_n(x) = E\{\widehat{\lambda}_{n,X|S}(x) - \lambda_{J_n,K_n}(x)\}^2 + \{\lambda_{X|S}(x) - \lambda_{J_n,K_n}(x)\}^2$$

where the first term is the variance $V_n(x)$ of $\widehat{\lambda}_{n,X|S}(x)$, it is a $O(n^{-1}J_n^2K_n^2)$. As $E\|\widehat{\lambda}_{n,X|S}(x) - \lambda_{J_n,K_n}(x)\|_2^2 = O(n^{-1}J_nK_n)$, the variance of $n^{\frac{1}{2}}J_n^{-1}K_n^{-1}(\widehat{\lambda}_{n,X|S} - \lambda_{X|S})$ converges to zero almost everywhere. For a conditional hazard function $\lambda_{X|S}$ in the class W_{2s}, defined by (2.20), the estimator $\widehat{\lambda}_{n,X|S}$ has the optimal rate $n^{-\frac{s}{2s+1}}$ as $J_nK_n = O(n^{\frac{1}{2s+1}})$.

Proposition 6.13. *Under the conditions (6.31) and with a uniformly bounded basis of functions, the process $(J_n^{-1}K_n^{-1}n)^{\frac{1}{2}}(\widehat{\lambda}_{n,X|S} - \lambda_{J_n,K_n})$ converges weakly to a centered Gaussian process with variance the limit of $n^{-1}V_n$.*

Proof. The function $J_n^{-1}K_n^{-1}n\int V_n(x)\,dx$ converges to a finite limit under the condition and as n tends to infinity, then the process $(K^{-1}n)^{\frac{1}{2}}(\widehat{\lambda}_{n,X|S} - \lambda_K)$ converges weakly to a Gaussian process, by the arguments of Proposition (2.3) and by the weak convergence of the variables $n^{\frac{1}{2}}(\widehat{a}_{kn} - a_k)$ to Gaussian variables. □

Replacing the scalar parameter β of the model (6.30) by a nonparametric function depending on a time variable S, X or $S + X$, the function β is expanded on an orthonormal basis of the space $L^2(\mathbb{R}_+)$ related to the time variable. The coefficients of the expansion are estimated like previous the regression parameter in (6.30), the regressor of the new model are the product of the covariate Z and estimated functions of the basis in the same way as in Section 6.4, the properties of the estimators are similar.

6.7 Goodness of fit tests

Let $(T_{ij})_{i=1,\ldots,I,j=1,\ldots,n_i}$ be random times right-censored by a independent variables $(C_{ij})_{i=1,\ldots,I,j=1,\ldots,n_i}$, in I sub-samples with a total size $n = \sum_{i=1}^{I} n_i$. The censored observations are the variables $X_{ij} = T_{ij} \wedge C_{ij}$ and the counting processes $N_{ni}(t)$ and $Y_{ni}(t)$ defined in Section 6.1, for every $i = 1,\ldots,I$. Let $\lambda_i(t)$ be the hazard function of the variables T_{ij}, for every $j = 1,\ldots,n_i$, it is estimated by the series estimator $\widehat{\lambda}_{ni}(t)$ given by (6.3). Tests for the comparison of the hazard functions λ_i are equivalent to tests of comparison of the estimated cumulative hazard functions Λ_i (Pons 2014). In the same way, a goodness-of-fit test for the validation of a regular parametric model $\mathcal{L}_\Theta = \{\lambda_\theta \in C^2(\Theta), \theta \in \Theta\}$, where Θ is a bounded open set, for the hazard function of a n sample of right-censored time variables is performed by a comparison of the estimated hazard function is performed by a comparison of the estimated cumulative hazard functions $\widehat{\Lambda}_n$ defined by (6.2) and the estimator $\Lambda_{\widehat{\theta}_n}$ in $\mathcal{L}$, using the maximum likelihood estimator $\widehat{\theta}_n$ of the parameter θ in Θ.

We consider now goodness-of-fit tests for the nonparametric or semi-parametric exponent of a proportional hazard model. Let

$$\mathcal{M}_\Theta = \{r(z) = r_\theta(z), \theta \in \Theta, z \in \mathcal{I}_Z, r_\theta \in C^2(\Theta)\} \tag{6.32}$$

be a parametric model for the regression of the hazard function (6.8), with Θ a bounded open subset of $\mathbb{R}^d$. A likelihood ratio test of the hypothesis H_0 of a proportional hazards model (6.6), against an alternative of a parametric model $\mathcal{M}_\Theta$, relies on the statistic

$$T_n = 2\sum_{i=1}^n \Big\{ r_{\widehat{\theta}_n}(Z_i(T_i) - \widehat{\beta}_n^T Z_i(T_i)\} - \log \frac{S_{2n}(T_i, \widehat{\theta}_n)}{S_{1n}(T_i, \widehat{\beta}_n)} \Big\}, \tag{6.33}$$

depending on the maximum likelihood estimators $\widehat{\theta}_n$ in model (6.8) and $\widehat{\beta}_n$ in model (6.6), with the sums $S_{1n}(t,\beta) = \sum_{i=1}^n 1_{\{X_i \geq t\}} e^{\beta^T Z_i(t)}$ defined in model (6.6) and $S_{2n}(t,\theta) = \sum_{i=1}^n 1_{\{X_i \geq t\}} e^{r_\theta(Z_i(t))}$ in model (6.8).

Proposition 6.14. *Under the hypothesis H_0, the statistic T_n converges weakly to the integral of a squared centered Gaussian process T_0 with respect to Λ_0. Under fixed alternatives such that $r_{\theta_0}(z)$ and $e^{\beta_0^T z}$ differ, T_n diverges.*

Proof. Under the hypothesis H_0, $r_{\theta_0}(z) = e^{\beta_0^T z}$ for every z in $\mathbb{Z}$, the estimators of the parameters are consistent and they have the expansions

$$n^{\frac{1}{2}}(\widehat{\beta}_n - \beta_0) = I_1^{-1} U_{1n}(\beta_0) + o_p(1),$$
$$n^{\frac{1}{2}}(\widehat{\theta}_n - \theta_0) = I_2^{-1} U_{2n}(\theta_0) + o_p(1)$$

with the score function U_{jn} and information matrices I_j defined as

$$U_{1n}(\beta_0) = n^{-\frac{1}{2}} \sum_{i=1}^n \int_0^T Z_i \, dN_i - n^{-\frac{1}{2}} \int_0^T \frac{S_{1n}^{(1)}(s,\beta)}{S_{1n}^{(0)}(s,\beta)} \, d\bar{N}(s),$$

$$U_{2n}(\beta_0) = n^{-\frac{1}{2}} \sum_{i=1}^n \int_0^T \dot{r}_\theta(Z_i) \, dN_i - n^{-\frac{1}{2}} \int_0^T \frac{S_{2n}^{(1)}(s,\beta)}{S_{2n}^{(0)}(s,\beta)} \, d\bar{N}(s),$$

$$I_{j0} = \int_0^T s_{j0}^{(1)2} s_{j0}^{(0)-1} \, d\Lambda_0,$$

where $s_{j0}^{(k)}$ is the limit of $n^{-1} S_{jn}^{(k)}$ at the parameter value under the hypothesis, for $j = 1, 2$. It follows that for every z in $\mathbb{Z}$, the process

$$\begin{aligned} W_n(z) &= n^{\frac{1}{2}}\{r_{\widehat{\theta}_n}(z) - \widehat{\beta}_n^T z\} \\ &= n^{\frac{1}{2}}(\widehat{\theta}_n - \theta_0)^T \dot{r}_{\theta_0}(z) - n^{\frac{1}{2}}(\widehat{\beta}_n - \beta_0)^T z + o_p(1) \end{aligned}$$

and it converges weakly to a centered Gaussian process W under H_0. Let η be the parameter with components those of θ and β, and let $S_n^{(2)}$ be the vector with components those of $S_{1n}^{(2)}$ and $S_{1n}^{(2)}$, the variable $\widehat{\eta}_n$ and the statistic T_n have the expansions

$$n^{\frac{1}{2}}(\widehat{\eta}_n - \eta_0) = I_0^{-1}U_n(\eta_0) + o_p(1)$$

and

$$\begin{aligned}\frac{1}{2}T_n &= n^{-\frac{1}{2}}\sum_{i=1}^{n}\int_0^{\mathcal{T}} W_n(Z_i(t))\,dN_i(t)\\ &\quad - \int_0^{\mathcal{T}} \frac{(\widehat{\theta}_n - \theta_0)^T S_{2n}^{(1)}(t,\theta_0) - (\widehat{\beta}_n - \beta_0)^T S_{1n}^{(1)}(t,\beta_0)}{S_{1n}^{(0)}(t,\theta_0)}\,d\bar{N}_n(t)\\ &\quad + \frac{1}{2}\int_0^{\mathcal{T}} \frac{(\widehat{\eta}_n - \eta_0)^T S_n^{(2)}(t,\eta_0)(\widehat{\eta}_n - \eta_0)}{S_{1n}^{(02)}(t,\theta_0)}\,d\bar{N}_n(t) + o_p(1).\end{aligned}$$

The martingales $\bar{M}_n(t) = n^{-\frac{1}{2}}\{\bar{N}_n(t) - \int_0^t S_n^{(0)}(s,\beta_0)\,d\Lambda_0(s)\}$ converge weakly to a centered Gaussian process B with independent increments and variance $\int_0^t s^{(0)}(s,\beta_0)\,d\Lambda_0(s)\}$, and the expansion of the statistic becomes

$$\begin{aligned}\frac{1}{2}T_n &= n^{-\frac{1}{2}}\sum_{i=1}^{n}\int_0^{\mathcal{T}} W_n(Z_i(t))\,dM_i(t)\\ &\quad - \int_0^{\mathcal{T}} n^{\frac{1}{2}}\frac{(\widehat{\theta}_n - \theta_0)^T S_{2n}^{(1)}(t,\theta_0) - (\widehat{\beta}_n - \beta_0)^T S_{1n}^{(1)}(t,\beta_0)}{S_{1n}^{(0)}(t,\theta_0)}\,d\bar{M}_n(t)\\ &\quad + \frac{1}{2}\int_0^{\mathcal{T}} \frac{(\widehat{\eta}_n - \eta_0)^T S_n^{(2)}(t,\eta_0)(\widehat{\eta}_n - \eta_0)}{S_{1n}^{(0)}(t,\theta_0)}\,d\Lambda_0(t) + o_p(1).\end{aligned}$$

The weak converge of the processes ends the proof. Under fixed alternatives such that $r_{\theta_0}(z)$ and $e^{\beta_0^T z}$ differ, $n^{\frac{1}{2}}\{r_{\theta_0}(z) - e^{\beta_0^T z}\}$ diverges so the statistic T_n diverges. □

Let A_n be a local alternative defined by parameters

$$\theta_n = \theta_0 + n^{-\frac{1}{2}}h_{1n},$$
$$\beta_n = \beta_0 + n^{-\frac{1}{2}}h_{2n},$$

where the sequences $(h_{jn})_n$ converge to nonzero limits h_j, $j = 1, 2$, such that $h_1^T \dot{r}_0(z)$ differs from $h_2^T z$ for every z in $\mathbb{Z}$.

Proposition 6.15. *Under a sequence of local alternatives such that sequences of parameters converge to θ_0 with the rate $n^{-\frac{1}{2}}$, the statistic T_n converges weakly to the integral of a squared uncentered Gaussian process with respect to Λ_0.*

Proof. Under the alternative A_n, the difference $n^{\frac{1}{2}}\{e^{r_{\theta_n}(z)} - e^{\beta_n^T z}\}$ converges to $\{h_1^T \dot{r}_0(z) - h_2^T z\}e^{r_0(z)}$. Moreover, $n^{\frac{1}{2}}(\widehat{\theta}_n - \theta_0)$, and respectively $n^{\frac{1}{2}}(\widehat{\beta}_n - \beta_0)$, converge under the alternative A_n to Gaussian variables with finite variances and with mean h_1 , and respectively h_2. By the same argument, the processes W_n is uncentered. □

The validation of a parametric model $\mathcal{M}_\Theta$ for a hazard function is achieved by a likelihood ratio test of $\mathcal{M}_\Theta$ against nonparametric model, following the same arguments as for the test of model (6.6) against a parametric model. Due to the lower convergence rate $n^{-\frac{1}{2}}K_n$ of the nonparametric estimator $\widehat{r}_n$ of the regression function r, the test statistic is normalized by K_n^{-2}, with $K_n^2 = \sum_{j=1}^d K_{jn}^2$ defined in Section 6.2. Let $S_{3n}(t,r) = \sum_{i=1}^n 1_{\{X_i \geq t\}} e^{r(Z_i)}$ in the nonparametric model, the test statistic is defined as

$$T_{2n} = 2K_n^{-2} \sum_{i=1}^n \int_0^\tau \Big\{\widehat{r}_n(Z_i) - r_{\widehat{\theta}_n}(Z_i\} - \log \frac{S_{3n}(t,\widehat{r}_n)}{S_{2n}(t,\widehat{\theta}_n)}\Big\}\, dN_i(t). \quad (6.34)$$

Let $r_0 = r_{\theta_0}$ be the true regression function of the hazards under the hypothesis H_0 of a parametric model $\mathcal{M}_\Theta$. By Proposition (6.4), the process $W_{2n} = n^{-\frac{1}{2}}K_n^{-1}(\widehat{r}_n - r_{\widehat{\theta}_n})$ has, under H_0, the approximation

$$W_{2n} = n^{\frac{1}{2}}K_n^{-1}(\widehat{r}_n - r_0) + o_p(1)$$

and it converges weakly to a Gaussian process W_2, by Proposition 6.3.

Proposition 6.16. *Under the hypothesis H_0, the statistic T_{2n} converges weakly to a limit $T_2 = \frac{1}{2}\int_0^\tau s_{2,r_0}^{(0)-1} U_2^2\, d\Lambda_0$. Under fixed alternatives such that r_{θ_0} and r_0 differ, the statistic T_{2n} diverges.*

Proof. Under the hypothesis H_0, the statistic $2T_{2n}$ has the expansion

$$n^{-\frac{1}{2}}K_n^{-1} \sum_{i=1}^n \int_0^\tau W_{2n}(Z_i)\, dN_i - K_n^{-2} \int_0^\tau \frac{S_{3n}^{(0)}(\widehat{r}_n) - S_{2n}^{(0)}(r_{\widehat{\theta}_n})}{S_{2n}^{(0)}(r_{\widehat{\theta}_n})}\, d\bar{N}_n$$
$$+ \frac{K_n^{-2}}{2} \int_0^\tau \frac{\{S_{3n}^{(0)}(\widehat{r}_n) - S_{2n}^{(0)}(r_{\widehat{\theta}_n})\}^2}{S_{2n}^{(0)2}(r_{\widehat{\theta}_n})}\, d\bar{N}_n + o_p(1)$$

where the process $U_{2n} = n^{-\frac{1}{2}}K_n^{-1}\{S_{3n}^{(0)}(\widehat{r}_n) - S_{2n}^{(0)}(r_{\widehat{\theta}_n})\}$ has the expansion

$$U_{2n}(t) = n^{-1} \sum_{i=1}^n Y_i(t) W_{2n}(Z_i) e^{r_0}(Z_i) + o_p(1)$$

and it converges weakly to $U_2 = E\{Y_i(t)W_2(Z_i)e^{r_0}(Z_i)\}$, as n tends to infinity.

The expansion of the statistic T_{2n} under H_0 becomes

$$\begin{aligned}
\frac{1}{2}T_{2n} &= n^{-\frac{1}{2}}K_n^{-1}\sum_{i=1}^n \int_0^\tau W_{2n}(Z_i)\,dM_i - K_n^{-1}\int_0^\tau \frac{U_{2n}}{S_{2n}^{(0)}(r_0)}\,d\bar{M}_n \\
&\quad + \frac{n^{\frac{3}{2}}}{2}\int_0^\tau \frac{U_{2n}^2}{S_{2n}^{(0)2}(r_0)}\,d\bar{M}_n \\
&\quad + \frac{K_n^{-2}}{2}\int_0^\tau \frac{\{S_{3n}^{(0)}(\widehat{r}_n) - S_{2n}^{(0)}(r_{\widehat{\theta}_n})\}^2}{S_{2n}^{(0)}(r_0)}\,d\Lambda_0 + o_p(1) \\
&= n^{-\frac{1}{2}}K_n^{-1}\sum_{i=1}^n \int_0^\tau W_{2n}(Z_i)\,dM_i - K_n^{-1}\int_0^\tau \frac{U_{2n}}{S_{2n}^{(0)}(r_0)}\,d\bar{M}_n \\
&\quad + \frac{n^{\frac{3}{2}}}{2}\int_0^\tau \frac{U_{2n}^2}{S_{2n}^{(0)2}(r_{\widehat{\theta}_n})}\,d\bar{M}_n + \frac{1}{2}\int_0^\tau \frac{U_{2n}^2}{n^{-1}S_{2n}^{(0)}(r_0)}\,d\Lambda_0 + o_p(1)
\end{aligned}$$

the first three terms converge in probability to zero and the last term converges in probability to $\frac{1}{2}\int_0^\tau s_{2,r_0}^{(0)-1}U_2^2\,d\Lambda_0$.

Under fixed alternatives such that r_{θ_0} and r_0 differ, the processes $n^{\frac{1}{2}}K_n^{-1}\{r_{\theta_0}(z) - r_0\}$ and U_{2n} diverge, hence the test statistic diverges. □

Let A_n be local alternatives defined by a sequence of parameters

$$\theta_n = \theta_0 + n^{-\frac{1}{2}}h_n,$$

where the sequence $(h_n)_n$ converges to a nonzero limit h.

Proposition 6.17. *Under the alternatives A_n, the statistic T_{2n} converges weakly to a limit different from T_2.*

A Kolmogorov–Smirnov type test for the validation of the model $\mathcal{M}_\theta$ is defined by the statistic

$$\widetilde{T}_{2n} = n^{\frac{1}{2}}K_n^{-1}\sup_{x\in\mathcal{I}_X}|\widehat{r}_n(x) - r_{\widehat{\theta}_n}(x)|$$

it has the same asymptotic behavior as a test statistic in regression models under the hypothesis and local alternatives, its limits are given by Propositions 3.17 and 3.18.

In a two-stage sampling, I independent sub-populations are sampled with a sub-sample of size n_i in the ith sub-population on an interval $[0, \tau_i]$, for $i = 1, \ldots, I$. We consider the semi-parametric proportional hazards models

$$\lambda_{ij}(t \mid Z_{ij}) = \lambda_0(t) e^{r_{\theta_i}(Z_{ij}(t))},\ i = 1, \ldots, n_i, i = 1, \ldots, I, \tag{6.35}$$

with baseline a hazard function λ_0 and a regression function r in a parametric class of functions $\mathcal{M}_\Theta$ defined by (6.32) with an open and bounded set Θ and distinct parameters $\theta_1, \ldots, \theta_I$. The parameter values under the probability P_0 of the observations are denoted θ_{0i}.

Let $N_{ij}(t) = \delta_{ij} 1_{\{T_{ij} \leq t\}}$ and $\bar{N}_{ni}(t) = \sum_{j=1}^{n_i} N_j(t)$ be the counting processes of the uncensored time variables in the population and let $S_{ni}^{(0)}(t, \theta_i) = \sum_{j=1}^{n_i} 1_{\{X_{ij} \geq t\}} e^{r_{\theta_i}(Z_i(t))}$. The maximum likelihood estimator $\widehat{\theta}_{ni}$ of the parameterθ_i in the model (6.35) maximizes the likelihood ratio

$$l_{ni}(\theta_i) = \sum_{j=1}^{n_i} \int_0^{\tau_i} \left[r_{\theta_i}(Z_{ij}(t)) - r_{\theta_{0i}}(Z_{ij}(t)) - \log \frac{S_{ni}^{(0)}(t, \theta_i)}{S_{ni}^{(0)}(t, \theta_{0i})} \right] dN_{ij}(t),$$

the estimators are $n_i^{\frac{1}{2}}$-consistent and asymptotically Gaussian. The cumulative hazard function $\Lambda_0(t) = \int_0^t \lambda_0(s)\, ds$ is estimated using the whole sample, by the process

$$\widehat{\Lambda}_n(t) = \int_0^t 1_{\{\sum_{i=1}^I S_{n_i}^{(0)}(s, \widehat{\theta}_{n_i}) > 0\}} \frac{d\bar{N}_{n_i}(s)}{\sum_{i=1}^I S_{n_i}^{(0)-1}(s, \widehat{\theta}_{n_i})},$$

$\widehat{\Lambda}_n$ is $n^{\frac{1}{2}}$-consistent and asymptotically Gaussian. Let $\widehat{r}_{n_i}$ be a series estimators of the functions r_i of the ith sample using K_{n_i} estimated function of an orthonormal basis and let $K_n^2 = \sum_{i=1}^I K_{n_i}^2$ such that $K_n^{-1} K_{n_i}$ converges to a strictly positive limit for every i.

A test for the validation of the model (6.35) may be defined by summing the statistics (6.34) of the sub-samples, let

$$T_{3n} = K_n^{-2} \sum_{i=1}^I \sum_{j=1}^{n_i} \int_0^{\tau_i} \left\{ \widehat{r}_{n_i}(Z_{ij}) - r_{\widehat{\theta}_{n_i}}(Z_{ij}) - \log \frac{S_{3n_i}(t, \widehat{r}_{n_i})}{S_{2n_i}(t, \widehat{\theta}_{n_i})} \right\} dN_{ij}(t).$$

Under the hypothesis H_0 of the model (6.35) and local alternatives, the limits of the statistic T_{3n} are weighted sums of the limits defined by Propositions 6.16 and 6.17.

Tests for the Cox model against the alternative of a time-varying coefficient model (6.21) or a model (6.24) with coefficients varying nonparametrically with a random variable are achieved with a Kolmogorov–Smirnov type test comparing the nonparametric and the parametric proportional hazard regression models, like in Section 3.12 for conditional expectations, or with a likelihood ratio test as previously.

6.8 Right-censoring of a nonparametric regression

Let (Y, X) be regression variables with a distribution function $F_{X,Y}$ on a subset $\mathcal{I}_{XY}$ of $\mathbb{R}^{d+1}$, and with marginal distribution functions F_X on $\mathcal{I}_X$ and F_Y on $\mathcal{I}_Y$. Let

$$Y = m(X) + \varepsilon, \tag{6.36}$$

where $E(\varepsilon \mid X) = 0$ and $E(\varepsilon^2 \mid X)$ is finite. The variable X has a density f_X on I_X, the variable Y has the conditional distribution function $F_{Y|X}$ on $I_Y \times I_X$, where $I_Y =]\tau_1, \tau_2[$, and the density $m(X)$, it is right-censored by a random variable C independent of (X, Y), the observations are

$$\delta = 1\{Y \leq C\},\ Y^c = Y \wedge C.$$

The conditional mean of the observed Y is

$$m^*(x) = E(Y\delta|X = x) = \int_{I_Y} y\bar{F}_C(y)\, F_{Y|X}(dy; x).$$

The conditional distribution and survival functions of the variable Y^c are

$$F^{nc}(y; x) = P(Y \leq y|\delta = 1, X = x) = \int_y^{\tau_2} \bar{F}_C(z)\, F_{Y|X}(dz; x),$$

$$\bar{F}^c(y; x) = P(Y \wedge C \geq y|X = x) = \bar{F}_{Y|X}(y; x)\bar{F}_C(y), \tag{6.37}$$

they have expansions on an orthonormal basis $(\psi_k)_{k\geq 0}$ of $L^2(\mathcal{I}_X, F_X)$ as

$$F^{nc}(y; x) = \sum_{k\geq 0} p_k(y)\psi_k(x), \quad \bar{F}^c(y; x) = \sum_{k\geq 0} q_k(y)\psi_k(x),$$

with the coefficients

$$p_k(y) = E\{\delta 1_{\{Y\leq y\}}\psi_k(X)\}, \quad q_k(y) = E\{1_{\{Y^c\geq y\}}\psi_k(X)\}.$$

The functions $p_k(y)$ and $q_k(y)$ are estimated with a n-sample $(X_i, Y_i^c, \delta_i)_{i\leq n}$, as

$$\widehat{p}_{nk}(y) = n^{-1} \sum_{i=1}^n \delta_i 1_{\{Y\_i\leq y\}} \widehat{\psi}_k(X_i),$$

$$\widehat{q}_{nk}(y) = n^{-1} \sum_{i=1}^n 1_{\{Y_i^c\geq y\}} \widehat{\psi}_k(X_i),$$

using the estimators $\widehat{\psi}_k$ of (3.2). Then the conditional empirical distributions $F^{nc}(y; x)$ and $\bar{F}^c(y; x)$ are estimated as

$$\widehat{F}_n^{nc}(y; x) = \sum_{k=0}^{K_n} \widehat{p}_{nk}(y)\widehat{\psi}_k(x),$$

$$\widehat{\bar{F}}{}_n^c(y; x) = \sum_{k=0}^{K_n} \widehat{q}_{nk}(y)\widehat{\psi}_k(x),$$

with $K_n = o(n)$ tending to infinity as n tends to infinity. By Lemmas 3.1 and 3.2, the estimators $\widehat{F}_n^{nc}$ and $\widehat{\bar{F}}_n^c$ are a.s. uniformly consistent and for every integer $k \leq K_n$, the processes $n^{\frac{1}{2}}(\widehat{p}_{nk} - p_k)$ and $n^{\frac{1}{2}}(\widehat{q}_{nk} - q_k)$ converge weakly to centered Gaussian variables with strictly positive variances. The MISE errors of the estimators $\widehat{F}_n^{nc}$ and $\widehat{\bar{F}}_n^c$ are $O(nK_n^2) + o(1)$ and we assume the condition (3.7) is fulfilled.

Proposition 6.18. *Under the condition (3.7), the processes $K_n^{-1} n^{\frac{1}{2}}(\widehat{F}_n^{nc} - F^{nc})$ and $K_n^{-1} n^{\frac{1}{2}}(\widehat{F}_n^c - F^c)$ converge weakly to Gaussian processes.*

The conditional hazard function of Y given X satisfies

$$\Lambda_{Y|X}(y;x) = \int_{\tau_1}^{y} \bar{F}_{Y|X}^{-1}(z;x)\, F_{Y|X}(dz;x) = \int_{\tau_1}^{y} \{\bar{F}_n^c(z;x)\}^{-1} F^{nc}(dz;x)$$

and it is estimated using the conditional empirical distributions as

$$\widehat{\Lambda}_{n,Y|X}(y;x) = \int_{\tau_1}^{y} \{\widehat{\bar{F}}_n^c(z;x)\}^{-1} \widehat{F}_n^{nc}(dz;x).$$

Proposition 6.19. *Under the condition (3.7), on every subset I of its support, the process $K_n^{-1} n^{\frac{1}{2}}(\widehat{\Lambda}_{n,Y|X} - \Lambda_{Y|X})$ converges weakly to a Gaussian process.*

The proof is similar to the proof of convergence of the hazard function of a right-censored time variable, using Proposition 6.18. Let $\Delta H(t)$ is the jump of a function H at t, the conditional product-limit estimator of $F_{Y|X}(y;x)$ is deduced from $\widehat{\Lambda}_{n,Y|X}$ as

$$\widehat{F}_{nY|X}(y;x) = \prod_{Y_i \leq y} \left\{1 - \Delta\widehat{\Lambda}_{n,Y|X}(Y_i)\right\}^{\delta_i}.$$

Proposition 6.20. *Under the condition (3.7) and on every bounded subset I of its support, the estimator $\widehat{F}_{nY|X}$ is uniformly consistent on I and the process $K_n^{-1} n^{\frac{1}{2}}(\widehat{F}_{nY|X} - F_{Y|X})$ converges weakly to a Gaussian process.*

An estimator of the regression function is obtained from $\widehat{F}_{nY|X}$ on every set where the estimator $\widehat{\bar{F}}_n^c$ is strictly positive

$$\widehat{m}_n(x) = \sum_{i=1}^{n} Y_i \delta_i \Delta\widehat{F}_{Y|X,n}(Y_i;x), \tag{6.38}$$

and the function $\bar{F}_Y(y)$ is estimated as

$$\widehat{F}_{nY}(y) = \int_{\mathcal{I}_X} \widehat{F}_{nY|X}(y;x)\, d\widehat{F}_{nX}(x),$$

with the empirical distribution function $\widehat{F}_{nX}$ of F_X. They are uniformly consistent on every bounded sub-interval of support of $F_{Y|X}$ and asymptotically Gaussian, with the convergence rate $K_n n^{-\frac{1}{2}}$.

Proposition 6.21. *The optimal convergence rate $n^{-\frac{s}{2s+1}}$ of the estimator (6.38) for a regression function m in W_{4s}, defined by (3.8), is reached for $K_n = O(n^{\frac{1}{2(2s+1)}})$.*

6.9 Left-truncation of a nonparametric regression

Let X and Y be the variables of the nonparametric regression model (6.36) observed conditionally on a minimal or maximal random threshold for Y, that is a truncation of the variable Y. With a left-truncation of Y by a variable T such that Y and T are independent conditionally on X, with distribution functions F_T and $F_{T|X}$, the variables Y and T are observed conditionally on the event $\{Y \geq T\}$ and none of the two variables is observed if $Y < T$.

Let $F_{X,Y}$ be the distribution function of (Y, X) on $\mathcal{I}_{XY}$, with marginal distribution functions F_X on $\mathcal{I}_X$ and F_Y on $\mathcal{I}_Y$, and with conditional distribution function $F_{Y|X}(y;x) = F_\varepsilon(y - m(x))$, where F_ε is the distribution function of the error ε independent of X. The probability of an observation conditionally on $\{X = x\}$ is

$$\alpha(x) = P(T \leq Y|X = x) = \int_{\mathcal{I}_Y} \bar{F}_{Y|X}(y;x)\, dF_{T|X}(t;x).$$

When $\alpha(x) > 0$, the distributions of the variables truncated conditionally on $\{X = x\}$ are

$$\begin{aligned} A(y;x) &= P(Y \leq y|X = x, T \leq Y) \\ &= \alpha^{-1}(x) \int_{-\infty}^{y} F_{T|X}(y;x)\, F_{Y|X}(dy;x) \qquad (6.39) \\ B(y;x) &= P(T \leq y \leq Y|X = x, T \leq Y) \\ &= \alpha^{-1}(x) F_{T|X}(y;x) \bar{F}_{Y|X}(y;x). \qquad (6.40) \end{aligned}$$

Under the left-truncation, the conditional mean $m(x) = E(Y|X = x)$ is biased and the apparent mean is

$$m^*(x) = E(Y|X = x, T \leq Y) = \alpha^{-1}(x) \int_{\mathcal{I}_Y} y F_{T|X}(y;x)\, F_{Y|X}(dy;x).$$

The problem is the estimation of the conditional distribution $F_{Y|X}$ and the mean $m(x)$, instead of the mean $m^*(x)$ of the observed variables. Under left-truncation, the distribution of Y conditionally on X is expressed from A and B by equations (6.39) and (6.40)

$$\begin{aligned} F_{Y|X}(y;x) &= \exp\{-\Lambda_{Y|X}(y;x)\}, \\ \Lambda_{Y|X}(y;x) &= \int_{-\infty}^{y} 1_{\{\bar{F}_{Y|X}(s;x)>0\}} \{\bar{F}_{Y|X}(s;x)\}^{-1} F_{Y|X}(ds;x) \\ &= \int_{-\infty}^{y} 1_{\{B(s;x)>0\}} \{B(s;x)\}^{-1} A(ds;x). \end{aligned}$$

The conditional sub-distribution functions

$$\begin{aligned} H_1(y;x) &= P(T \le Y \le y \mid X = x), \\ H_2(y;x) &= P(T \le y \le Y \mid X = x) \end{aligned}$$

have expansions on an orthonormal basis $(\psi_k)_{k\ge0}$ of $L^2(\mathcal{I}_X, F_X)$ as $H_1(y;x) = \sum_{k\ge0} p_k(y)\psi_k(x)$ and $H_2(y;x) = \sum_{k\ge0} h_k(y)\psi_k(x)$ with the coefficients

$$\begin{aligned} p_k(y) &= E\{H_1(y;X)\psi_k(X)\} = E\{1_{\{T\le Y\le y\}}\psi_k(X)\}, \\ h_k(y) &= E\{H_2(y;X)\psi_k(X)\} = E\{1_{\{T\le y\le Y\}}\psi_k(X)\}. \end{aligned}$$

The functions ψ_k have the estimators $\widehat{\psi}_k$ defined as (3.2) from a n-sample $(X_i)_{i\le n}$ and the functions $p_k(y)$, $h_k(y)$ and $b_k(y)$ are estimated from the observation of a n-sample $(X_i, Y_i 1_{\{T_i\ge Y_i\}})_{i\le n}$

$$\begin{aligned} \widehat{p}_{nk}(y) &= n^{-1} \sum_{i=1}^{n} 1_{\{T_i\ge Y_i\le y\}} \widehat{\psi}_k(X_i), \\ \widehat{h}_{nk}(y) &= n^{-1} \sum_{i=1}^{n} 1_{\{T_i\le y\le Y_i\}} \widehat{\psi}_k(X_i), \end{aligned}$$

then the functions H_1 and H_2 have the estimators

$$\begin{aligned} \widehat{H}_{1n}(y;x) &= \sum_{k=0}^{K_n} \widehat{p}_{nk}(y)\widehat{\psi}_k(x), \\ \widehat{H}_{2n}(y;x) &= \sum_{k=0}^{K_n} \widehat{h}_{nk}(y)\widehat{\psi}_k(x) \end{aligned}$$

and the function $\alpha(x)$ is estimated by $\widehat{H}_{1n}(\infty;x)$. Estimators of the conditional functions A and B are the ratios

$$\widehat{A}_n(y;x) = \widehat{\alpha}_n^{-1}(x)\widehat{H}_{1n}(y;x), \quad \widehat{B}_n(y;x) = \widehat{\alpha}_n^{-1}(x)\widehat{H}_{2n}(y;x).$$

The MISE errors of the estimators $\widehat{H}_{jn}$, $j = 1, 2$, are $O(nK_n^2) + o(1)$, it is the same for $\widehat{A}_n$ and $\widehat{B}_n$.

Proposition 6.22. *Under the condition (3.7), the processes $K_n^{-1} n^{\frac{1}{2}}(\widehat{A}_n - A)$ and $K_n^{-1} n^{\frac{1}{2}}(\widehat{B}_n - B)$ converge weakly to Gaussian processes on every subset of their support where the function α is strictly positive.*

The conditional hazard function $\Lambda_{Y|X}$ is then estimated on every interval where the function α is strictly positive as

$$\widehat{\Lambda}_{nY|X}(y;x) = \int_{-\infty}^{y} 1_{\{\widehat{B}_n(s;x)>0\}} \frac{\widehat{A}_n(dy;x)}{\widehat{B}_n(s;x)}$$

and the product-limit estimator of $F_{Y|X}(y;x)$ is

$$\widehat{F}_{nY|X}(y;x) = \prod_{Y_i \le y} \left\{ 1 - 1_{\{\widehat{B}_n(s;x)>0\}} \frac{\widehat{A}_n(dy;x)}{\widehat{B}_n(s;x)} \right\}.$$

Let I be a bounded subset of the support of $\Lambda_{Y|X}$ where the function α is strictly positive.

Proposition 6.23. *Under the condition (3.7), the estimator $\widehat{\Lambda}_{nY|X}$ is uniformly consistent on I and the process $K_n^{-1} n^{\frac{1}{2}}(\widehat{\Lambda}_{nY|X} - \Lambda_{Y|X})$ converges weakly to a Gaussian process on I.*

As a product-limit estimator based on $\widehat{\Lambda}_{nY|X}$, the estimator $\widehat{F}_{nY|X}$ has the same properties.

Proposition 6.24. *Under the condition (3.7), the estimator $\widehat{F}_{nY|X}$ is uniformly consistent on I and the process $K_n^{-1} n^{\frac{1}{2}}(\widehat{F}_{nY|X} - F_{Y|X})$ converges weakly to a Gaussian process on I.*

An estimator of the regression function is deduced from $\widehat{F}_{nY|X}$ as

$$\begin{aligned} \widehat{m}_n(x) &= \int y \, \widehat{F}_{Y|X,n}(dy;x), \\ &= \sum_{i=1}^{n} Y_i 1\{T_i \le Y_i\}\{\widehat{F}_{Y|X,n}(Y_i;x) - \widehat{F}_{nY|X}(Y_i^-;x)\} \qquad (6.41) \end{aligned}$$

and the function $\bar{F}_Y(y)$ is estimated by

$$\widehat{F}_{nY}(y) = \int_{\mathcal{I}_X} \widehat{F}_{nY|X}(y;x) \, d\widehat{F}_{nX}(x),$$

with the empirical distribution function $\widehat{F}_{nX}$ of the sample $(X_i)_{i \le n}$. The optimal convergence rate of the estimator $\widehat{m}_n$ for a regression function m in W_{4s} is again $n^{\frac{s}{2s+1}}$.

For an error variable ε independent of X, an estimator of its distribution function F_ε is deduced from estimators of $F_{Y|X}$, F_X and m as

$$\widehat{F}_{n\varepsilon}(s) = n^{-1} \sum_{1\leq i\leq n} \widehat{F}_{nY|X}(s + \widehat{m}_n(X_i); X_i).$$

Proposition 6.25. *Under the condition (3.7) and with a basis of functions uniformly bounded on I, the estimator $\widehat{m}_n$ and $\widehat{F}_{\varepsilon,n}$ are uniformly consistent on I. The processes $K_n^{-1}n^{\frac{1}{2}}(\widehat{m}_n - m)$ and $K_n^{-1}n^{\frac{1}{2}}(\widehat{F}_{n\varepsilon} - F_\varepsilon)$ converge weakly to Gaussian processes on I.*

6.10 Truncation and censoring of a nonparametric regression

Let X and Y be the variables of the nonparametric regression model (6.36), the variable Y is supposed randomly left-truncated by an unobserved variable T and right-censored by a variable C, and such that T and C are mutually independent and independent of (X, Y). The notations α and those of the joint and marginal distribution function of X, Y and T are in like in Section 6.9 and F_C is the distribution function of C. The observations are $\delta = 1_{\{Y\leq C\}}$, and $Y \wedge C$, conditionally on $Y \wedge C \geq T$. Let

$$\alpha(x) = P(T \leq Y|X = x) = \int_{\tau_1}^{\tau_2} F_T(y)\, F_{Y|X}(dy; x).$$

The conditional sub-distributions of the observed variables are

$$\begin{aligned}
A(y; x) &= P(Y \leq y \wedge C|X = x, T \leq Y) \\
&= \alpha^{-1}(x) \int_{\tau_1}^{y} F_T(v)\bar{F}_C(v)\, F_{Y|X}(dv; x) \\
B(y; x) &= P(T \leq y \leq Y \wedge C|X = x, T \leq Y) \\
&= \alpha^{-1}(x) F_T(y)\bar{F}_C(y)\bar{F}_{Y|X}(y; x).
\end{aligned}$$

The conditional hazard function $\Lambda_{Y|X}$ is defined for every $y < \tau_2$ as

$$\begin{aligned}
\Lambda_{Y|X}(y; x) &= \int_{\tau_1}^{y} \bar{F}_{Y|X}^{-1}(v; x)\, F_{Y|X}(dv; x) \\
&= \int_{\tau_1}^{y} 1_{\{B(v;x)>0\}} B^{-1}(v; x) A(dv; x)
\end{aligned}$$

and the conditional survival function is $\bar{F}_{Y|X}(y; x) = \exp\{-\Lambda_{Y|X}(y; x)\}$. The conditional sub-distribution functions

$$\begin{aligned}
H_1(y; x) &= P(T \leq Y \leq y \wedge C \mid X = x), \\
H_2(y; x) &= P(T \leq y \leq Y \wedge C \mid X = x)
\end{aligned}$$

and the function $\alpha(x)$ have expansions on an orthonormal basis $(\psi_k)_{k\geq 0}$ of $L^2(\mathcal{I}_X, F_X)$ as $\alpha(x) = \sum_{k\geq 0} a_k(y)\psi_k(x)$, $H_1(y;x) = \sum_{k\geq 0} h_{1k}(y)\psi_k(x)$ and $H_2(y;x) = \sum_{k\geq 0} h_{2k}(y)\psi_k(x)$ with the coefficients

$$\begin{aligned}
\alpha(x) &= E\{1_{\{T\leq Y\}}\psi_k(X)\},\\
h_{1k}(y) &= E\{H_1(y;X)\psi_k(X)\} = E\{1_{\{T\leq Y\leq y\wedge C\}}\psi_k(X)\},\\
h_{2k}(y) &= E\{H_2(y;X)\psi_k(X)\} = E\{1_{\{T\leq y\leq Y\wedge C\}}\psi_k(X)\}.
\end{aligned}$$

Let $\widehat{\psi}_k$ be the estimators (3.2 of the functions $p_k(y)$, the coefficients a_k, $h_{1k}(y)$ and $h_{2k}(y)$ are estimated from the censored and truncated observations $(X_i, (Y_i \wedge C_i)1_{\{T_i\geq Y_i\}})_{i\leq n}$

$$\begin{aligned}
\widehat{a}_{nk}(y) &= n^{-1}\sum_{i=1}^n 1_{\{T_i\geq Y_i\}}\widehat{\psi}_k(X_i),\\
\widehat{h}_{1nk}(y) &= n^{-1}\sum_{i=1}^n 1_{\{T_i\geq Y_i\leq y\wedge C_i\}}\widehat{\psi}_k(X_i),\\
\widehat{h}_{2nk}(y) &= n^{-1}\sum_{i=1}^n 1_{\{T_i\leq y\leq Y_i\wedge C_i\}}\widehat{\psi}_k(X_i),
\end{aligned}$$

then the functions α, H_1 and H_2 have the estimators

$$\begin{aligned}
\widehat{\alpha}_n(y;x) &= \sum_{k=0}^{K_n} \widehat{a}_{nk}(y)\widehat{\psi}_k(x),\\
\widehat{H}_{1n}(y;x) &= \sum_{k=0}^{K_n} \widehat{h}_{1nk}(y)\widehat{\psi}_k(x),\\
\widehat{H}_{2n}(y;x) &= \sum_{k=0}^{K_n} \widehat{h}_{2nk}(y)\widehat{\psi}_k(x).
\end{aligned}$$

The conditional sub-distributions A and B are estimated by

$$\begin{aligned}
\widehat{A}_n(y;x) &= \widehat{\alpha}_n^{-1}(x)\widehat{H}_{1n}(y;x),\\
\widehat{B}_n(y;x) &= \widehat{\alpha}_n^{-1}(x)\widehat{H}_{2n}(y;x),
\end{aligned}$$

they are uniformly consistent.

Proposition 6.26. *Under the condition (3.7), the processes $K_n^{-1}n^{\frac{1}{2}}(\widehat{A}_n - A)$ and $K_n^{-1}n^{\frac{1}{2}}(\widehat{B}_n - B)$ converge weakly to Gaussian processes on every subset of their support where the function α is strictly positive.*

The estimators $\widehat{\Lambda}_{nY|X}$, and respectively $\widehat{F}_{nY|X}$, of the conditional hazard and distribution functions, have the same form as in Section 6.9

$$\widehat{\Lambda}_{nY|X}(y;x) = \int_{-\infty}^{y} 1_{\{\widehat{B}_n(s;x)>0\}} \frac{\widehat{A}_n(dy;x)}{\widehat{B}_n(s;x)},$$
$$\widehat{F}_{nY|X}(y;x) = \prod_{Y_i \le y} \left\{1 - \Delta\widehat{\Lambda}_{nY|X}(Y_i;x)\right\}^{\delta_i}.$$

Let I be a bounded subset of the support of $\Lambda_{Y|X}$ where the function α is strictly positive.

Proposition 6.27. *Under the condition (3.7), the estimators $\widehat{\Lambda}_{nY|X}$ and $\widehat{F}_{nY|X}$ are uniformly consistent on I, the process $K_n^{-1}n^{\frac{1}{2}}(\widehat{\Lambda}_{nY|X} - \Lambda_{Y|X})$ and $K_n^{-1}n^{\frac{1}{2}}(\widehat{F}_{nY|X} - F_{Y|X})$ converge weakly to Gaussian processes on I.*

An estimator of the regression function is deduced from the conditional distribution functions

$$\begin{aligned}\widehat{m}_n(x) &= \int y\,\widehat{F}_{Y|X,n}(dy;x),\\ &= \sum_{i=1}^{n} Y_i 1_{\{T_i \le Y_i\}} \Delta\widehat{F}_{Y|X,n}(Y_i;x),\end{aligned}$$

it satisfies Proposition 6.25.

Chapter 7

Nonparametric diffusion processes

7.1 Introduction

We consider several models of diffusion processes with nonparametric autoregressive or time varying drift and variance functions. The parametric estimation of diffusion processes has mainly been studied using a discretization of their sample paths, here we define nonparametric estimators for continuously observed diffusions and for their discretization, and we study their asymptotic behavior. The simpler diffusion processes are defined with time dependent drift α and variance β by the stochastic differential equation observed on an interval $[0, T]$

$$dX_t = \alpha(t)\,dt + \beta(t)\,dB_t, \tag{7.1}$$

with $X_0 = 0$, where B is the standard Brownian motion, α and β^2 are functions of the space $L^2(\mathbb{R}_+)$ defined with respect to the Lebesgue measure on $\mathbb{R}_+$. The Brownian process $(B_t)_{t\geq 0}$ is a martingale with respect to the filtration generated by the $(B_u)_{u<t}$, it satisfies $E(B_t - B_s \mid B_s) = 0$, for every $0 < s < t$, and its moments are the values at zero of the derivatives of its Laplace transform $L_t(x) = \exp(\frac{1}{2}x^2 t)$, so $EB_t^{2k+1} = 0$ for every integer k, $EB_t^2 = t$ and $EB_t^4 = 3t^2$.

The process X_t is a Gaussian process with the expectation and the variance functions

$$EX_t = A_t := \int_0^t \alpha(s)\,ds, \quad \operatorname{Var} X_t = V_t := \int_0^t \beta^2(s)\,ds.$$

Its Laplace transform is $L_t(x) = \exp(\frac{1}{2}x^2 V_t - xA_t)$, and its first moments are $E(X_t^2) = A_t^2 + V_t$, $E(X_t^3) = -A_t(A_t^2 + 3V_t)$, $E(X_t^4) = 3V_t^2 + 6A_t^2 V_t + A_t^4$ and they are finite as A_t and V_t are finite. The Brownian motion is a process with independent increments, it is a stationary Markov process with transition probabilities

$$\begin{aligned}\pi_{A'}(A; t-s) &= P(B_t \in A \mid B_s \in A') \\ &= \int 1_{\{y\in A\}} 1_{\{x\in A'\}} P(B_{t-s} = y - x)\, dx\, dy \\ &= \int 1_{\{y\in A\}} 1_{\{x\in A'\}} f_{t-s}(y-x)\, dx\, dy,\end{aligned}$$

where $f_t(x)$ is the Gaussian density with mean zero and variance t at x. Let $F_B(x)$ be the distribution function with density

$$f_B(x) = \lim_{T\to\infty} \frac{2}{\sqrt{2\pi T}} \int_0^T e^{-\frac{x^2}{2u^2}}\, du$$

with respect to Lebesgue's measure on $\mathbb{R}$.

Ergodic properties of the Brownian motion follow from the definition of f_B, for every real function ψ integrable with respect to F_B we have

$$\begin{aligned}\lim_{T\to\infty} ET^{-1} \int_T \psi(B_t)\, dt &= \lim_{T\to\infty} T^{-1} \int_{\mathbb{R}} \int_T \psi(x) P(B_t = x)\, dt\, dx \\ &= \int_{\mathbb{R}} \psi(x)\, dF_B(x). \end{aligned} \tag{7.2}$$

On $\mathbb{R}^2$, we define the measure

$$\pi \otimes F_B(A; A') = \int_{\mathbb{R}_+^2} \int 1_{\{y\in A\}} 1_{\{x\in A'\}} \pi_x(dy; t-s) f_s(x)\, dx\, ds\, dt,$$

for every function ψ of $L^1(\pi\otimes F_B)$ we have

$$E\int_{[0,T]^2} \psi(B_s, B_t)\, ds\, dt = \int_{\mathbb{R}^2} \int_{[0,T]^2} \psi(x,y) f_{t-s}(y-x) f_s(x)\, dx\, dy\, ds\, dt,$$

$$\lim_{T\to\infty} ET^{-2} \int_{[0,T]^2} \psi(B_s, B_t)\, ds\, dt = \int_{\mathbb{R}^2} \psi(x,y)\, \pi\otimes F_B(dy; dx). \tag{7.3}$$

For every ψ of $L^2(\pi\otimes F_B)$, the variable

$$T^{-\frac{1}{2}} \int_0^T \Big\{\psi(B_t) - T^{-1} E \int_0^T \psi(B_t)\, dt\Big\}\, dt \tag{7.4}$$

converges weakly, as T tends to infinity, to a centered Gaussian variable with variance

$$v^2 = \int_{\mathbb{R}^2} \psi(x)\psi(y)\, \pi\otimes F_B(dy; dx) - \Big\{\int_{\mathbb{R}} \psi(x)\, dF_B(x)\Big\}^2.$$

In a parametric diffusion model (7.1), let $\alpha(t,\theta)$ and $\beta^2(t,\theta)$ be the drift and the variance of the diffusion depending on a parameter θ of an open and bounded subset Θ of $\mathbb{R}^d$, such that for every t the functions $\alpha(t,\cdot)$ and $\beta^2(t,\cdot)$ belong to $C^2(\Theta)$ and $L^2(\Theta)$. With observations at times t_i such that $t_{i+1}-t_i=\Delta$ for every i, the variables $Y_i=X_{t_{i+1}}-X_{t_i}$ have the approximation

$$Y_i=\Delta\alpha(t_i,\theta)+\beta(t_i,\theta)\,\varepsilon_i$$

where the variables $\varepsilon_i=B_{t_{i+1}}-B_{t_i}$ are independent centered Gaussian variables with variance Δ, and ε_i is independent of X_{t_i}. The density of Y_i is

$$\phi_\theta(Y_i)=\frac{1}{\beta(t_i,\theta)\sqrt{2\pi\Delta}}\exp\Big\{-\frac{(Y_i-\Delta\alpha(t_i,\theta))^2}{2\Delta\beta^2(t_i,\theta)}\Big\}, \tag{7.5}$$

for $i=1,\ldots,n=\Delta^{-1}T$. The log-likelihood of the sample $(Y_i)_{i=1,\ldots,n}$ with density (7.5) is

$$l_n(\theta)=-\sum_{i=1}^n\frac{(Y_i-\Delta\alpha(t_i,\theta))^2}{2\Delta\beta^2(t_i,\theta)}-\sum_{i=1}^n\log\beta(t_i,\theta) \tag{7.6}$$

up to constants. The process $-n^{-1}l_n(\theta)$ converges in probability under P_0 to the limit $-l(\theta)$ of $n^{-1}\sum_{i=1,\ldots,n}\log\beta(t_i,\theta)$ uniformly on Θ, and $-l(\theta)$ is minimum at the true parameter value θ_0, its second order derivative is the Fisher information matrix $I(\theta)$. Under the probability P_0 of the sample $(Y_i,X_{t_i})_{i=1,\ldots,n}$, the maximum likelihood estimator $\widehat{\theta}_n$ is therefore a.s. consistent and the variable $n^{\frac12}(\widehat{\theta}_n-\theta_0)$ converges to a centered Gaussian variable with variance I_0^{-1} where $I_0=I(\theta_0)$.

In the nonparametric diffusion model (7.1), the functions α and β^2 are estimated by projection on the transformed Laguerre orthonormal basis of functions $(\varphi_k)_{k\geq0}$ of $L^2(\mathbb{R}_+)$ defined by (2.8), such that the expansions $\alpha(t)=\sum_{k\geq0}a_k\varphi_k(t)$ and $\beta^2(t)=\sum_{k\geq0}b_k\varphi_k(t)$ are convergent, with the coefficients

$$a_k=\int_0^\infty\varphi_k(t)\alpha(t)\,dt,\quad b_k=\int_0^\infty\varphi_k(t)\beta^2(t)\,dt.$$

The series expansions of α and β^2 are approximated by

$$\alpha_T(t)=\sum_{k=0}^{m_T}a_k\varphi_k(t),$$

$$\beta_T^2(t)=\sum_{k=0}^{n_T}b_k\varphi_k(t),$$

where m_T and n_T tend to infinity as T tends to infinity.

The function α is consistently estimated on $[0,T]$ by

$$\widehat{\alpha}_T(t) = \sum_{k=0}^{m_T} \widehat{a}_{Tk}\varphi_k(t)$$

with the estimators of the coefficients

$$\widehat{a}_{Tk} = \int_0^T \varphi_k(t)\, dX_t. \tag{7.7}$$

Lemma 7.1. *The estimator $\widehat{a}_{Tk}$ is a.s. consistent under P_0.*

Proof. The estimator $\widehat{a}_{Tk}$ converges a.s. to the limits of its expectation

$$a_{Tk} = E\widehat{a}_{Tk} = \int_0^T \alpha(t)\varphi_k(t)\, dt$$

and it converges to a_k, for every integer k. □

The centered diffusion process

$$Y_t = X_t - \int_0^t \alpha(s)\, ds = \int_0^t \beta(s)\, dB(s)$$

has the variance $EY_t^2 = \int_0^t \beta^2(s)\, ds$ and it is estimated by

$$\begin{aligned}\widehat{Y}_T(t) = X_t - \int_0^t \widehat{\alpha}_T(s)\, ds &= \int_0^t \{\alpha(s) - \widehat{\alpha}_T(s)\}\, ds + \int_0^t \beta(s)\, dB(s)\\ &= Y_t + o_{as}(1).\end{aligned}$$

The coefficients b_k and the function $\beta^2(t)$ are consistently estimated by

$$\widehat{b}_{Tk} = 2\int_0^T \varphi_k(t)\, \widehat{Y}_T(t)\, d\widehat{Y}_T(t) \tag{7.8}$$

and $\widehat{\beta}^2_T(t) = \sum_{k=0}^{n_T} \widehat{b}_{Tk}\varphi_k(t)$.

Lemma 7.2. *The estimator $\widehat{b}_{Tk}$ is a.s. consistent under P_0.*

Proof. The approximation of the process $\widehat{Y}_T$ implies

$$E\{\widehat{Y}_T^2(t)\} = \int_0^t \beta^2(s)\, ds + o(1),$$

$$b_{Tk} = E\widehat{b}_{Tk} = \int_0^T \varphi_k(t)\beta^2(t)\, dt + o(1),$$

and it converges to b_k as T tends to infinity. □

Lemmas 7.1 and 7.2 imply that the processes $\widehat{\alpha}_T(t)$, and respectively $\widehat{\beta}^2_T(t)$, converge uniformly on every compact interval of $\mathbb{R}_+$ to the functions $\alpha(t)$, and respectively $\beta^2(t)$.

The variable $\widehat{a}_{Tk} - a_{Tk} = \int_0^T \varphi_k(t)\beta(t)\,dB(t) = \int_0^T \varphi_k(t)\,dY_t$ is a centered Gaussian variable and

$$\begin{aligned}
\widehat{b}_{Tk} - b_{Tk} &= \int_0^T \varphi_k(t)\,d\{\widehat{Y}^2_T(t) - Y^2_t\} \\
&= \int_0^T \varphi_k(t)\Big[\Big\{2Y_t + \int_0^t \{\alpha(s) - \widehat{\alpha}_T(s)\}\,ds\Big\}\{\alpha(t) - \widehat{\alpha}_T(t)\}\,dt \\
&\qquad + \{2\,dY_t + \alpha(t)\,dt - \widehat{\alpha}_T(t)\,dt\}\int_0^t \{\alpha(s) - \widehat{\alpha}_T(s)\}\,ds\Big].
\end{aligned}$$

Their variances are

$$\begin{aligned}
\text{Var}\,\widehat{a}_{Tk} &= \int_0^T \varphi^2_k(t)\beta^2(t)\,dt, \\
\text{Var}\,\widehat{b}_{Tk} &= E\Big[\int_0^T \varphi_k(s)\,d\Big\{\int_0^s \beta(u)\,dB(u)\Big\}^2\Big]^2 - b^2_{Tk} + o(1) \\
&= 2\int_0^T \varphi^2_k(s)\Big[\beta^2(s)\Big\{3s\beta^2(s) + \int_0^s \beta^2(u)\,du\Big\}\,ds\Big] + o(1)
\end{aligned}$$

where the moments of the Brownian motion imply

$$E\Big\{\int_0^s \beta(u)\,dB(u)\Big\}^4 = \Big\{\int_0^s \beta^2(u)\,du\Big\}^2 + 6\int_0^s u\beta^4(u)\,du.$$

Lemma 7.3. *Under the condition of bounded integrals $\int_0^\infty \varphi^2_k(t)\beta^2(t)\,dt$, $\int_0^\infty t\varphi^2_k(t)\beta^4(t)\,dt$ and $\int_0^\infty \varphi^2_k(t)\beta^2(t)\Big\{\int_0^t \beta^2(s)\,ds\Big\}\,dt$, the variables $\widehat{a}_{Tk} - a_k$ and $\widehat{b}_{Tk} - b_k$ converge weakly to centered Gaussian variables with finite variances.*

The mean integrated squared errors in $L^2(\mathbb{R}_+)$ of the estimators are the sums of the squared error of estimation and approximation

$$\begin{aligned}
\text{MISE}_T(\alpha) &= E\|\widehat{\alpha}_T - \alpha_T\|^2_2 + \|\alpha_T - \alpha\|^2_2 \\
&= \int_0^\infty \Big\{\sum_{k=0}^{m_T} \varphi^2_k(t)\Big\}\beta^2(t)\,dt + \sum_{k>m_T} a^2_k, \\
\text{MISE}_T(\beta_0) &= E\|\widehat{\beta}_T - \beta_T\|^2_2 + \|\beta_T - \beta\|^2_2 \\
&= \sum_{k=0}^{n_T} \text{Var}\,\widehat{b}_{Tk} + \sum_{k>n_T} b^2_k,
\end{aligned}$$

they converge under the conditions

$$\sigma_\alpha^2 = \int_0^\infty \sum_{k\geq 0} \varphi_k^2(t)\beta^2(t)\,dt < \infty, \quad \sigma_\beta^2 = \sum_{k\geq 0} \operatorname{Var} \widehat{b}_{Tk} < \infty. \tag{7.9}$$

From the Lemmas and the expression of the variances of the estimated coefficients, we deduce the weak convergence of the estimators $\widehat{\alpha}_T$ and $\widehat{\beta}_T$.

Proposition 7.1. *Under the conditions (7.9), the processes $\widehat{\alpha}_T - \alpha_T$ and $\widehat{\beta}_T - \beta_T$ converge weakly to centered Gaussian processes.*

The optimal numbers n_T and m_T may be chosen by cross-validation like for the series estimator of a density.

7.2 Auto-regressive diffusions

Let α and β be continuous functions on a metric space $(\mathbb{X}, \|\cdot\|_2)$, and let B be the standard Brownian motion. A diffusion process is defined by a stochastic differential equation

$$dX_t = \alpha(X_t)\,dt + \beta(X_t)\,dB_t, \tag{7.10}$$

for t in a time interval $[0,T]$, with an initial value X_0 such that $E(X_0^2)$ is finite. The process X_t is solution of the implicit equation

$$X_t = X_0 + \int_0^t \alpha(X_s)\,ds + \int_0^t \beta(X_s)\,dB_s.$$

By discretization on n sub-intervals of length Δ_n which tends to zero as n tends to infinity, the equation has the approximation

$$Y_i = X_{t_{i+1}} - X_{t_i} = \Delta_n \alpha(X_{t_i}) + \beta(X_{t_i})\varepsilon_i, \tag{7.11}$$

where $\varepsilon_i = B_{t_{i+1}} - B_{t_i}$ is a Gaussian variable with mean zero and variance Δ_n conditionally on the σ-algebra $\mathcal{F}_{t_i}$ generated by the sample-paths of X up to t_i, then $E\{\varepsilon_i \mid X_{t_i}\} = 0$

$$\begin{aligned}
\operatorname{Var}(Y_i \mid X_{t_i}) &= \Delta_n \beta^2(X_{t_i}),\\
EY_i &= \Delta_n E\{\alpha(X_{t_i})\},\\
\operatorname{Var} Y_i &= \Delta_n^2 \operatorname{Var}\{\alpha(X_{t_i})\} + \Delta_n E\{\beta^2(X_{t_i})\}\\
&= \Delta_n E\{\beta^2(X_{t_i})\} + o(\Delta_n).
\end{aligned}$$

Let α and β be Lipschitz functions such that for $i = 0, \ldots, n-1$ and for every t in the interval $I_i =]t_i, t_{i+1}]$

$$|\alpha(X_t) - \alpha(X_{t_i})| \leq k_\alpha \Delta_n,$$
$$|\beta(X_t) - \beta(X_{t_i})| \leq k_\beta \Delta_n,$$

then the approximation error

$$X_t - X_{t_i} = \int_{t_i}^{t} \{\alpha(X_s) - \alpha(X_{t_i})\}\, ds + \int_{t_i}^{t} \{\beta(X_s) - \beta(X_{t_i})\}\, dB_s$$

of the process X_t by X_{t_i} on I_i satisfies $E|X_t - X_{t_i}| = O(\Delta_n^2)$ and its variance is bounded by

$$\int_{t_i}^{t} E\{\alpha(X_s) - \alpha(X_{t_i})\}^2 ds + \int_{t_i}^{t} E\{\beta(X_s) - \beta(X_{t_i})\}^2 ds = O(\Delta_n^3),$$

they have smaller orders than the expectation and variance of Y_i.

Let $P_0 = P_{\alpha_0, \beta_0}$ the probability of the observed process and let $X_i = X_{t_i}$. The sequence $(X_i)_{i \geq 1}$ is a Markov chain with the conditional expectation $E_0(X_{i+1} \mid X_i) = X_i + \Delta_n \alpha_0(X_i)$ and

$$\lim_{n \to \infty} E_0(X_{i+1} \mid X_i) = X_i,$$

it is a sub-martingale if $\alpha_0 \geq 0$ and a super-martingale if $\alpha_0 \leq 0$. For subsets A and A' of $\mathcal{I}_X$, the transition probabilities of the chain are $\pi(A', A; i) = P(X_{i+1} \in A \mid X_i \in A')$ and its invariant distribution function is defined by the limit

$$F_X(A) = \lim_{n \to \infty} n^{-1} \sum_{i=1}^{n} \int_A \pi(\mathcal{I}_X, dy; i).$$

For every real function ψ integrable with respect to F_X, the Markov chain satisfies the ergodic property

$$\begin{aligned} \lim_{n \to \infty} n^{-1} \sum_{i=1}^{n} \psi(X_i) &= \lim_{n \to \infty} n^{-1} \sum_{i=1}^{n} \int_{\mathcal{I}_X} \psi(x)\, dP(X_i = x) \\ &= \int_{\mathcal{I}_X} \psi(x)\, dF_X(x). \end{aligned} \tag{7.12}$$

For every subset A of $\mathcal{I}_X$, let $\pi \otimes F_X$ be the transition distribution such that

$$\pi \otimes F_X(A', A) = \lim_{n \to \infty} n^{-1} \sum_{i=1}^{n} P(X_i \in A', X_{i+1} \in A),$$

every function ψ of $L^1(\mathcal{I}_X, \pi \otimes F_X)$ satisfies

$$\lim_{n\to\infty} n^{-1} \sum_{i=1}^{n} \psi(X_i, X_{i+1}) = \lim_{n\to\infty} n^{-1} \sum_{i=1}^{n} \int_{\mathcal{I}_X^2} \psi(x,y)\, \pi(x, dy; i)\, dP(X_i = x)$$
$$= \int_{\mathcal{I}_X^2} \psi(x,y) \pi \otimes F_X(dx, dy). \tag{7.13}$$

Let $\pi\psi(x) = \int_{\mathcal{I}_X} \psi(x,y)\, \pi(x, dy)$, for every function ψ of $L^2(\mathcal{I}_X, \pi \otimes F_X)$, the chain has the property that

$$n^{-\frac{1}{2}} \sum_{i=1}^{n} \Big\{ \psi(X_i, X_{i+1}) - \pi\psi(X_i) \Big\} \tag{7.14}$$

converges weakly under P_0 to a centered Gaussian variable with variance

$$v^2 = \int_{\mathcal{I}_X} \psi^2(x,y)\, \pi(x, dy)\, dF_X(x) - \int_{\mathcal{I}_X} (\pi\psi)^2(x)\, dF_X(x).$$

The variables Y_i follows a nonparametric heteroscedastic regression model and we define estimators of its conditional expectation and variance based on projections on an orthonormal basis. They are extended to estimators defined with the continuous diffusion process X_t. Like in Chapter 3.1, the coefficients are expressed as the scalar product of the functions of the basis and the diffusion process, their convergence does not require differentiability conditions.

7.3 Estimators of a discrete diffusion

In a parametric auto-regressive model (7.10), let $\alpha(x,\theta)$ and $\beta^2(x,\theta)$ be the drift and the variance of the diffusion depending on a parameter θ of an open and bounded subset Θ of $\mathbb{R}^d$, such that the functions $\alpha(x,\cdot)$ and $\beta^2(x,\cdot)$ belong to $C^2(\Theta)$ and $L^2(\Theta)$, uniformly in $\mathcal{I}_X$. With observations at times t_i such that $t_{i+1} - t_i = \Delta$ for every i, the variables $Y_i = X_{t_{i+1}} - X_{t_i}$ have the approximation

$$Y_i = \Delta\alpha(X_{t_i}, \theta) + \beta(X_{t_i}, \theta)\, \varepsilon_i$$

where the variables $\varepsilon_i = B_{t_{i+1}} - B_{t_i}$ are independent centered Gaussian variables with variance Δ, and ε_i is independent of X_{t_i}. The density of Y_i conditionally on X_{t_i} is

$$\phi_\theta(Y_i) = \frac{1}{\beta(X_{t_i}, \theta)\sqrt{2\pi\Delta}} \exp\Big\{ -\frac{(Y_i - \Delta\alpha(X_{t_i}, \theta))^2}{2\Delta\beta^2(X_{t_i}, \theta)} \Big\}. \tag{7.15}$$

for $i = 1, \ldots, n = \Delta^{-1}T$. The log-likelihood of the sample $(Y_i)_{i=1,\ldots,n}$ with density (7.15) conditionally on X_{t_i} is

$$l_n(\theta) = -\sum_{i=1,\ldots,n} \frac{(Y_i - \Delta\alpha(X_{t_i}, \theta))^2}{2\Delta\beta^2(X_{t_i}, \theta)} - \sum_{i=1,\ldots,n} \log \beta(X_{t_i}, \theta) \tag{7.16}$$

up to constants. From the expression of the conditional expectation of the variables Y_i, the maximum likelihood estimator $\widehat{\theta}_n$ of the parameter θ minimizes the conditional mean variance of the variables Y_i. The process $n^{-1}l_n(\theta)$ converges in probability under P_0 to

$$l(\theta) = -\int_{\mathcal{I}_X} \log \beta(x, \theta)\, dF_X(x)$$

uniformly on Θ, it is minimum at the true parameter value θ_0 and its second order derivative is the Fisher information matrix $I(\theta)$. Under the probability P_0 of the sample $(Y_i, X_{t_i})_{i=1,\ldots,n}$, the maximum likelihood estimator $\widehat{\theta}_n$ is a.s. consistent and by the weak convergence property (7.14) with respect to the invariant distribution function F_X, the variable $n^{\frac{1}{2}}(\widehat{\theta}_n - \theta_0)$ converges to a centered Gaussian variable with variance I_0^{-1} where $I_0 = I(\theta_0)$.

For a discretized nonparametric diffusion (7.11), the ergodic distribution function F_X defined by (7.12) is estimated by

$$\widehat{F}_{nX}(x) = n^{-1} \sum_{i=1}^{n} 1_{\{X_i \le x\}}, \tag{7.17}$$

for every x of $\mathcal{I}_X$, with $X_i = X_{t_i}$. The a.s. convergence of $\widehat{F}_n(x)$ is a consequence of the ergodicity (7.12), and its uniform consistency on $\mathcal{I}_X$ is proved like for the empirical distribution function of a sample of independent and identically distributed variables.

Proposition 7.2. *The empirical process $\nu_{nX}(x) = n^{\frac{1}{2}}(\widehat{F}_{nX} - F_X)(x)$ converges weakly to the Brownian bridge $G_{\pi \otimes F_X}$ with covariance function*

$$v(x, y) = \pi \otimes F_X(x, y) - F_X(x)F_X(y).$$

The expression of the covariance v is deduced from (7.13).

Let $(\varphi_k)_{k \ge 0}$ be the orthonormal basis of functions (3.1) of $L^2(\mathcal{I}_X, F_X)$ defined by the ergodic distribution function F_X, they are estimated on subsets of $\mathcal{I}_X$ where $\widehat{F}_{nX} < 1$ using transformed Laguerre polynomials, let

$$\widehat{\varphi}_{nk}(x) = L_k \circ \widehat{H}_n(x) \tag{7.18}$$

with the estimator $\widehat{H}_n$ of the cumulative hazard function of the variables X_i, given by (3.3), for every integer k. Lemma 3.1 extends to the estimators $\widehat{\psi}_{nk}$.

Lemma 7.4. *For every integer $k \geq 1$, on every sub-interval of $\mathcal{I}_X$ where $\widehat{F}_{nX} < 1$, the estimator $\widehat{\psi}_{nk}$ is a.s. uniformly consistent and the process $n^{\frac{1}{2}}(\widehat{\psi}_{nk} - \psi_k)$ converges weakly to a centered Gaussian process with a bounded covariance function.*

The drift function $\alpha(x)$ is approximated as $\alpha_n(x) = \sum_{k=0}^{K_n} a_k \varphi_k(x)$ with K_n tending to infinity as n tends to infinity and coefficients $a_0 = E_0\alpha(X)$ and

$$a_k = \int_{\mathcal{I}_X} \alpha(x)\varphi_k(x)\, dF_X(x).$$

They are estimated by

$$\widehat{a}_{n,k} = (n\Delta_n)^{-1} \sum_{i=1}^{n} Y_i \widehat{\varphi}_{nik}(X_i),\ k \geq 0,$$

where the observation X_i is omitted in th estimator $\widehat{\varphi}_{nik}$, then

$$\widehat{\alpha}_n(x) = \sum_{k=0}^{K_n} \widehat{a}_{n,k} \widehat{\varphi}_{nk}(x).$$

In model (7.11), by the ergodic property (7.12) and Lemma 7.4, the estimators $\widehat{a}_{n,k}$ converge a.s. under P_0 to the limit of

$$\Delta_n^{-1} E_0\{Y \widehat{\varphi}_{nk}(X)\} = E_0\{\alpha(X)\widehat{\varphi}_{nk}(X)\} = a_k + o(1).$$

The variance of

$$\begin{aligned}
\widehat{a}_{n,k} &= \Delta_n^{-1} \int_{\mathcal{I}_{XY}} y \widehat{\varphi}_{nk}(x)\, d\widehat{F}_{n,XY}(x,y) \\
&= \int_{\mathbb{R}} \int_{\mathcal{I}_X} \{\alpha(x) + \Delta_n^{-1}\beta(x)\varepsilon\} \widehat{\varphi}_{nk}(x)\, d\widehat{F}_{n,X}(x)\, d\widehat{F}_{n,\varepsilon}(\varepsilon)
\end{aligned}$$

is the variance of

$$\begin{aligned}
&\Delta_n^{-1} \int_{\mathbb{R}} \int_{\mathcal{I}_X} \beta(x)\varepsilon \widehat{\varphi}_{nk}(x)\, d\widehat{F}_{n,X}\, d\widehat{F}_{n,\varepsilon}(\varepsilon) \\
&+ \int_{\mathcal{I}_X} \alpha(x)\{\widehat{\varphi}_{nk}(x)\, d\widehat{F}_{n,X}(x) - \varphi_k(x)\, dF_X\},
\end{aligned}$$

the variance of the second term of this sum is a $O(n^{-1})$, the expectation of the first term is zero and its variance is

$$n^{-1} \sum_{i=1}^{n} E\{\beta^2(X_i)\widehat{\varphi}_{nik}^2(X_i)\},$$

it converges to a finite limit, by (7.12). Then the variables $n^{\frac{1}{2}}(\widehat{a}_{nk}-a_k)$ converge weakly to centered Gaussian variables with bounded covariances, for $k=0,\dots,K_n$. By Lemma 7.4, the processes $n^{\frac{1}{2}}(\widehat{a}_{nk}\widehat{\psi}_{nk}-a_k\psi_k)$ converge weakly to centered Gaussian processes with finite and non degenerated covariance functions.

Let

$$\begin{aligned}Z_i &= Y_i - \Delta_n\widehat{\alpha}_n(X_i)\\ &= \Delta_n(\alpha_0-\widehat{\alpha}_n)(X_i)+\beta_0(X_i)\varepsilon_i,\end{aligned}$$

the expectation of Z_i under F_X is $E_0Z_i = o(\Delta_n)$ and its variance is $\Delta_n E_0\{\beta_0^2(X)\}$. The function $\beta_0^2(x)$ has an approximation $\beta_n^2(x) = \sum_{k=1}^{L_n} b_k\varphi_k(x)$ with L_n tending to infinity as n tends to infinity, and with the coefficients $b_0 = E_0\beta_0^2(X)$ and

$$b_k = \int_{\mathcal{I}_X}\beta_0^2(x)\varphi_k(x)\,dF_X(x),$$

for every integer $k\geq 1$. They are estimated by

$$\widehat{b}_{n,k} = (n\Delta_n)^{-1}\sum_{i=1}^n Z_i^2\widehat{\varphi}_{nik}(X_i),\ k\geq 0,$$

and $\widehat{\beta}_n(x) = \sum_{k=0}^{L_n}\widehat{b}_{n,k}\widehat{\varphi}_{nk}(x)$. By the ergodic property (7.12) and Lemma 7.4, the estimators $\widehat{b}_{n,k}$ converge a.s. under P_0 to b_k. Like for the estimation of the function α

$$E(\widehat{b}_{n,k}-b_k)^2 = O(n^{-1})$$

and the variables $n^{\frac{1}{2}}(\widehat{b}_{nk}-b_k)$ converge weakly to centered Gaussian variables with bounded covariances, for $k=0,\dots,L_n$. By Lemma 7.4, the processes $n^{\frac{1}{2}}(\widehat{b}_{nk}\widehat{\psi}_{nk}-b_k\psi_k)$ converge weakly to centered Gaussian processes with finite and non degenerated covariance functions.

The mean integrated squared errors in $L^2(\mathcal{I}_X, F_X)$ of the estimators are the sums of the squared error of estimation and approximation

$$\begin{aligned}\mathrm{MISE}_n(\alpha_0) &= E\|\widehat{\alpha}_n-\alpha_n\|_2^2+\|\alpha_0-\alpha_n\|_2^2,\\ \mathrm{MISE}_n(\beta_0) &= E\|\widehat{\beta}_n-\beta_n\|_2^2+\|\beta_0-\beta_n\|_2^2.\end{aligned}$$

The estimation errors have the same orders as in Lemma 3.3 for the regression models with independent variables.

Let α and β^2 be functions of the space W_{4s} defined by (3.8) for $s>1$, then there exists a constant C such that the squared approximation errors are $\|\alpha_0 - \alpha_n\|_2^2 = O(K_n^{-4s})$ and $\|\beta_0 - \beta_n\|_2^2 = O(L_n^{-4s})$. As n tends to infinity, the size K_n and L_n defining the approximating function α_n and β_n^2 are optimum as the errors have the same order, they are chosen so that under the probability P_0 of the observations

$$\begin{aligned} K_n &= o(n^{\frac{1}{2}}), \quad \|\alpha_0 - \alpha_n\|_2 = O(n^{-\frac{1}{2}} K_n), \\ L_n &= o(n^{\frac{1}{2}}), \quad \|\beta_0 - \beta_n\|_2 = O(n^{-\frac{1}{2}} K_n). \end{aligned} \tag{7.19}$$

Under the conditions (7.19), Propositions 3.3 and 3.4 are still valid, the asymptotic variances of the estimators are given by the convergence of a variable written as (7.14) for the ergodic measure $\pi \otimes F_X$.

Lemma 7.5. *Under the conditions (7.19), the estimators $\widehat{\alpha}_n$ and $\widehat{\beta}_n$ satisfy $E\|\widehat{\alpha}_n - \alpha_0\|_2 = O(n^{-\frac{1}{2}} K_n)$ and $E\|\widehat{\beta}_n - \beta_0\|_2 = O(n^{-\frac{1}{2}} L_n)$, the norms converge to zero under P_0 as K_n and L_n are $o(n^{\frac{1}{2}})$.*

Proposition 7.3. *The optimal convergence rate of the estimators $\widehat{\alpha}_n$ and $\widehat{\beta}_n$ for functions α and β in W_{4s} is $n^{\frac{s}{2s+1}}$ as K_n and $L_n = O(n^{\frac{1}{2(2s+1)}})$.*

The weak convergence of the processes $K_n^{-1} n^{\frac{1}{2}}(\widehat{\alpha}_n - \alpha_n)$ and $L_n^{-1} n^{\frac{1}{2}}(\widehat{\beta}_n - \beta_n)$ to centered Gaussian processes is deduced from Lemma 7.5 and from the weak convergence of the series of variables $n^{\frac{1}{2}}(\widehat{a}_{nk}\widehat{\psi}_{nk} - a_k\psi_k)$ and $n^{\frac{1}{2}}(\widehat{b}_{nk}\widehat{\psi}_{nk} - b_k\psi_k)$.

Proposition 7.4. *Under P_0 and conditions (7.19), the variables $K_n^{-1} n^{\frac{1}{2}}(\widehat{\alpha}_n - \alpha_0)$ and $L_n^{-1} n^{\frac{1}{2}}(\widehat{\beta}_n - \beta_0)$ converge weakly to Gaussian processes.*

The numbers K_n, and respectively L_n, may be estimated by minimization of the integrated squared error $I_{n\alpha}(K_n)$ of the estimators $\widehat{\alpha}_n$, and respectively $\widehat{\beta}_n$ with respect to the invariant distribution function F_X. Let

$$I_{n\alpha}(K_n) = \int_{\mathcal{I}_X} \{\widehat{\alpha}_n^2(x) - 2\alpha(x)\widehat{\alpha}_n(x) + \alpha^2(x)\} dx,$$

its minimization with respect to K_n is equivalent to the minimization of $I_{n\alpha}(K_n) - \int_{\mathcal{I}_X} \alpha^2 \, dF_X$. Like in Section 3.4, the orthogonality of the functions of the basis implies that for $\widehat{\alpha}_n$ defined with their estimators we have

$$I_{n\alpha}(K_n) - \int_{\mathcal{I}_X} \alpha^2 \, dF_X = \sum_{k=0}^{K_n} \widehat{a}_{kn}^2 - 2\sum_{k=0}^{K_n} a_k \widehat{a}_{kn} + O_p(n^{-\frac{1}{2}}).$$

Let $\widehat{\alpha}_{n,j}$ be the estimator of the drift function defined with the sub-sample without the observation (Y_j, X_j), the difference $I_{n\alpha}(K_n) - \int_{\mathcal{I}_X} \alpha^2 \, dF_X$ has the estimator

$$J_{n\alpha}(K_n) = \sum_{k=0}^{K_n} \widehat{a}_{kn}^2 - \frac{2}{\Delta_n^2 n(n-1)} \sum_{k=0}^{K_n} \sum_{j \neq i=1}^{n} Y_i Y_j \widehat{\phi}_{nk,i}(X_i) \widehat{\phi}_{nk,j}(X_j),$$

where $\widehat{\phi}_{nk,i}$ is the estimator of ϕ_k without the observation X_i, and the expectation of J_n is $EJ_{n\alpha}(K_n) = EI_n(K_n) - \int_{\mathcal{I}_X} \alpha^2 \, dF_X + o(1)$. It follows that the minima of I_n and J_n are asymptotically equivalent in probability and the estimator $\widehat{\alpha}_{\widehat{K}_n}$ satisfies

$$\lim_{n\to\infty} \frac{I_{n\alpha}(\widehat{K}_n)}{\inf_K I_{n\alpha}(K)} = 1$$

in probability. The optimal number $\widehat{L}_n$ for the estimation of the variance of the diffusion is determined in the same way.

Weighted series estimators of the functions α and β^2 which do not depend on K_n and L_n are defined like (3.11) for a regression function. They have the same optimal convergence rates as $\widehat{\alpha}_n$ and $\widehat{\beta}_n$.

7.4 Estimation for a continuous diffusion

Let $(X_t)_{t\in[0,T]}$ be a diffusion process defined on a probability space $(\Omega, \mathcal{A}, P_{\alpha,\beta})$ by (7.10) and the initial value $X_0 = 0$, with continuous functions α and β on a metric space $(\mathbb{X}, \|\cdot\|_2)$, and the standard Brownian motion B. The uniform norms of the functions α and β on a bounded subsets $\mathcal{J}$ of I_X are denoted $\|\alpha\|_{\mathcal{J}}$, $\|\beta\|_{\mathcal{J}}$, and their L^2-norms $\|\alpha^2(X)\|_{L^2}$ and $\|\beta^2(X)\|_{L^2}$ are supposed to be finite. Under the probability P_0 of the observed process, the diffusion is defined by functions α_0 and β_0.

Equation (7.10) has a unique solution of X_t, it is a continuous Gaussian process. Under P_0, its expectation is $E_0 X_t = \int_0^t E\alpha_0(X_s)\, ds$ and the variance of the process $Y_t = \int_0^t \beta_0(X_s) dB_s$ is $E_0 \int_0^t \beta_0^2(X_s)\, ds$ and

$$\mathrm{Var}_0 X_t = \mathrm{Var}_0 \int_0^t \alpha_0(X_s)\, ds + E_0 \int_0^t \beta_0^2(X_s)\, ds.$$

The process Y_t is a square integrable centered martingale and (Pons 2017) for every $x > 0$

$$P_0(\sup_{0\le t\le T} Y_t > x) \le \exp\Big\{-\frac{x^2}{2\sup_{0\le t\le T} E_0|\beta_0(X_t)|^2}\Big\}. \qquad (7.20)$$

The diffusion process (7.10) is a Markov process, its transition probabilities $P(X_t = y \mid X_s = x)$, for $s < t$, define a distribution function $\pi_x(dy; s, t)$ on $\mathcal{I}_X \times \mathcal{I}_X$ and there exists an invariant distribution function F_X on $\mathcal{I}_X$ such that for every function ψ of $L^1(\mathcal{I}_X, F_X)$

$$\lim_{T\to\infty} T^{-1} \int_{[0,T]} \psi(X_t)\, dt = \lim_{T\to\infty} T^{-1} \int_{[0,T]} \int_{\mathcal{I}_X} \psi(x)\, P(X_t \in (x, x+dx))\, dt$$
$$= \int_{\mathcal{I}_X} \psi(x)\, dF_X(x). \tag{7.21}$$

On $\mathcal{I}_X^2$, the measure $\pi \otimes F_X$ is defined as

$$\pi \otimes F_X(x, y) = \lim_{T\to\infty} T^{-1} \int_{[0,T]^{\otimes 2}} dP(X_s \le x, X_t \le y)\, ds\, dt,$$

for every function ψ of $L^1(\mathcal{I}_X^2, \pi \otimes F_X)$ the ergodicity property implies

$$\lim_{T\to\infty} T^{-1} \int_{[0,T]^{\otimes 2}} \psi(X_s, X_t)\, ds\, dt$$
$$= \int_{\mathcal{I}_X^{\otimes 2}} \psi(x, y)\, d\pi \otimes F_X(x, y), \tag{7.22}$$

the limit is also denoted $\pi \otimes F_X(\psi)$. The property (7.21) implies that the Laplace transform of the distribution function F_X is

$$L_{F_X}(\lambda) = \lim_{T\to\infty} ET^{-1} \int_{[0,T]} e^{-\lambda X_t}\, dt = \int_{\mathcal{I}_X} e^{-\lambda x}\, dF_X(x),$$

and it has the moments

$$E_{F_X} X^k = \lim_{T\to\infty} ET^{-1} \int_{[0,T]} X_t^k\, dt.$$

For every function ψ of $L^2(\mathcal{I}_X, \pi \otimes F_X)$, the variable

$$T^{\frac{1}{2}} \Big\{ T^{-1} \int_{[0,T]^2} \psi(X_s, X_t)\, ds\, dt - \pi \otimes F_X(\psi) \Big\} \tag{7.23}$$

converges weakly under P_0 to a centered Gaussian variable with variance

$$v^2 = \pi \otimes F_X(\psi^2) - \{\pi \otimes F_X(\psi)(x)\}^2.$$

The ergodic distribution function F_X is estimated by

$$\widehat{F}_{T,X}(x) = T^{-1} \int_0^T 1_{\{X_t \le x\}}\, dt, \tag{7.24}$$

for every x, the a.s. convergence of $\widehat{F}_T(x)$ is a consequence of the ergodicity (7.21), and the uniform consistency is proved like for the empirical distribution function of a sample of independent and identically distributed variables.

Proposition 7.5. *The empirical process $\nu_{T,X} = T^{\frac{1}{2}}(\widehat{F}_{T,X} - F_X)$ converges weakly to the Brownian bridge $G_{\pi\otimes F_X}$ with covariance function*

$$v(x,y) = \pi \otimes F_X(x,y) - F_X(x)F_X(y).$$

The expression of the covariance of the process ν_{TX} is deduced from (7.21).

The functions α and β^2 have been estimated by smoothing the expectation of X and the variance of Y with a symmetric kernel density K under integrability and second order differentiability conditions (Pons 2011), and with a bandwidth $h = h_T$ such that $h_T = O(T^{-\frac{1}{5}})$. The kernel estimators are uniformly a.s. consistent on compact subsets of I_X, their convergence rate is $(Th)^{-\frac{1}{2}}$ and they converge weakly on compact subsets of $\mathcal{I}_X$ to Gaussian processes with finite variance functions depending on the density of the invariant distribution function F_X. We now define series estimators by projection on an estimated basis of functions of $L^2(\mathcal{I}_X, F_X)$.

Let $(\varphi_k)_{k\geq 0}$ be the orthonormal basis of functions (3.1) of $L^2(\mathcal{I}_X, F_X)$ defined by the ergodic distribution function F_X, they are estimated like for nonparametric regression functions as

$$\widehat{\varphi}_{Tk}(x) = L_k \circ \widehat{H}_T(x), \tag{7.25}$$

for every integer k, using the estimator of the cumulative hazard function H_X related to F_X

$$\widehat{H}_{T,X}(x) = \int_0^x 1_{\{\widehat{F}_{T,X}(y)<1\}} \frac{d\widehat{F}_{T,X}(y)}{1 - \widehat{F}_{T,X}(y)}.$$

The process $T^{\frac{1}{2}}(\widehat{H}_T - H_X)$ converges weakly on every interval where $\widehat{F}_{T,X} < 1$ to a centered Gaussian process with covariance function

$$c_H(x,y) = \int_0^{x\wedge y} 1_{\{F_X(y)<1\}} \frac{dF_X(y)}{\{1 - F_X(y)\}^2}.$$

Lemma 3.1 extends to the estimators $\widehat{\psi}_{Tk}$.

Lemma 7.6. *For every integer $k \geq 1$, on every sub-interval of $\mathcal{I}_X$ where $F_X(x) < 1$, the estimator $\widehat{\varphi}_{Tk}$ is a.s. uniformly consistent and the process $W_{T\varphi} = T^{\frac{1}{2}}(\widehat{\varphi}_{Tk} - \varphi_k)$ converges weakly to a centered Gaussian process with a bounded covariance function.*

Proof. By the differentiability of the polynomials L_k we have

$$T^{\frac{1}{2}}\{\widehat{\varphi}_{Tk}(x) - \varphi_k(x)\} = T^{\frac{1}{2}}\{\widehat{H}_T(x) - H_X(x)\}\, L'_k \circ \varphi_k(x)\{1 + o_{as}(1)\}$$

and the result is deduced from the weak convergence of the process $T^{\frac{1}{2}}(\widehat{H}_T - H_X)$ to a centered Gaussian process. □

The drift function α is approximated by $\alpha_T(x) = \sum_{k=0}^{K_T} a_k\varphi_k(x)$ where K_T tends to infinity as T tends to infinity, the coefficients

$$a_k = \int_{\mathcal{I}_X} \alpha_0(x)\varphi_k(x)\, dF_X(x) = \lim_{T\to\infty} T^{-1}\int_0^T \varphi_k(X_t)\, dX_t$$

are estimated by

$$\widehat{a}_{Tk} = T^{-1}\int_0^T \widehat{\varphi}_{Tk}(X_t)\, dX_t$$

and the function α is estimated by

$$\widehat{\alpha}_T = \sum_{k=0}^{K_T} \widehat{a}_{Tk}\widehat{\varphi}_{Tk}(x).$$

Lemma 7.7. *The estimators $\widehat{a}_{Tk}$ converges a.s. to α_k under P_0, for every $k \geq 0$, and the variables $T^{\frac{1}{2}}(\widehat{a}_{Tk} - a_k)$ converge weakly to centered Gaussian variables with strictly positive variances v_{ak}.*

Proof. The estimator converges a.s. to the limit of its expectation which reduces to

$$\begin{aligned}
\lim_{T\to\infty} E\widehat{a}_{Tk}(x) &= \lim_{T\to\infty} T^{-1}E\int_0^T \alpha_0(X_t)\varphi_k(X_t)\, dt \\
&= \int_{\mathcal{I}_X} \alpha_0(x)\varphi_k(x)\, dF_X(x) = a_k,
\end{aligned}$$

since $E\int_0^T \beta(X_t)\, dB_t = 0$. For $k = 0$, the estimator $\widehat{a}_{T,0} = T^{-1}X_T$ converges a.s. to $a_0 = E\alpha_0(X)$. The variable $T^{\frac{1}{2}}(\widehat{a}_{Tk} - a_k)$ is written as

$$\begin{aligned}
T^{\frac{1}{2}}(\widehat{a}_{Tk} - a_k) &= T^{\frac{1}{2}}\Big\{\int_{\mathcal{I}_X} \alpha_0(x)\widehat{\varphi}_{Tk}(x)\, d\widehat{F}_{T,X}(x) - \int_{\mathcal{I}_X} \alpha_0(x)\varphi_k(x)\, dF_X(x)\Big\} \\
&= \int_{\mathcal{I}_X} \alpha_0(x)\widehat{\varphi}_{Tk}(x)\, d\nu_{T,X}(x) \\
&\quad + T^{\frac{1}{2}}\int_{\mathcal{I}_X} \{\widehat{\varphi}_{Tk}(x) - \varphi_k(x)\}\, dF_X(x),
\end{aligned}$$

its weak convergence follows from the weak convergence of the empirical process $\nu_{T,X}$ and from Lemma 7.6. □

Centering X_t with the estimator of its expectation yields the process

$$Z_t = X_t - \int_0^t \widehat{\alpha}_T(X_s)\,ds$$
$$= \int_0^t (\alpha_0 - \widehat{\alpha}_T)(X_s)\,ds + \int_0^t \beta_0(X_s)\,dB_s.$$

As t tends to infinity, the expectation of the process $t^{-1}Z_t$ under P_0 converges to $E\int_{I_X}(\alpha_0-\widehat{\alpha}_T)(x)\,dF_X(x)$ and by the consistency of the estimator $\widehat{\alpha}_T$, Z_t has the asymptotic expansion

$$Z_t = \int_0^t \beta_0(X_s)\,dB_s + o_p(1).$$

The variance under P_0 of $t^{-1}Z_t$ converges to $v_0 = \int_{I_X}\beta_0^2(x)\,dF_X(x)$. The function

$$\beta_0^2(x) = E_0\{\beta_0^2(X_s) \mid X_s = x\}$$

is approximated by $\beta_T^2(x) = \sum_{k=1}^{L_T} b_k\varphi_k(x)$ where L_T tends to infinity as T tends to infinity, and the coefficients are

$$b_k = \int_{\mathcal{I}_X} \beta_0^2(x)\varphi_k(x)\,dF_X(x),$$

and $b_0 = E\beta_T^2(X)$. They are estimated by

$$\widehat{b}_{Tk} = 2T^{-1}\int_0^T Z_t\widehat{\varphi}_{Tk}(X_t)\,dZ_t$$

and the function β is estimated by

$$\widehat{\beta}_{T,X}(x) = \sum_{k=1}^{K_T} \widehat{b}_{Tk}\widehat{\varphi}_{Tk}(x).$$

Lemma 7.8. *The estimator $\widehat{b}_{Tk}$ converge a.s. under P_0 to β_k on every sub-interval of $\mathcal{I}_X$ where $\widehat{F}_{T,X}$ is strictly lower than 1, for $k \geq 1$. Under the condition $\|\beta_0^2(X)\|^4$ finite, the variables $T^{\frac{1}{2}}(\widehat{b}_{Tk} - b_k)$ converge weakly to centered Gaussian variables with strictly positive variances v_{bk}.*

Proof. The expectation of the process Z is a $o(1)$ by Lemma 7.7 and its variance is estimated by Z_t^2 hence

$$2E_0\int_0^T Z_t\varphi_k(X_t)\,dZ_t = E_0\int_{I_X}\beta_0^2(x)\varphi_k(x)\,dF_X(x).$$

The ergodic property (7.21) and the expression of the limit v_0 of the variance of Z imply that $E\widehat{b}_{Tk}$ converges to b_k, the a.s. uniformly convergence of $\widehat{b}_{Tk}$ to its expectation proves the consistency.

The variable $T^{\frac{1}{2}}(\widehat{b}_{Tk} - b_k)$ is asymptotically equivalent to

$$T^{-\frac{1}{2}}\Big\{\int_0^T \widehat{\varphi}_{Tk}(X_t)\,d(Z_t^2) - \int_0^T \beta_0^2(X_t)\varphi_k(X_t)\,dt\Big\}$$
$$= T^{-\frac{1}{2}}\int_0^T \widehat{\varphi}_{Tk}(X_t)\,\{dZ_t^2 - \beta_0^2(X_t)\,dt\}$$
$$+T^{-\frac{1}{2}}\int_0^T \beta_0^2(X_t)\{\widehat{\varphi}_{Tk}(X_t) - \varphi_k(X_t)\}\,dt,$$

the second term of this sum converges weakly to $\int_{\mathcal{I}_X} \beta_0^2(x)W_{T\varphi}(x)\,dF_X(x)$ as T tends to infinity. The variable $T^{-1}\int_0^T \{Z_t^2 - \beta_0^2(X_t)\}\varphi_k(X_t)\,dt$ converges in probability to zero and, by the ergodic weak convergence (7.23), the variable $T^{-\frac{1}{2}}\int_0^T \{Z_t^2 - \beta_0^2(X_t)\}\varphi_k(X_t)\,dt$ converges weakly to a centered Gaussian variable with finite variance, the result follows from the uniform consistency of the estimator of the function φ_k. □

The weak convergence of the process $T^{\frac{1}{2}}(\widehat{\psi}_{Tk} - \psi_k)_{k\geq 0}$ given in Lemma 7.6, and Lemmas 7.7 and 7.8 imply the weak convergence of the series of variables $T^{\frac{1}{2}}(\widehat{a}_{Tk}\widehat{\varphi}_{Tk} - a_k\varphi_k)_{k\geq 0}$ and $T^{\frac{1}{2}}(\widehat{b}_{Tk}\widehat{\varphi}_{Tk} - b_k\varphi_k)_{k\geq 0}$ to centered Gaussian processes, their variances are bounded.

The mean integrated squared errors in $L^2(\mathcal{I}_X, F_X)$ of the estimators $\widehat{\alpha}_T$ and $\widehat{\beta}_T^2$ are the sums

$$\mathrm{MISE}_T(\alpha_0) = E\|\widehat{\alpha}_T - \alpha_T\|_2^2 + \|\alpha_0 - \alpha_T\|_2^2,$$
$$\mathrm{MISE}_T(\beta_0^2) = E\|\widehat{\beta}_T^2 - \beta_T^2\|_2^2 + \|\beta_0^2 - \beta_T^2\|_2^2.$$

The squared estimation errors

$$E\|\widehat{\alpha}_T - \alpha_T\|_2^2 = E\int_{\mathcal{I}_X}\Big\{\sum_{k\leq K_T}(\widehat{\alpha}_{Tk}\widehat{\varphi}_{Tk} - a_k\varphi_k)\Big\}^2 dF_X$$

and $E\|\widehat{\beta}_T - \beta_T\|_2^2$ depend on the variances of the estimators and the squared approximation errors depend on the basis $(\varphi_k)_{k\geq 0}$, their orders are similar to the orders of the estimators for discretized diffusions.

Lemma 7.9. *Let K_T and L_T tend to infinity as T tends to infinity, then $E\|\widehat{\alpha}_T - \alpha_0\|_2 = O(T^{-\frac{1}{2}}K_T)$ and $E\|\widehat{\beta}_T^2 - \beta_0^2\|_2 = O(T^{-\frac{1}{2}}L_T)$, they converge to zero as K_T and L_T are $o(T^{\frac{1}{2}})$.*

Proof. The squared estimation error of the function α_T develops as

$$E\|\widehat{\alpha}_T - \alpha_T\|_2^2 = E \sum_{k \le K_T} \Big\{ (\widehat{a}_{Tk} - a_k)^2 + \int_{\mathcal{I}_X} \widehat{a}_{Tk}^2 (\widehat{\varphi}_{Tk} - \varphi_k)^2 \, dF_X$$
$$+ 2(\widehat{a}_{Tk} - a_k)\widehat{a}_{Tk} \int_{\mathcal{I}_X} (\widehat{\varphi}_{Tk} - \varphi_k)\varphi_k \, dF_X \Big\}$$
$$+ 2E \sum_{k \ne k' \le K_T} (\widehat{a}_{Tk} - a_k)\widehat{a}_{Tk'} \int_{\mathcal{I}_X} \widehat{\varphi}_{Tk'}\varphi_k \, dF_X$$
$$+ E \sum_{k \ne k' \le K_T} \widehat{a}_{Tk}\widehat{a}_{Tk'} \int_{\mathcal{I}_X} (\widehat{\varphi}_{Tk} - \varphi_k)(\widehat{\varphi}_{Tk'} - \varphi_{k'}) \, dF_X.$$

For every $k \le K_T$, $\widehat{a}_{Tk} - a_k = O_p(T^{-\frac{1}{2}})$ and $E(\widehat{a}_{Tk} - a_k)^2$ is a $O(T^{-1})$ by Lemmas 7.7 and 7.8. By the weak convergence of Lemma 7.6,

$$\int_{\mathcal{I}_X} (\widehat{\varphi}_{Tk} - \varphi_k)(\widehat{\varphi}_{Tk'} - \varphi_{k'}) \, dF_X = O_p(T^{-1})$$

and the expectation of its limit is a $O(T^{-1})$ for all integers k and k'. Then $\int_{\mathcal{I}_X} (\widehat{\varphi}_{Tk} - \varphi_k)^2 \, dF_X = O_p(T^{-1})$ and, from the orthogonality of the functions φ_k and $\varphi_{k'}$, for $k' \ne k$

$$\int_{\mathcal{I}_X} \widehat{\varphi}_{Tk'}\varphi_k \, dF_X = \int_{\mathcal{I}_X} (\widehat{\varphi}_{Tk'} - \varphi_{k'})\varphi_k \, dF_X = O_p(T^{-\frac{1}{2}}),$$

it follows that $E\|\widehat{\alpha}_T - \alpha_T\|_2^2 = O(K_T^2 T^{-1})$. The proof is the same for the estimator $\widehat{\beta}_T^2$. □

As T tends to infinity, the sizes K_T and L_T defining the approximating functions α_T and β_Y are optimum as the errors have the same order, we assume the following conditions

$$K_T = o(T^{\frac{1}{2}}), \quad \|\alpha - \alpha_T\|_2 = O(T^{-\frac{1}{2}} K_T) \tag{7.26}$$
$$L_T = o(T^{\frac{1}{2}}), \quad \|\beta^2 - \beta_T^2\|_2 = O(T^{-\frac{1}{2}} L_T). \tag{7.27}$$

Proposition 7.6. *Under the conditions (7.27), the convergence rate of the MISE error for the estimators $\widehat{\alpha}_T$ and $\widehat{\beta}_T^2$ are the optimal convergence rates of kernel estimators for α and β^2 in C^s, $s \ge 2$, for K_T and $L_T = O(T^{\frac{1}{2(2s+1)}})$.*

Proposition 7.7. *Under the conditions (7.27), the processes $K_T^{-1} T^{\frac{1}{2}} (\widehat{\alpha}_T - \alpha)$ and $L_T^{-1} T^{\frac{1}{2}} (\widehat{\beta}_T^2 - \beta^2)$ converge weakly to Gaussian processes with bounded expectation and variance functions.*

The weak convergence is deduced from Lemma 7.9 and from the weak convergence of the series of variables $T^{\frac{1}{2}}(\widehat{a}_{Tk}\widehat{\varphi}_{Tk}-a_k\varphi_k)_{k\geq 0}$ and $T^{\frac{1}{2}}(\widehat{a}_{Tk}\widehat{\varphi}_{Tk}-b_k\varphi_k)_{k\geq 0}$.

The number K_T of functions of the basis to estimate the function α may be estimated by minimization of the integrated squared error of the estimator $\widehat{\alpha}_T$ with respect to the invariant distribution function F_X defined by (7.21). The integrated squared error for the estimator of α is

$$I_{T\alpha}(K_T)=\int_{\mathcal{I}_X}\{\widehat{\alpha}_T(x)-\alpha_0(x)\}^2\,dF_X(x),$$

its minimization is equivalent to the minimization of $I_{T\alpha}(K_T)-\int_{\mathcal{I}_X}\alpha_0^2\,dF_X$. Like in Section 3.4, the orthogonality of the functions of the basis implies that for their estimators we have

$$I_{T\alpha}(K_T)-\int_{\mathcal{I}_X}\alpha_0^2\,dF_X=\sum_{k=0}^{K_T}\widehat{a}_{Tk}^2-2\sum_{k=0}^{K_T}a_k\widehat{a}_{Tk}+O_p(n^{-\frac{1}{2}}).$$

The difference $I_{T\alpha}(K_T)-\int_{\mathcal{I}_X}\alpha_0^2\,dF_X$ has the estimator

$$J_{T\alpha}(K_T)=\sum_{k=0}^{K_T}\widehat{a}_{Tk}^2-\frac{2}{T}\int_0^T\widehat{\alpha}_T(X_t)\,dX_t.$$

The minima of $I_{T\alpha}$ and $J_{T\alpha}$ are asymptotically equivalent in probability, which entails the convergence in probability

$$\lim_{T\to\infty}\frac{I_{T\alpha}(\widehat{K}_T)}{\inf_K I_T(K)}=1.$$

The optimal number $\widehat{L}_T$ for the estimation of the variance function $\beta^2(x)$ of the diffusion is determined in the same way by minimization of

$$\begin{aligned}I_{T\beta}(L_T)-\int_{\mathcal{I}_X}\beta_0^2\,dF_X&=\int_{\mathcal{I}_X}\widehat{\beta}_{Tk}^4\,dF_X-2\int_{\mathcal{I}_X}\beta_0^2\widehat{\beta}^2\,dF_X\\&=\sum_{k=0}^{L_T}\widehat{b}_{kT}^2-2\sum_{k=0}^{L_T}b_k\widehat{b}_{kT}+o_p(1)\end{aligned}$$

or, equivalently by minimization of

$$J_{T\beta}(L_T)=\sum_{k=0}^{L_T}\widehat{b}_{Tk}^2-\frac{2}{T}\int_0^T\widehat{\beta}_T^2(X_t)\,d(Z_t^2).$$

Weighted series estimators of the functions α and β^2 which do not depend on K_T and L_T are defined like (3.11) for a regression function. Their optimal convergence rates are the same as $\widehat{\alpha}_T$ and $\widehat{\beta}_T$.

Let X be a d dimensional vector with components satisfying multivariate diffusion models

$$dX_{jt} = \alpha_j(X_t)dt + \beta_j(X_t)dB_{jt}, \tag{7.28}$$

with functions α_j and β_j depending on the vector X_t, for $j = 1, \ldots, d$ and t in a time interval $[0, T]$. The estimators of the functions α_j and β_j for the discretized diffusion are the same as for multivariate regression models, with orthogonalization of the components of the process X, and they are extended to the continuous diffusion, like above. The estimators are generalized to additive diffusion models and additive models with interactions, like the nonparametric regression functions.

Diffusion processes are generalized as long-range dependence Markov processes such as

$$\begin{aligned} dX_t &= \alpha(Y_t)\,dt + \beta(Y_t)\,dB_t, \\ Y_t &= g(X_t - X_{t-u}), \end{aligned}$$

for t in $[0, T]$, with a known duration u and a known function g so the distribution of Y_t depends on the increment $X_t - X_{t-u}$ of the past history of X_t. By discretization of the intervals $[0, T]$ as I sub-intervals with length δ and $[t_i, t_i - u]$ as J sub-intervals of the same length δ, the variable Y_{t_i} and the variations of the process X_t on $]t_i, t_{i+1}]$ satisfy the equations

$$\begin{aligned} X_{t_{i+1}} - X_{t_i} &= \delta\alpha(Y_{t_i}) + \beta(Y_{t_i})\varepsilon_i, \\ Y_{t_i} &= g\Big(\sum_{t-u \le t_j \le t} \{\delta\alpha(Y_{t_j}) + \beta(Y_{t_j})\varepsilon_j\}\Big), \end{aligned}$$

with $\varepsilon_i = B_{t_{i+1}} - B_{t_i}$ independent of Y_{t_i}. The sequence of variables $(X_{t_i})_i$ is asymptotically a Markov chain as δ tends to zero and the sequence of variables $(Y_{t_i})_i$ is a Markov chain with range J, the properties of the diffusion extend to X_t and the drift and variance functions are estimated by projections on the estimated basis defined from the empirical distribution of the process Y_t.

7.5 Diffusions with stochastic volatility

A model nonparametric of diffusion with a hidden stochastic volatility is defined by the differential equations for an observed real centered Gaussian process X_t and an unobserved real markovian process V_t defined on $\mathbb{R}_+$ as

$$\begin{aligned} dX_t &= \sigma_t\,dB_t, \\ dV_t &= \alpha(V_t)\,dt + \beta(V_t)\,dW_t, \end{aligned} \tag{7.29}$$

with functions α and β of $L^2(\mathbb{R}_+)$, with $V_t = \sigma_t^2 > 0$, $X_0 = 0$ and $V_0 = \eta$ a random variable independent of the two dimensional standard Brownian motion (B_t, W_t). Assuming that B_t and W_t are independent, the process X_t is centered and its variance is

$$\mathrm{Var}_0 X_t = E_0 \int_0^t V_s \, ds.$$

The process V_t is a Markov process and it satisfies the ergodic property (7.21) with an invariant distribution function F_V on $\mathbb{R}_+$, it is estimated like $\beta^2(t)$ in the diffusion model (7.1), by projection on Laguerre's orthonormal basis of functions $(\psi_k)_{k\geq 0}$ of $L^2(\mathbb{R}_+)$, let

$$\widehat{b}_{Tk} = 2 \int_0^T \psi_k(t) \, X_t \, dX_t, \tag{7.30}$$

then

$$\widehat{V}_{Tt} = \sum_{k=0}^{n_T} \widehat{b}_{Tk} \psi_k(t)$$

is an a.s. consistent estimator of V_t, by Lemma 7.2. The ergodic distribution function F_V is estimated by

$$\widehat{F}_{T,\widehat{V}_T}(x) = T^{-1} \int_0^T 1_{\{\widehat{V}_{Tt} \leq x\}} \, dt, \tag{7.31}$$

and an orthonormal basis of functions $(\varphi_k)_{k\geq 0}$ of $L^2(F_V)$, is estimated by the processes $(\widehat{\varphi}_{Tk})_{k\geq 0}$ defined with the estimated ergodic distribution (7.31). The weak convergence of the empirical process

$$\nu_{T,\widehat{V}_T}(x) = T^{\frac{1}{2}} (\widehat{F}_{T,\widehat{V}_T} - F_{\widehat{V}_T})(x)$$

relies on the following lemma.

Let $\nu_{n,P} = n^{\frac{1}{2}}(\widehat{P}_n - P)$ be the empirical process of a n-sample on a probability space $(\Omega, \mathcal{A}, P)$ and let G_P be the Brownian bridge related to P, they are considered as defined on a family $\mathcal{F}$ of functions on $\mathcal{X}$, included in L_P^2, be the relationship $P(f) = \int f \, dP$, the uniform norm of probabilities on $\mathcal{F}$ is denoted $\|P\|_{\mathcal{F}}$.

Lemma 7.10. *Let P_n and P be probabilities on a complete Borelian metric space $(\mathcal{X}, \mathcal{B})$ and let $\mathcal{F}$ be a subset of $\bigcap_n L^2_{P_n} \bigcap L^2_P$ on $\mathcal{X}$, with an envelop F in $\bigcap_n L^2_{P_n} \bigcap L^2_P$ and with a finite entropy dimension, such that* $\lim \|P_n - P\|_{\mathcal{F}} = 0$ *and* $\lim \|P_n - P\|_{\mathcal{F}^2} = 0$. *Then for every n there exist uniformly continuous versions of the Brownian bridges related to P and P_n such that for every $\varepsilon > 0$,* $\lim_n P\left\{\|G_{P_n}^{(n)} - G_P^{(n)}\|_{\mathcal{F}} > \varepsilon\right\} = 0$.

Proposition 7.8. *The empirical process*

$$\nu_{T,\widehat{V}_T} = T^{\frac{1}{2}}(\widehat{F}_{T,\widehat{V}_T} - F_{\widehat{V}_T})$$

converges weakly to the Brownian bridge G_{F_V} related to the ergodic distribution function F_V, and G_{F_V} is the limit of $\nu_{T,V}(x) = T^{\frac{1}{2}}(\widehat{F}_{T,V} - F_V)$.

The equation (7.29) defining the model of the auto-regressive diffusion V_t is similar to (7.10) previously defined for the process X_t, the asymptotic distribution G_{F_V} of $\nu_{T,V}$ related to the distribution function F_V is therefore given by Proposition 7.5. The asymptotic properties of the orthonormal basis estimated using the empirical ergodic distribution $\widehat{F}_{T,\widehat{V}_T}$ defined by (7.31) are then similar to the limit of $(\widehat{\psi}_{Tk})_{k\geq 0}$ given in Lemma 7.6. The estimators of the drift and variance functions of the diffusion equations (7.10) and (7.29) are similar and their asymptotic properties deduced from the empirical processes $\nu_{T,X}$, and respectively $\nu_{T,\widehat{V}_T}$, are the same.

Let X be the process defined by the model (7.1) as

$$dX_t = \mu_t\,dt + \sigma_t\,dB_t$$

or by model (7.10) as

$$dX_t = \alpha(X_t)\,dt + \sigma(X_t)\,dB_t,$$

with a stochastic volatility $\sigma^2 = V$ following the equation (7.1), or respectively (7.29). The estimators of the function V defined in Section 7.1 or 7.4 are used to estimate the orthonormal basis of $L^2(F_V)$ as above, estimators of the drift and variance functions of the diffusion process X_t and their asymptotic properties follow. In the same way, the results obtained with discrete observation, in Section 7.3, for the estimator of the variance of the process X apply to the estimation of the drift and variance functions of the diffusion equation (7.29) for V with discrete observations.

7.6 The Langevin diffusion equation

The auto-regressive diffusion of Section 7.2 is extended with a drift $\alpha(x,t)$ of $L^2(\mathbb{X} \times \mathbb{R}_+)$ and a variance function $\beta(t)$ of $L^2(\mathbb{R}_+)$ as

$$dX_t = \alpha(X_t, t)\,dt + \beta(X_t)\,dB_t, \tag{7.32}$$

on a time interval $[0,T]$, with an initial value x_0. The process X_t is solution of the equation

$$X_t = x_0 + \int_0^t \alpha(X_s, s)\,ds + \int_0^t \beta(X_s)\,dB_s,$$

it is a Markov process with transition functions on $\mathcal{I}_X \times \mathcal{I}_X$

$$P(X_t \le y \mid X_s = x) = \int_{x_0}^{x} \pi_x(dy; s, t),$$

for $s < t$. There exists an invariant distribution function F_X on $\mathcal{I}_X$ satisfying (7.21) and the transition function $\pi \otimes F_X$ satisfies (7.22). The estimator $\widehat{F}_{T,X}$ of F_X on $[0,T]$ defines an empirical process $\nu_{T,X}$ which converges weakly to the Brownian bridge $G_{\pi\otimes F_X}$. Moreover, for every function ψ of $L^1(\mathcal{I}_X \times \mathbb{R}_+, F_X \times \mu)$, with the Lebesgue measure μ, we have

$$\begin{aligned} T^{-1}E\int_{[0,T]} \psi(X_t,t)\,dt &= T^{-1}\int_{[0,T]}\int_{\mathcal{I}_X} \psi(x)\,P(X_t \in (x,x+dx)),\,dt \\ &\to \int_{\mathbb{R}_+}\int_{\mathcal{I}_X} \psi(x,t)\,dF_X(x)\,dt. \end{aligned} \tag{7.33}$$

The function β is estimated by projections on an orthonormal basis of functions $(\phi_k)_{k\ge 0}$ of $L^2(\mathbb{X}, F_X)$ which is estimated by $(\widehat{\phi}_k)_{k\ge 0}$ defined by (7.25), like in Section 7.4. Let $(\varphi_k)_{k\ge 0}$ be the transformed Laguerre polynomials (2.8), the function α is now estimated by projections on the orthonormal basis of functions $(\phi_k \times \varphi_l)_{k,l\ge 0}$ of $L^2(\mathbb{X}\times\mathbb{R}_+, F_X \times \mu)$. The function α has the approximation

$$\alpha_T(x,t) = \sum_{k=0}^{K_T}\sum_{l=0}^{L_T} a_{kl}\phi_k(x)\varphi_l(t)$$

where K_T and L_T tend to infinity as T tends to infinity. For $k, l \ge 0$, we have

$$\begin{aligned} \lim_{T\to\infty} T^{-1}E\int_0^T \phi_k(X_t)\varphi_l(t)\,dX_t &= \lim_{T\to\infty} T^{-1}E\int_0^T \phi_k(X_t)\varphi_l(t)\alpha(X_t,t)\,dt \\ &= \int_{\mathbb{R}_+}\int_{\mathcal{I}_X} \phi_k(x)\varphi_l(t)\alpha(x,t)\,dF_X(x)\,dt, \end{aligned}$$

and the coefficients

$$a_{kl} = \int_{\mathbb{R}_+}\int_{\mathcal{I}_X} \alpha(x,t)\phi_k(x,t)\varphi_l(t)\,dF_X(x)\,dt$$

are consistently estimated as

$$\widehat{a}_{Tkl} = T^{-1}\int_0^T \widehat{\phi}_{Tk}(X_t)\varphi_l(t)\,dX_t.$$

On every interval $[0,a]$ such that $a < X_T$, the function α is uniformly consistently estimated as

$$\widehat{\alpha}_T = \sum_{k=0}^{K_T}\sum_{l=0}^{L_T} \widehat{a}_{Tkl}\widehat{\phi}_{Tk}(x)\varphi_l(t).$$

Lemma 7.11. *For k and $l \ge 0$, the variables $T^{\frac{1}{2}}(\widehat{a}_{Tkl} - a_{kl})$ converge weakly to centered Gaussian variables with strictly positive variances v_{akl}.*

Proof. The variable $T^{\frac{1}{2}}(\widehat{a}_{Tkl} - a_{kl})$ is written as

$$\int_{\mathcal{I}_X}\int_0^T \alpha(x,t)\widehat{\phi}_{Tk}(x)\varphi_l(t)\,d\nu_{TX}(x)\,dt$$
$$+T^{\frac{1}{2}}\int_{\mathcal{I}_X}\int_0^T \alpha(x,t)\{\widehat{\phi}_{Tk}(x) - \phi_k(x)\}\varphi_l(t)\,dF_X(x)\,dt,$$

it converges weakly to a centered Gaussian process with a bounded covariance function, by the weak convergence of the empirical process $\nu_{T,X}$ and Lemma 7.6. □

The mean integrated squared errors in $L^2(\mathcal{I}_X, F_X)$ of the estimator $\widehat{\alpha}_T$ is the sum $\mathrm{MISE}_T(\alpha_0) = E\|\widehat{\alpha}_T - \alpha_T\|_2^2 + \|\alpha_0 - \alpha_T\|_2^2$. The squared approximation error $\|\alpha_0 - \alpha_T\|_2^2$ depends on the basis $(\phi_k)_{k\geq 0}$ and the squared estimation error is

$$E\|\widehat{\alpha}_T - \alpha_T\|_2^2 = \sum_{l=0}^{\varphi_T}\int_{\mathcal{I}_X} E\Big\{\sum_{k\leq K_T}(\widehat{\alpha}_{Tkl}\widehat{\phi}_{Tk} - a_{kl}\phi_k)\Big\}^2\,dF_X.$$

Lemma 7.12. *Let K_T and L_T tend to infinity as T tends to infinity, then $E\|\widehat{\alpha}_T - \alpha_0\|_2 = O(T^{-\frac{1}{2}}K_T L_T^{\frac{1}{2}})$ converges to zero as $K_T^2 L_T = o(T)$.*

Proof. The estimation error $\int_{\mathcal{I}_X} E\Big\{\sum_{k\leq K_T}(\widehat{\alpha}_{Tkl}\widehat{\phi}_{Tk} - a_{kl}\phi_k)\Big\}^2 dF_X$ develops like in Lemma 7.9, it is a $O(K_T^2 T^{-1})$ for every $l \geq 0$. Their sum for $l \leq L_T$ is multiplied by L_T. □

We assume the following conditions

$$K_T L_T^{\frac{1}{2}} = o(T^{\frac{1}{2}}), \quad \|\alpha - \alpha_T\|_2 = O(T^{-\frac{1}{2}} L_T^{\frac{1}{2}} K_T). \tag{7.34}$$

Proposition 7.9. *Under the conditions (7.34), the convergence rate of the MISE error for the estimator $\widehat{\alpha}_T$ is the optimal convergence rates of kernel estimators for α in C^s, $s \geq 2$, for $K_T L_T^{\frac{1}{2}} = O(T^{\frac{1}{2(2s+1)}})$.*

Proposition 7.10. *Under the conditions (7.34), $K_T^{-1} L_T^{-\frac{1}{2}} T^{\frac{1}{2}}(\widehat{\alpha}_T - \alpha)$ converges weakly to a Gaussian process with bounded expectation and variance functions.*

The Langevin diffusion equations for processes X_t and V_t define a model of second order differentiability

$$dX_t = V_t\,dt,$$
$$dV_t = K(X_t, t)\,dt - \alpha_t V_t\,dt + \beta_t\,dB_t, \tag{7.35}$$

with initial values X_0 and V_0, and with a Brownian motion B. In the usual model, the function K is the product $K_t X_t$ with a time function K_t, α and β are constants, the process V is solution of the differential equation

$$dV_t = (X_t - X_0)K(t)\,dt - \alpha V_t\,dt + \beta\,dB_t. \tag{7.36}$$

Assuming that $dV_t = 0$ in (7.35), the process X_t is solution of a diffusion equation of the form (7.32)

$$dX_t = \alpha_t^{-1}\{K(X_t,t)\,dt + \beta_t\,dB_t\}$$

where only the functions $\alpha_t^{-1}K(X_t,t)$ and $\alpha_t^{-1}\beta_t$ are identifiable.

The unobserved process V_t is estimated from the first equation of (7.35) like the drift function of nonparametric diffusion model (7.1), the projection $Y_t = \sum_{k\geq 0} y_k\varphi_k(t)$ of $Y_t = EV_t$ on the transformed Laguerre basis (2.8) is estimated from the observation of the process X on an interval $[0,T]$ as

$$\widehat{V}_T(t) = \sum_{k=0}^{K_T} \widehat{v}_{Tk}\varphi_k(t)$$

with $\widehat{v}_{Tk} = \int_0^T \varphi_k(t)dX_t$, and the estimator $\widehat{V}_T(t)$ is consistent for V_t, by Lemma 7.1. The equation (7.36) with constants can be solved with exponential solutions and the parameters α and K are estimated by maximum likelihood in a Gaussian model. The variance of V_t conditionally on the past of the processes up to t is function $\int_0^t \beta_s^2\,ds$, the function β_t has the same estimator as in model (7.1) where X_t is repaced by $\widehat{V}_T(t)$.

7.7 Random sampling of a diffusion process

Let $(X_t)_{t\in[0,T]}$ be the auto-regressive diffusion process (7.10), we assume it is observed at the random times T_i, $i \geq 1$, of a Poisson process N with distribution $\mathcal{P}(\lambda)$ and independent of the process X. The duration times $S_i = T_{i+1} - T_i$ are independent and they have an exponential distribution with parameter λ^{-1}. Observing the process X at times T_i provides a random discretization of its sample paths and the diffusion equation (7.10) has the approximation

$$\begin{aligned} Y_i &= X_{T_{i+1}} - X_{T_i} \\ &= S_i\,\alpha(X_{T_i}) + \beta(X_{T_i})\,\varepsilon_i, \end{aligned} \tag{7.37}$$

where $\varepsilon_i = B_{T_{i+1}} - B_{T_i}$. By the independence of the variables S_i for a Poisson process, the variables ε_i are independent Gaussian variables with mean zero and variance $ES_i = \lambda$ conditionally on the σ-algebra $\mathcal{F}_{T_i}$ generated by the sample-paths of X and N up to T_i, then $E\{\varepsilon_i \mid X_{T_i}\} = 0$, $E\{\varepsilon_i \mid S_i\} = 0$ and $\text{Var}(\varepsilon_i \mid S_i) = S_i$.

The conditional expectation and variance of Y_i are

$$E(Y_i \mid \mathcal{F}_{T_i}) = \lambda\alpha(X_{T_i}),$$
$$\text{Var}(Y_i \mid \mathcal{F}_{T_i}) = \lambda\beta^2(X_{T_i}),$$

it follows that $EY_i = \lambda E\{\alpha(X_{T_i})\}$ and

$$\text{Var}\, Y_i = \lambda^2 \, \text{Var}\{\alpha(X_{T_i})\} + \lambda E\{\beta^2(X_{T_i})\}.$$

The sequence $(X_{T_i})_{i\geq 1}$ is a Markov chain with the conditional expectation

$$E_0(X_{T_{i+1}} \mid \mathcal{F}_{T_i}) = X_{T_i} + \lambda\alpha(X_{T_i}).$$

There exists an invariant distribution function F_X such that the Markov chain satisfies

$$\lim_{T\to\infty} T^{-1} \sum_{i=1}^{N_T} \psi(X_{T_i}) = \lambda \int_{\mathcal{I}_X} \psi(x)\, dF_X(x), \tag{7.38}$$

for every function ψ of $L^1(\mathcal{I}_X, F_X)$.

The parameter λ is consistently estimated from the observations of the process N on a time interval $[0,T]$ by $\widehat{\lambda}_T = T^{-1}N_T$ and $T^{\frac{1}{2}}(\widehat{\lambda}_T - \lambda)$ converges weakly to a centered Gaussian variables with variance λ, as T tends to infinity.

The drift function $\alpha(x)$ is estimated like in Section 7.3 for a discretely observed diffusion, its expansion by projection on the orthonormal basis of functions (3.1) of $L^2(\mathcal{I}_X, F_X)$ is approximated by

$$\alpha_T(x) = \sum_{k=0}^{K_T} a_k \varphi_k(x)$$

where K_T tends to infinity as T tends to infinity. It is estimated by

$$\widehat{\alpha}_T(x) = \sum_{k=0}^{K_T} \widehat{a}_{Tk} \widehat{\varphi}_{Tk}(x),$$

with the estimated basis (7.18) and

$$\widehat{a}_{Tk} = (N_T\widehat{\lambda}_T)^{-1} \sum_{i=1}^{N_T-1} Y_{(i)} \widehat{\varphi}_{Tki}(X_{(N_T:i)}).$$

As T tends to infinity, the estimators $\widehat{a}_{Tk}$ are asymptotically equivalent to

$$N_T^{-1} \sum_{i=1}^{N_T-1} \alpha(X_{(N_T:i)})\widehat{\varphi}_{Tki}(X_{(N_T:i)})$$

and by (7.38) it converges a.s. to $a_k = \int_{\mathcal{I}_X} \alpha(x)\varphi_k(x)\,dF_X(x)$. The properties of the estimated basis given in Lemma 7.4 and (7.38) imply the weak convergence of the variable $N_T^{\frac{1}{2}}(\widehat{a}_{Tk} - a_k)$ to a centered Gaussian variable, for every integer k. Let

$$Z_i = Y_i - \widehat{\lambda}_T\widehat{\alpha}_T(X_{T_i}) = \lambda\alpha(X_{T_i}) - \widehat{\lambda}_T\widehat{\alpha}_T(X_{T_i}) + \beta(X_{T_i})\,\varepsilon_i,$$

its conditional expectation is

$$E(Z_i \mid \mathcal{F}_{T_i}) = E\{\lambda\alpha(X_{T_i}) - \widehat{\lambda}_T\widehat{\alpha}_T(X_{T_i}) \mid \mathcal{F}_{T_i}) = o_p(1)$$

and its conditional variance is $\lambda\beta^2(X_{T_i}) + o_p(1)$. The function $\beta^2(x)$ has an approximation $\beta_T^2(x) = \sum_{k=1}^{L_T} b_k\varphi_k(x)$ with L_T tending to infinity as T tends to infinity, and with the coefficients $b_k = \int_{\mathcal{I}_X} \beta^2(x)\varphi_k(x)\,dF_X(x)$. They have the estimators

$$\widehat{b}_{Tk} = (N_T\widehat{\lambda}_T)^{-1} \sum_{i=1}^{N_T-1} Z_i^2\widehat{\varphi}_{Tki}(X_{(N_T:i)}),\ k \le L_T,$$

the estimator $\widehat{\beta}_T(x) = \sum_{k=0}^{L_T} \widehat{b}_{Tk}\widehat{\varphi}_{Tk}(x)$ is a.s. consistent and $N_T^{\frac{1}{2}}(\widehat{b}_{Tk} - b_k)$ converges weakly to a centered Gaussian variable, for every integer k.

The mean integrated squared errors in $L^2(\mathcal{I}_X, F_X)$ of the estimators are the sums of the squared error of estimation and approximation

$$\text{MISE}_T(\alpha) = E\|\widehat{\alpha}_T - \alpha_T\|_2^2 + \|\alpha_T - \alpha_0\|_2^2,$$
$$\text{MISE}_T(\beta) = E\|\widehat{\beta}_T - \beta_T\|_2^2 + \|\beta_T - \beta_0\|_2^2.$$

The estimation errors have orders similar to those of the estimators in Section 7.3 and Lemma 7.5 applies under similar conditions

$$K_T = o(T^{\frac{1}{2}}), \quad \|\alpha_T - \alpha_0\|_2 = O(T^{-\frac{1}{2}}K_T),$$
$$L_T = o(T^{\frac{1}{2}}), \quad \|\beta_T - \beta_0\|_2 = O(T^{-\frac{1}{2}}K_T). \tag{7.39}$$

Lemma 7.13. *Under the conditions (7.39), the estimators have the MISE errors* $E\|\widehat{\alpha}_T - \alpha_0\|_2 = O(T^{-\frac{1}{2}}K_T)$ *and* $E\|\widehat{\beta}_T - \beta_0\|_2 = O(T^{-\frac{1}{2}}L_T)$*, they converge to zero as* T *tends to infinity.*

The weak convergence of the processes $K_T^{-1}T^{\frac{1}{2}}(\widehat{\alpha}_T - \alpha_0)$ and $L_T^{-1}T^{\frac{1}{2}}(\widehat{\beta}_T - \beta_0)$ to Gaussian processes is deduced from Lemma 7.13 and from the weak convergence of the series of variables $T^{\frac{1}{2}}(\widehat{a}_{Tk}\widehat{\psi}_{Tk} - a_k\psi_k)$ and $T^{\frac{1}{2}}(\widehat{b}_{Tk}\widehat{\psi}_{Tk} - b_k\psi_k)$, $k \ge 0$.

7.8 Estimation of processes

Let $Y(t)$, $t \geq 0$, be a process of $L^2(\mathcal{I}_Y)$, with expectation $\mu(t)$ and covariance function $v(s,t) = E[\{Y(t) - \mu(t)\}\{Y(s) - \mu(s)\}]$ such that

$$Y(t) = \mu(t) + \varepsilon(t), \tag{7.40}$$

with a centered process ε having independent increments so the covariance of $\varepsilon(t)$ and $\varepsilon(s)$ is $v(s,t) = v(s \wedge t)$, (7.40) generalizes the diffusion (7.1).

Let $\mathcal{F}_t$ be the σ-algebra generated by the sample paths of Y on $[0,t]$, the properties of the process ε imply that Y is a Markov process and $E\{Y(t) - Y(s) \mid \mathcal{F}_s\} = \mu(t) - \mu(s)$. The transition probabilities $P(Y_t = y \mid Y_s = x)$, for $s < t$, define a distribution function $\pi(x, dy; s, t)$ on $\mathcal{I}_Y \times \mathcal{I}_Y$ and there exists an invariant distribution function F_Y on $\mathcal{I}_Y$ such that

$$F_Y(A) = \int_{\mathbb{R}_+^2} \int_A \pi(\mathcal{I}_Y, dy; s, t)\, ds\, dt,$$

the ergodicity properties (7.2). The estimation of a discretized sample path of the process proceeds of the same methods as the discrete diffusions in Section 7.1.

By projection of the mean and variance functions of $L^2(\mathbb{R})$ in an orthonormal basis of functions $(\phi_k)_{k\geq 0}$ of $L^2(\mathbb{R})$ they have expansions

$$\mu(t) = \sum_{k\geq 0} a_k \phi_k(t),$$
$$v(t) = \sum_{k\geq 0} b_k \phi_k(t),$$

with the coefficients $a_k = \int_{\mathbb{R}_+} \mu(t)\phi_k(t)\, dt$ and $b_k = \int_{\mathbb{R}_+} v(t)\phi_k(t)\, dt$. The functions μ and v are estimated from the continuous observation of the processes Y in an increasing interval $[0,T]$ by

$$\widehat{\mu}_T(t) = \sum_{k=0}^{K_T} \widehat{a}_{Tk}\phi_k(t), \quad \widehat{v}_T(s,t) = \sum_{k=0}^{L_T} \widehat{b}_{Tl}\phi_l(t)$$

with the estimated coefficients

$$\widehat{a}_{Tk} = \int_0^T Y(t)\phi_k(t)\, dt,$$
$$\widehat{b}_{Tl} = \int_0^T \{Y(t) - \widehat{\mu}_T(t)\}^2 \phi_l(t)\, dt,$$

for $k \leq K_T, l \leq L_T$ such that K_T and $L_T = o(T)$ and they tend to infinity as T tends to infinity. The estimators $\widehat{\mu}_T$ and $\widehat{v}_T$ are unbiased and a.s.

consistent by ergodicity, the covariance of $\widehat{\mu}_T(s)$ and $\widehat{\mu}_T(t)$ is $v(s,t)$ and the variance of $\widehat{v}_T(s,t)$ is finite if $E(Y^4)$ is finite.

Lemma 7.14. *If $E(Y^4)$ is finite and the basis is uniformly bounded, the variables $\widehat{a}_{Tk} - a_k$ and $\widehat{b}_{Tk} - b_k$ converge weakly to centered Gaussian variables with finite variances.*

Proof. The variance of $\widehat{a}_{Tk}$ converges to b_{kk} and it is finite by assumption. As the variable Y is Gaussian, the limit of $\widehat{a}_{Tk} - a_k$ is a consequence of the consistency. The variance of $\widehat{b}_{Tk}$ is

$$\left(E \int_{[0,T]^2} [\{Y(t) - \widehat{\mu}_T(t)\}\{Y(s) - \widehat{\mu}_T(s)\} - v(s,t)]\phi_k(s)\phi_k(t)\, ds\, dt\right)^2$$

and its limit as T tends to infinity is bounded. □

Proposition 7.11. *Under the conditions (7.9), the processes $\widehat{\mu}_T - \mu_T$ and $\widehat{v}_T - v_T$ converge weakly to centered Gaussian processes.*

Under an independent Poisson random sampling of the process Y, we assume that Y is observed at the jump times T_i, $i = 1, \ldots, N_T$, of a Poisson process N with distribution $\mathcal{P}(\lambda)$ and independent of the process Y. The duration times $S_i = T_{i+1} - T_i$ are independent and have an exponential distribution with parameter λ^{-1} and $P(T_i \le t) = \lambda t$ for $i = 1, \ldots, N_T$. The sequence $(Y_i)_{i\ge1} = (Y_{T_i})_{i\ge1}$ is a Markov chain with the conditional expectation

$$E(Y_{i+1} \mid \mathcal{F}_i) = Y_i + E\{\mu(T_{i+1}) - \mu(T_i)\}$$

and for every function μ of $C_b^1(\mathbb{R}_+)$, there exists a constant C such that

$$|E(Y_{i+1} - Y_i \mid \mathcal{F}_{T_i})| \le \lambda C.$$

By the properties of the Markov chain, there exists an invariant distribution function F_Y satisfying the ergodicity property.

The functions μ and v have expansions by projection on an orthonormal basis $(\phi_k)_{k\ge0}$ of $L^2(\mathbb{R}_+)$, let $\mu(t) = \sum_{k\ge0} a_k\phi_k(t)$ and $v(t) = \sum_{k\ge0} b_k\phi_k(t)$ with the coefficients $a_k = \int_{\mathbb{R}_+} \mu(t)\phi_k(t)\, dt$ and $b_k = \int_{\mathbb{R}_+} v(t)\phi_k(t)\, dt$. The estimators $\widehat{\mu}_T$ and $\widehat{v}_T$ are defined by the estimators of the coefficients

$$\widehat{a}_{Tk} = \widehat{\lambda}_T^{-1} \int_0^T \phi_k(t)Y(t)\, dN(t) = \widehat{\lambda}_T^{-1} \sum_{i=1}^{N_T} Y_i\phi_k(T_i),$$

$$\begin{aligned}\widehat{b}_{Tk} &= \widehat{\lambda}_T^{-1} \int_0^T \{Y(t) - \widehat{\mu}_T(t)\}^2\phi_k(t)\, dN(t)\\ &= \widehat{\lambda}_T^{-1} \sum_{i=1}^{N_T} \{Y_i - \widehat{\mu}_T(T_i)\}^2\phi_k(T_i).\end{aligned}$$

By the independence of the processes N and Y, we have

$$E\int_0^T \phi_k(t)Y(t)\,dN(t) = E\int_0^T \phi_k(t)\mu(t)\,dN(t) = \lambda\int_0^T \phi_k(t)\mu(t)\,dt$$

and it converges to λa_k as T tends to infinity, therefore the estimator $\widehat{\mu}_T(t) = \sum_{k=0}^{K_T} \widehat{a}_{Tk}\phi_k(t)$ is consistent, with $K_T = o(T)$ tending to infinity. The equality $E\{N(t)-\lambda t\}\{N(s)-\lambda s\} = \lambda(s\wedge t)$ implies

$$\begin{aligned}E(\widehat{\lambda}_T\widehat{b}_{Tk}) &= E\int_0^T \{\widehat{\mu}_T(t)-\mu(t)\}^2\phi_k(t)\,dN(t)\\ &\quad+E\int_0^T \{Y(t)-\mu(t)\}^2\phi_k(t)\,dN(t)\\ &= \lambda b_{Tk} + o(1)\end{aligned}$$

and the estimator

$$\widehat{v}_T(t) = \sum_{l=0}^{L_T} \widehat{b}_{Tk}\phi_l(t)$$

of the variance function v is consistent as $L_T = o(T)$ tends to infinity.

Lemma 7.15. *If $E(Y^2)$ is finite and the basis is uniformly bounded, the variables $T^{-\frac{1}{2}}(\widehat{a}_{Tk}-a_k)$ converge weakly to centered Gaussian variables with finite variances.*

Proof. The variance of $\widehat{\lambda}_T\widehat{a}_{Tk}$ is the sum

$$\mathrm{Var}(\widehat{a}_{Tk})\,E(\widehat{\lambda}_T^2) + E(\widehat{a}_{Tk}^2)\,\mathrm{Var}(\widehat{\lambda}_T)$$

where $\mathrm{Var}(\widehat{\lambda}_T) = \lambda T^{-1}$ and $E(\widehat{\lambda}_T^2) = \lambda T^{-1}\{1+o(1)\}$ therefore

$$\mathrm{Var}(\widehat{a}_{Tk}) = \{\lambda^{-1}T\ \mathrm{Var}(\widehat{\lambda}_T\widehat{a}_{Tk}) - a_{Tk}^2\{1+o(1)\}.$$

The expectation of $\widehat{\lambda}_T\widehat{a}_{Tk}$ is $\lambda a_{Tk} = \lambda\int_0^T \phi_k(t)\mu(t)\,dt$ and its variance is the variance of $\int_0^T \phi_k(t)\mu(t)\,\{dN(t)-\lambda\,dt\} + \int_0^T \phi_k(t)\varepsilon(t)\,dN(t)$ which is equal to $\lambda\int_0^T \phi_k^2(t)\{\mu^2(t)+v(t)\}\,dt$. It follows

$$\mathrm{Var}(\widehat{a}_{Tk}) = T\int_0^T \phi_k^2(t)\{\mu^2(t)+v(t)\}\,dt\{1+o(1)\} - a_{Tk}^2,$$

and $T^{-1}\,\mathrm{Var}(\widehat{a}_{Tk})$ converges to $\sigma_{a,k}^2 = \int_{\mathbb{R}_+}\{\mu^2(t)+v(t)\}\phi_k^2(t)\,dt$, its limit is finite by assumption.

As the variable Y is Gaussian, the limit of $T^{-\frac{1}{2}}(\widehat{a}_{Tk}-a_k)$ follows. □

Proposition 7.12. *Under the conditions of Lemma 7.3, the process $K_T^{-\frac{1}{2}}T^{-\frac{1}{2}}(\widehat{\mu}_T-\mu)$ converges weakly to a centered Gaussian process.*

Proof. The expectation of $\widehat{\mu}_T$ is $\mu_T(t) = \sum_{k=0}^{K_T} \phi_k(t) \int_0^T \mu(s)\phi_k(s)\,ds$ and it converges uniformly to $\mu(t)$. The covariance of $\widehat{a}_{Tk}$ and $\widehat{a}_{Tk'}$ is bounded and it is asymptotically equivalent to

$$T \int_{\mathbb{R}_+^2} \{\mu(s)\mu(t) + v(s,t)\}\phi_k(s)\phi_{k'}(t)\,ds\,dt - \widehat{a}_k\widehat{a}_{k'}$$

by the same arguments as for the limit of the variance of $\widehat{a}_{Tk}$. Moreover the mean integrated squared estimation error of $\widehat{\mu}_T$ is

$$E\int_{\mathbb{R}_+} (\widehat{\mu}_T - \mu_T)^2\,dt = \sum_{k=0}^{K_T} \mathrm{Var}(\widehat{a}_{Tk}) = O(TK_T),$$

the weak convergence of the process $T^{-\frac{1}{2}}K_T^{-\frac{1}{2}}(\widehat{\mu}_T - \mu)$ to a centered Gaussian process is then a consequence of Lemma 7.15. □

Lemma 7.16. *If $E(Y^4)$ is finite and the basis is uniformly bounded, the variables $T^{-\frac{1}{2}}(\widehat{b}_{Tk} - b_k)$ converge weakly to centered Gaussian variables with finite variances.*

Proof. The variable $\widehat{\varepsilon}_T(t) = Y(t) - \widehat{\mu}_T(t) = \varepsilon(t) + \mu(t) - \widehat{\mu}_T(t)$ has a finite moment of order 4 if $E(Y^4)$ is finite and with an uniformly bounded basis, the variance of

$$\widehat{\lambda}_T\widehat{b}_{Tk} = \int_0^T \widehat{\varepsilon}_T^2(t)\phi_k(t)\,dN(t)$$

is bounded. It follows that the variance of $\widehat{b}_{Tk}$ is is a $O(T)$. □

Proposition 7.13. *Under the conditions of Lemma 7.3, $E(Y^4)$ finite and with an uniformly bounded basis,the process $T^{-\frac{1}{2}}L_T^{-\frac{1}{2}}(\widehat{v}_T - v_T)$ converges weakly to a centered Gaussian process.*

Proof. The mean integrated squared estimation error of $\widehat{v}_T$ is

$$E\int_0^T \{\widehat{v}_T(t) - v_T(t)\}^2\,ds\,dt = \sum_{k=0}^{L_T} \mathrm{Var}(\widehat{b}_{Tk}) = O(TL_T),$$

the weak convergence of the process $\widehat{v}_T - v$ to a centered Gaussian process is a consequence of Lemma 7.16. □

7.9 Tests for diffusions

Let X_t be a regular parametric diffusion model (7.1) with drift function α and variance β belonging to a space

$$\mathcal{M}_\Theta = \{\alpha_\theta(t), \beta_\theta(t) \in C^2(\Theta), \theta \in \Theta, t \in \mathbb{R}_+\},$$

where Θ is a bounded open subset of $\mathbb{R}^d$. A goodness-of-fit test for the validation of the model $\mathcal{M}_\Theta$ is performed by a comparison of their nonparametric estimators $\widehat{\alpha}_T$ and $\widehat{\beta}_T$ defined in Section (7.1) with the parametric estimators of the drift and variance functions defined by the maximum likelihood estimator $\widehat{\theta}_n$ of the parameter θ in Θ.

With a discretization of the samples of the diffusion process, the model is equivalent to a regression model and the goodness of fit tests for a parametric model may be used. A continuously observed diffusion on an interval $[0, T]$ has the conditional log-likelihood

$$l_T(\theta) = -\frac{\{X_T - \int_0^T \alpha_\theta(s)\,ds\}^2}{2\int_0^T \beta_\theta^2(s)\,ds} - \log \int_0^T \beta_\theta^2(s)\,ds \tag{7.41}$$

up to constants. The process $T^{-1} l_T(\theta)$ converges in probability under the probability P_0 of the observations to the function $l(\theta) = E l_\infty(\theta)$. Let α_0 and β_0 be the drift and variance functions of X_t under P_0. The first derivative of l_T defines the process

$$\begin{aligned}
l_T'(\theta) &= \frac{\{\int_0^T \dot\alpha_\theta(s)\,ds\}\{X_T - \int_0^T \alpha_\theta(s)\,ds\}}{\int_0^T \beta_\theta^2(s)\,ds} - \frac{\int_0^T \dot\beta_\theta(s)\beta_\theta\,ds}{\int_0^T \beta_\theta^2(s)\,ds} \\
&\quad + \frac{\{X_T - \int_0^T \alpha_\theta(s)\,ds\}^2\{\int_0^T \dot\beta_\theta(s)\beta_\theta(s)\,ds\}}{\{\int_0^T \beta_\theta^2(s)\,ds\}^2},
\end{aligned}$$

its expectation under P_0 is zero at θ_0. The expectation of the second derivative of $l_T(\theta)$ under P_0 is

$$E_0 l_T''(\theta_0) = -\frac{\{\int_0^T \dot\alpha_0(s)\,ds\}^2}{\int_0^T \beta_0^2(s)\,ds} - 2\Big\{\frac{\int_0^T \dot\beta_\theta(s)\beta_\theta\,ds}{\int_0^T \beta_\theta^2(s)\,ds}\Big\}^2,$$

and the limit I_0 of $-E_0 l_T''(\theta_0)$ as T tends to infinity is a positive definite matrix. The variable

$$\begin{aligned}
l_T'(\theta_0) &= \frac{\{\int_0^T \dot\alpha_0(s)\,ds\}\{\int_0^T \beta_0(s)\,dB(s)\}}{\int_0^T \beta_0^2(s)\,ds} \\
&\quad + \Big[\Big\{\int_0^T \beta_0(s)\,dB(s)\Big\}^2 - \int_0^T \beta_0^2(s)\,ds\Big] \frac{\int_0^T \dot\beta_0(s)\beta_0(s)\,ds}{\{\int_0^T \beta_0^2(t)\,ds\}^2}
\end{aligned}$$

converges weakly under P_0 to a centered Gaussian variable U_0 with variance V_0, as T tends to infinity. The maximum likelihood estimator of the parameter has the approximation

$$\widehat{\theta}_T - \theta_0 = I_0^{-1} U_0 + o_p(1),$$

it follows that the variable $\widehat{\theta}_T - \theta_0$ converges weakly under P_0 to the centered Gaussian variable $I_0^{-1} U_0$ with variance $I_0^{-1} V_0 I_0^{-1}$, as T tends to infinity. Then the processes $\alpha_{\widehat{\theta}_T} - \alpha_0$ and $\beta_{\widehat{\theta}_T} - \beta_0$ converge weakly under P_0 to centered Gaussian processes.

A Kolmogorov–Smirnov type statistic for the validation of $\mathcal{M}_\Theta$ is

$$S_T = \sup_{t \in [0,T]} |\widehat{\alpha}_T(t) - \alpha_{\widehat{\theta}_T}(t)| + \sup_{t \in [0,T]} |\widehat{\beta}_T(t) - \beta_{\widehat{\theta}_T}(t)|.$$

Let $W_{T\alpha} = \widehat{\alpha}_T - \alpha_{\widehat{\theta}_T}$ and $W_{T\beta} = \widehat{\beta}_T - \beta_{\widehat{\theta}_T}$, under the conditions of Proposition 7.1, they converge weakly under H_0 to centered Gaussian processes W_α and respectively W_β.

Proposition 7.14. *Under H_0 and the conditions of Proposition 7.1, the statistic S_T converges weakly to*

$$S = \sup_{t \geq 0} |W_\alpha(t)| + \sup_{t \geq 0} |W_\beta(t)|.$$

Under fixed alternatives, the variable S_T diverges.

A Cramer-von Mises type statistic for H_0 is a $L^2(\mathbb{R}_+, \mu)$ distance between the estimators

$$\widetilde{S}_T = \int_0^T \{\widehat{\alpha}_T(t) - \alpha_{\widehat{\theta}_T}(t)\}^2 w_T(t)\, dt + \int_0^T \{\widehat{\beta}_T(t) - \beta_{\widehat{\theta}_T}(t)\}^2 w_T(t)\, dt,$$

where the sequence of weighting functions w_T converge uniformly to a function w on $\mathbb{R}_+$.

Proposition 7.15. *Under H_0 and the conditions of Proposition 7.1, the statistic $\widetilde{S}_T$ converges weakly to the variable*

$$\widetilde{S} = \int_{\mathbb{R}_+} W_\alpha^2(t) w(t)\, dt + \int_{\mathbb{R}_+} W_\beta^2(t) w(t)\, dt.$$

Under a fixed alternative, it tends to infinity.

We consider now a parametric auto-regressive diffusion process (7.10) defined by functions $\alpha_\theta(x)$ and $\beta_\theta(x)$ of $C^2(\Theta)$, where Θ is a bounded open subset of $\mathbb{R}^d$. Under the probability P_0 of the observed process, the parameter values are θ_0, α_0 and β_0. A continuously observed diffusion on an interval $[0,T]$ has the conditional log-likelihood

$$l_T(\theta) = -\frac{\{X_T - \int_0^T \alpha_\theta(X_s)\,ds\}^2}{2\int_0^T \beta_\theta^2(X_s)\,ds} - \frac{1}{2}\log\int_0^T \beta_\theta^2(X_s)\,ds \tag{7.42}$$

up to constants. The first derivative of l_T is

$$\begin{aligned} l_T'(\theta) = & \frac{\{\int_0^T \dot\alpha_\theta(X_s)\,ds\}\{X_T - \int_0^T \alpha_\theta(X_s)\,ds\}}{\int_0^T \beta_\theta^2(X_s)\,ds} - \frac{\int_0^T \dot\beta_\theta(X_s)\beta_\theta(X_s)\,ds}{\int_0^T \beta_\theta^2(X_s)\,ds} \\ & + \frac{\{X_T - \int_0^T \alpha_\theta(X_s)\,ds\}^2\{\int_0^T \dot\beta_\theta(X_s)\beta_\theta(X_s)\,ds\}}{\{\int_0^T \beta_\theta^2(X_s)\,ds\}^2}, \end{aligned}$$

and the expectation of $l_T'(\theta_0)$ is zero. The process $T^{-1}l_T(\theta)$ converges in probability under P_0 to the function

$$l(\theta) = -\frac{\{\int_{\mathcal{I}_X}(\alpha_0 - \alpha_\theta)\,dF_X\}^2}{2\int_{\mathcal{I}_X}\beta_\theta^2\,dF_X},$$

its first derivative is

$$\begin{aligned} l'(\theta) = & \frac{\{\int_{\mathcal{I}_X}(\alpha_0 - \alpha_\theta)\,dF_X\}\int_{\mathcal{I}_X}\dot\alpha_\theta\,dF_X}{\int_{\mathcal{I}_X}\beta_\theta^2\,dF_X} \\ & + \frac{\{\int_{\mathcal{I}_X}(\alpha_0 - \alpha_\theta)\,dF_X\}^2\{\int_{\mathcal{I}_X}\beta_\theta\dot\beta_\theta\,dF_X\}}{\{\int_{\mathcal{I}_X}\beta_\theta^2\,dF_X\}^2}, \end{aligned}$$

its value at θ_0 is zero and its second derivative at θ_0 is

$$l''(\theta_0) = -\frac{\{\int_{\mathcal{I}_X}\dot\alpha_0\,dF_X\}^2}{\int_{\mathcal{I}_X}\beta_0^2\,dF_X}.$$

As the matrix $I_0 = -l''(\theta_0)$ is positive definite, the function l is minimum at θ_0. The variable $U_T = T^{-\frac{1}{2}}l_T'(\theta_0)$ converges weakly under P_0 to a centered Gaussian variable U_0 with variance V_0, as T tends to infinity. The maximum likelihood estimator of the parameter has the approximation

$$T^{\frac{1}{2}}(\widehat\theta_T - \theta_0) = I_0^{-1}U_0 + o_p(1),$$

and it converges weakly under P_0 to the centered Gaussian variable $I_0^{-1}U_0$ with variance $I_0^{-1}V_0I_0^{-1}$. Then the processes $T^{\frac{1}{2}}(\alpha_{\widehat\theta_T} - \alpha_0)$ and $T^{\frac{1}{2}}(\beta_{\widehat\theta_T} - \beta_0)$ converge weakly under P_0 to centered Gaussian processes.

As the convergence rates of the nonparametric estimators $\widehat\alpha_T$ and $\widehat\beta_T$ depend on the number of terms of the basis they use, we have

$$\begin{aligned} W_{T\alpha} &= K_T^{-1}T^{\frac{1}{2}}(\widehat\alpha_T - \alpha_{\widehat\theta_T}) = K_T^{-1}T^{\frac{1}{2}}(\widehat\alpha_T - \alpha_0) + o_p(1), \\ W_{T\beta} &= L_T^{-1}T^{\frac{1}{2}}(\widehat\beta_T - \beta_{\widehat\theta_T}) = L_T^{-1}T^{\frac{1}{2}}(\widehat\beta_T - \beta_0) + o_p(1), \end{aligned}$$

under the conditions (7.27), they converge weakly to Gaussian processes W_α, and respectively W_β.

A Kolmogorov–Smirnov type statistic for the validation of a parametric family $\mathcal{M}_\Theta = \{(\alpha_\theta, \beta_\theta) \in C^1(\mathcal{I}_X) \times C^1(\mathcal{I}_X), \theta \in \Theta\}$ is

$$S_T = \sup_{x \in \mathcal{I}_X} |W_{T\alpha}(x)| + \sup_{x \in \mathcal{I}_X} |W_{T\beta}(x)|.$$

Proposition 7.16. *Under H_0 and the conditions (7.27), the statistic S_T converges to*

$$S = \sup_{x \in \mathcal{I}_X} |W_\alpha(x)| + \sup_{x \in \mathcal{I}_X} |W_\beta(x)|.$$

Under fixed alternatives, the variable S_T diverges.

A Cramer-von Mises type statistic for H_0 has the form

$$\begin{aligned} \widetilde{S}_T &= \int_0^T \{\widehat{\alpha}_T(X_t) - \alpha_{\widehat{\theta}_T}(X_t)\}^2 w_T(X_t)\, dt \\ &\quad + \int_0^T \{\widehat{\beta}_T(X_t) - \beta_{\widehat{\theta}_T}(X_t)\}^2 w_T(X_t)\, dt, \end{aligned}$$

where the sequence of weighting functions w_T converge uniformly to a function w on $\mathcal{I}_X$.

Proposition 7.17. *Under H_0 and the conditions (7.27), the statistic $\widetilde{S}_T$ converges weakly to the variable*

$$\widetilde{S} = \int_{\mathcal{I}_X} W_\alpha^2(x) w(x)\, dF_X(x) + \int_{\mathcal{I}_X} W_\beta^2(x) w(x)\, dF_X(x).$$

Under a fixed alternative, it tends to infinity.

Local alternatives A_n are defined by a sequence of parameters

$$\begin{aligned} \theta_T &= \theta_0 + T^{-\frac{1}{2}} h_T, \\ \alpha_T &= \alpha_0 + K_T T^{-\frac{1}{2}} a_T, \\ \beta_T &= \beta_0 + L_T T^{-\frac{1}{2}} b_T, \end{aligned}$$

where a_T and b_T converge uniformly to functions a and respectively b, h_T converges to a non zero limit h. Under the alternatives A_n, θ_T converges to θ_0 and the parametric estimators of the drift and the variance have the approximations

$$\begin{aligned} T^{\frac{1}{2}}(\alpha_{\widehat{\theta}_T} - \alpha_{\theta_T}) &= T^{\frac{1}{2}}(\alpha_{\widehat{\theta}_T} - \alpha_0) + \alpha_0' h_T + o(1), \\ T^{\frac{1}{2}}(\beta_{\widehat{\theta}_T} - \beta_{\theta_T}) &= T^{\frac{1}{2}}(\beta_{\widehat{\theta}_T} - \beta_0) + \beta_0' h_T + o(1), \end{aligned}$$

the processes $W_{T\alpha}$ and $W_{T\beta}$ have the approximations

$$W_{T\alpha} = K_T^{-1}T^{\frac{1}{2}}(\widehat{\alpha}_T - \alpha_{\widehat{\theta}_T}) = K_T^{-1}T^{\frac{1}{2}}(\widehat{\alpha}_T - \alpha_0) + a_T + o_p(1),$$
$$W_{T\beta} = L_T^{-1}T^{\frac{1}{2}}(\widehat{\beta}_T - \beta_{\widehat{\theta}_T}) = L_T^{-1}T^{\frac{1}{2}}(\widehat{\beta}_T - \beta_0) + b_T + o_p(1)$$

and they converges weakly to $W_\alpha + a$, and respectively $W_\beta + b$.

Proposition 7.18. *Under the local alternatives A_n and the conditions (7.27), the statistic $\widetilde{S}_T$ converges weakly to the variable*

$$\widetilde{S} = \int_{\mathcal{I}_X} (W_\alpha + a)^2(x)w(x)\,dF_X(x) + \int_{\mathcal{I}_X} (W_\beta + b)^2(x)w(x)\,dF_X(x).$$

Tests of comparison of time dependent drift or variance functions of $d \geq 2$ diffusion processes defined by (7.1), or auto-regressive diffusion processes (7.10), are performed with statistics similar to S_T or $\widetilde{S}_T$. The multivariate diffusion model (7.28) develops as an additive diffusion model

$$\alpha(X_t) = a + \sum_{j=1}^{d} \alpha_j(X_{jt})$$

and models with interactions, like for nonparametric regression functions. Let

$$\alpha_j(X_{jt}) = E\{\alpha(X_t) \mid X_{jt}\} - E\alpha(X_t),$$

and let $\beta_j(X_{jt})$ be defined conditionally on X_{jt}, in the same way for $j = 1, \ldots, d$, the estimators of the functions α_j and β_j are expressed as expectations of the multivariate estimators α_T and $\widehat{\beta}_T$ conditionally on the process X_j. Tests of sub-models in such additive models with functions α_j and β_j depending on the vector X_t, for $j = 1, \ldots, d$ and t in a time interval $[0, T]$ are performed with Kolmogorov–Smirnov and Cramer-von Mises type statistic, as above.

The model (7.32) extends the time dependent and the auto-regressive drift function of the previous sections. By Proposition 7.10, under the hypothesis H_0 of an auto-regressive drift, the approximation

$$K_T^{-1}L_T^{-\frac{1}{2}}T^{\frac{1}{2}}\{\widehat{\alpha}_T(x,t) - \widehat{\alpha}_T(x)\} = K_T^{-1}L_T^{-\frac{1}{2}}T^{\frac{1}{2}}\{\widehat{\alpha}_T(x,t) - \alpha_0(x)\} + o_p(1)$$

converges weakly to a Gaussian process. A test of the hypothesis H_0 is still performed with statistics similar to S_T or $\widetilde{S}_T$.

Chapter 8

Functional wavelet estimators

8.1 Introduction

The multiresolution analysis with wavelets is defined by two orthonormal bases derived from functions ϕ and ψ of $L^2(\mathbb{R})$ by scaling and shift

$$\phi_{j_0 k} = 2^{\frac{j_0}{2}} \phi(2^{j_0} x - k), \; k \in \mathbb{Z}, \tag{8.1}$$

$$\psi_{jk} = 2^{\frac{j}{2}} \psi(2^j x - k), \; k \in \mathbb{Z}, j \geq j_0. \tag{8.2}$$

The function ψ is a wavelet of class m (Meyer 1990) if

Condition 8.1.
1. For $m = 0$, the function ψ belongs to $L^\infty(\mathbb{R})$, for $m \geq 1$, the derivatives $\psi^{(k)}$ belong to $L^\infty(\mathbb{R})$ for every $k \leq m$,
2. the function ψ and its derivatives $\psi^{(k)}$, $k \leq m$, decrease rapidly as x tends to $\pm\infty$,
3. $\int_{\mathbb{R}} x^k \psi(x)\, dx = 0$ for $k = 0, \ldots, m$,
4. the functions $(\psi_{jk})_{j,k \in \mathbb{Z}}$ is an orthonormal basis of $L^2(\mathbb{R})$.

The function ψ_{jk} are concentrated on the interval $I_{jk} = [k2^{-j}, (k+1)2^{-j})$, by the condition 2. Meyer (1990) proved the existence of a function ψ such that the derivatives of ψ_{jk} have the bound

$$|\psi_{jk}^{(k)}| \leq C_N 2^{\frac{j}{2}} 2^{jk} (1 + |2^j x - k|)^{-N},$$

for all $k \leq m$ and $N \geq 1$, and $(\psi(x-k))_{k \in \mathbb{Z}}$ is an orthonormal basis of W_0. It follows that $(\psi_{jk})_{k \in \mathbb{Z}}$ is an orthonormal basis of W_j, for every j. The function ϕ is a density of $C^m(\mathbb{R})$ with $\int_{\mathbb{R}} x^k \psi(x)\, dx = 0$ for $k = 1, \ldots, m$, such that ϕ and its derivatives $\phi^{(k)}$ have a fast decay

$$|\phi^{(k)}(x)| \leq C_m (1 + |x|)^{-m}$$

for $k = 1, \ldots, m$. The function ψ_{jk} is concentrated on the interval I_{jk}. The differentiability conditions define m-regular wavelets, the use of indicator functions enables to estimate functions with discontinuities.

The spaces V_j generated by the set of functions $\{\phi_{jk}, k \in \mathbb{Z}\}$ are increasing such that $L^2(\mathbb{R}) = \cup_{j\in\mathbb{Z}} V_j$ and $\cap_{j\in\mathbb{Z}} V_j = \emptyset$. Let W_j be the spaces generated by the set of functions $\{\psi_{jk}, k \in \mathbb{Z}\}$, the set $(W_j)_{j\geq 0}$ is an orthonormal basis of functions of $L^2(\mathbb{R})$ and

$$L^2(\mathbb{R}) = V_{j_0} \oplus W_{j_0} \oplus W_{j_0+1} \oplus \cdots$$

The two-dimensional wavelets are defined by product as

$$\psi_{jkl} = 2^j \phi(2^j x - k, 2^j y - l), \; j, k, l \in \mathbb{Z},$$

they form a Hilbert basis of W_j, for all (k, l) in $\mathbb{Z}^2$ and the space $L^2(\mathbb{R}^2)$ is the orthogonal sum of the spaces generated by the two-dimensional bases of functions $(\phi_{j_0kl})_{kl}$ and $(\psi_{jkl})_{kl}$, for $j \geq j_0$.

Every function f of $L^2(\mathbb{R})$ has an expansion on the wavelet basis as

$$f(x) = \sum_{k\in\mathbb{Z}} \alpha_{j_0k}\phi_{j_0k}(x) + \sum_{j\geq j_0}\sum_{k\in\mathbb{Z}} \beta_{jk}\psi_{jk}(x) \tag{8.3}$$

with the coefficients

$$\alpha_{j_0k} = \int f(x)\phi_{j_0k}(x)\,dx,$$

$$\beta_{jk} = \int f(x)\phi_{jk}(x)\,dx,$$

the squared norm $\|f\|_{L^2}^2$ is the sum of $\|\alpha_{j_0}\|_{l^2} = \sum_{k\in\mathbb{Z}} \alpha_{j_0k}^2$ and $\sum_{j\geq j_0}\sum_{k\in\mathbb{Z}} \beta_{jk}^2$ and they are finite. Under the condition (2.22) for the bases $(\phi_{j_0k})_{k\geq 1}$ and $(\psi_{jk})_{j\geq j_0, k\geq 1}$, the norm $\|f\|_\infty$ is bounded (Section 2.6). For a function f of a space B_{pq}^σ, $\frac{1}{2} < \sigma < m$, the norm

$$J_{pq}^\sigma(f) = \|E_0 f\|_{L^p(\mathbb{R})} + \Big\{\sum_{j\geq 0} (2^{j\sigma}\|P_{W_j} f\|_{L^p(\mathbb{R})})^q\Big\}^{\frac{1}{q}}$$

is finite, where $P_{W_j} f = \sum_{k\in\mathbb{Z}} \beta_{jk}\psi_{jk}$ is the projection of f on W_j and $E_0 f = \sum_{k\in\mathbb{Z}} \alpha_{0k}\phi_{0k}$. Let

$$I_{pq}^\sigma(f) = \|\alpha_{j_0\cdot}\|_{l^p} + \Big\{\sum_{j\geq 0} (2^{j\sigma}\|\beta_{j\cdot}\|_{l^p})^q\Big\}^{\frac{1}{q}},$$

for functions ϕ and ψ in $L^p(\mathbb{R})$, the conditions $J_{pq}^\sigma(f)$ finite and $I_{pq}^\sigma(f)$ finite are equivalent.

The projections $P_j f = \sum_{k \in \mathbb{Z}} \alpha_{jk}\phi_{jk}$ of a function f of B^σ_{pq} on the spaces V_j, and its projections $\langle f, \psi_{jk}\rangle$ on the spaces W_{jk} satisfy

$$\|f - P_j f\|_{L^2(\mathbb{R})} = O(2^{-j\sigma\wedge m}),$$
$$|\langle f, \psi_{jk}\rangle| = O(2^{-j\{(\sigma\wedge m)+\frac{1}{2}\}}),$$

for $j \geq j_0$ in $\mathbb{Z}$ and $k \leq 2^j - 1$. The properties of estimators by projection on wavelets for densities in the class $\mathcal{D}_{pq\sigma} = \{f \in L^2(\mathbb{R}) : I^\sigma_{pq} \leq M\}$ have been widely studied, as the sizes j_n, k_n and k_{jn} are chosen by thresholding (Antoniadis 1994, 1997, Kerkyacharian and Picard 1993, Donoho *et al.* 1996).

Here the norm $J^\sigma_{22}(f)$ is modified by weighting the main wavelets of the expansion (8.3) like in $W_{2\sigma}$, we define the quadratic norm

$$N^\sigma_2(f) = \Big\{\sum_{k\in\mathbb{Z}} k^{2\sigma} a^2_{j_0k}\Big\}^{\frac{1}{2}} + \Big\{\sum_{j\geq 0}(2^{j\sigma}\|\beta_{j\cdot}\|_{l^2})^2\Big\}^{\frac{1}{2}} \tag{8.4}$$

and we study the wavelet estimator of functions in the space

$$\mathcal{F}^\sigma_2 = \{f \in L^2(\mathbb{R}) : N^\sigma_2(f) < \infty\}. \tag{8.5}$$

The norm N^σ_2 is extended to norms

$$N^\sigma_p(f) = \Big\{\sum_{k\in\mathbb{Z}} k^{2\sigma} a^p_{j_0k}\Big\}^{\frac{1}{p}} + \Big\{\sum_{j\geq 0}(2^{j\sigma}\|\beta_{j\cdot}\|_{l^p})^p\Big\}^{\frac{1}{p}}$$

for bases defined by functions ϕ and ψ of $L^p(\mathbb{R})$, their norms are omitted in the definition of $N^\sigma_{pq}(f)$ and this does not modify the order of the $L^p(\mathbb{R})$ norm of f, $p > 1$. The properties of the $L^2(\mathbb{R})$-error of estimators extend to the $L^p(\mathbb{R})$-errors when the pth order moments of the estimated coefficients are determined.

8.2 Wavelet density estimator

Let $(X_i)_{i=1,\dots,n}$ be a n sample with a density f of $L^2(\mathbb{R})$ having an expansion (8.3) on the wavelet bases, f has an approximation

$$f_n(x) = \sum_{|k|\leq k_n} \alpha_{j_0k}\phi_{j_0k}(x) + \sum_{j=j_0}^{j_n}\sum_{|k|\leq k_{jn}} \beta_{jk}\psi_{jk}(x)$$

with j_n, k_n and k_{jn} tending to infinity as n tends to infinity. The coefficients α_{j_0k} and β_{jk} are estimated from the sample $(X_i)_{i=1,\dots,n}$ by projection on the orthonormal basis as

$$\widehat{\alpha}_{n,j_0k} = n^{-1}\sum_{i=1}^{n}\phi_{j_0k}(X_i),$$

$$\widehat{\beta}_{n,jk} = n^{-1}\sum_{i=1}^{n}\psi_{jk}(X_i)$$

and the density f has the estimator

$$\widehat{f}_n(x) = \sum_{|k|\le k_n}\widehat{\alpha}_{n,j_0k}\phi_{j_0k}(x) + \sum_{j=j_0}^{j_n}\sum_{|k|\le k_{jn}}\widehat{\beta}_{n,jk}\psi_{jk}(x). \tag{8.6}$$

Due to the orthogonality of the spaces V_{j_0} and W_{j_0+k} the variables $n^{\frac{1}{2}}(\widehat{\alpha}_{n,j_0k} - \alpha_{j_0k})$ and $n^{\frac{1}{2}}(\widehat{\beta}_{n,jk} - \beta_{jk})$ converge weakly to independent centered Gaussian variables with variances $\sigma^2_{\alpha,j_0k} = \int_{\mathbb{R}}\phi^2_{j_0k}(x)\,d[F(x)\{1 - F(x)\}]$, and respectively $\sigma^2_{\beta,jk} = \int_{\mathbb{R}}\psi^2_{jk}(x)\,d[F(x)\{1 - F(x)\}]$, for all $|k| \le k_n$ and $j_0 \le j \le j_n$. The variance of $\widehat{f}_n$ is their sum

$$V_n = \sum_{|k|\le k_n}\sigma^2_{\alpha,j_0k} + \sum_{j=j_0}^{j_n}\sum_{|k|\le k_{jn}}\sigma^2_{\beta,jk}.$$

The mean integrated squared error of the estimator $\widehat{f}_n$ is

$$\begin{aligned}\mathrm{MISE}_n(f) = \|f - f_n\|^2_{L^2} &+ \sum_{|k|\le k_n}E\{(\widehat{\alpha}_{n,j_0k} - \alpha_{j_0k})^2\}\|\phi_{j_0k}\|^2_{L^2}\\ &+\sum_{j=j_0}^{j_n}\sum_{|k|\le k_{jn}}E\{(\widehat{\beta}_{n,jk} - \beta_{jk})^2\}\|\psi_{jk}\|^2_{L^2},\end{aligned}$$

the order of $\mathrm{MISE}_n(f)$ is therefore

$$\|f - f_n\|^2_{L^2} + O(n^{-1}(k_n + K_n))$$

with $K_n = \sum_{j=j_0}^{j_n}k_{jn}$ and the approximation error $\|f - f_n\|^2_{L^2}$ converges to zero as n tends to infinity. With $k_n + K_n = o(n)$, the estimation error converges to zero as n tends to infinity. The errors have the same order if

$$\|f - f_n\|^2_{L^2} = O(n^{-1}(k_n + K_n)). \tag{8.7}$$

For a density of the space $\mathcal{F}^\sigma_2$, $\frac{1}{2} < \sigma < m$, the approximation error has the norm

$$\begin{aligned}\|f - f_n\|^2_{L^2} &= \sum_{|k|\ge k_n}\alpha^2_{j_0k} + \sum_{j>j_n}\sum_{k\in\mathbb{Z}}\beta^2_{jk} + \sum_{j=j_0}^{j_n}\sum_{|k|\ge k_{jn}}\beta^2_{jk}\\ &= \sum_{|k|\ge k_n}\alpha^2_{j_0k} + \Big(\sum_{j>j_n}\sum_{k\in\mathbb{Z}}\beta^2_{jk}\Big)\{1 + o(1)\}\\ &= O(k_n^{-2\sigma}) + O(2^{-j_n\sigma}),\end{aligned}$$

Proposition 8.1. *Under the condition (8.5), the error* $E\|\widehat{f}_n - f\|_2$ *for the wavelet estimator (8.6) of a density* f *in* $\mathcal{F}_2^\sigma$ *defined by (8.5 is*

$$\mathrm{MISE}_n(f) = O(n^{-\frac{2\sigma}{2\sigma+1}}).$$

Proof. Assuming that $k_{jn} = O(2^{\frac{j}{2}})$ for every n, we obtain $K_n = \sum_{j=j_0}^{j_n} k_{jn} = O(2^{\frac{j_n}{2}})$. Under the condition (8.7) and with the order of the squared estimation error, it follows that for the optimal j_n and k_n we have

$$2^{-j_n\sigma} = O(n^{-1}2^{\frac{j_n}{2}}),$$
$$k_n^{-2\sigma} = O(n^{-1}k_n),$$

therefore $2^{\frac{j_n}{2}}$ and k_n have the order $n^{\frac{1}{2\sigma+1}}$, this implies

$$\mathrm{MISE}_n(f) = O(n^{-\frac{2\sigma}{2\sigma+1}})$$

and that is the minimax L^2-risk for a wavelet estimator of f in $\mathcal{F}_2^\sigma$. □

Proposition 8.2. *Under the condition (8.7), the process* $n^{\frac{\sigma}{2\sigma+1}}(\widehat{f}_n - f)$ *converges weakly to a centered Gaussian process with finite covariance function.*

The optimal numbers of the functions of the bases used for the estimation of a density f in W_σ can be chosen by cross-validation, under the conditions for their optimal orders. The minimization of the integrated squared error $I_n(j_n, k_n, k_{jn})$ is asymptotically equivalent to the minimization of $I_n - \int f^2$ which is estimated by

$$CV_n = n^{-1}\sum_{i=1}^n \int \widehat{f}_{n,i}^2(x)\,dx - 2n^{-1}\sum_{i=1}^n \widehat{f}_{n,i}(X_i),$$

where $\widehat{f}_{n,i}$ be the wavelet estimator of the density defined with the sub-sample without the observation X_i, for $i = 1, \ldots, n$, and CV_n is asymptotically equivalent to

$$\begin{aligned}
J_n = &\sum_{|k|\le k_n} \widehat{\alpha}_{n,j_0k}^2 + \sum_{j=j_0}^{j_n}\sum_{|k|\le k_{jn}} \widehat{\beta}_{n,jk}^2 \\
&- \frac{2}{n(n-1)} \sum_{|k|\le k_n} \widehat{\alpha}_{n,j_0k}^2 \sum_{i'\ne i=1}^n \phi_{j_0k}(X_i)\phi_{j_0k}(X_{i'}) \\
&- \frac{2}{n(n-1)} \sum_{j=j_0}^{j_n}\sum_{|k|\le k_{jn}} \widehat{\beta}_{n,jk}^2 \sum_{i'\ne i=1}^n \psi_{jk}(X_i)\psi_{jk}(X_{i'}).
\end{aligned}$$

The expectation of J_n is $EJ_n = EI_n - \int f^2(x)\,dx$ and the minimization of I_n and J_n are asymptotically equivalent in probability. Estimators $\widehat{k}_n$ of k_n, $\widehat{j}_n$ of j_n and $\widehat{k}_{jn}$ of k_{jn} which minimize J_n are asymptotically optimal i.e.

$$\frac{I_n(\widehat{k}_n, \widehat{j}_n, \widehat{k}_{jn})}{\inf_{k_n, j_n, k_{jn}} I_n(k_n, j_n, k_{jn})} \to 1$$

in probability as n tends to infinity.

A weighted wavelet density estimator is defined similarly to the estimator (2.23) for a density of the space $\mathcal{F}_2^\sigma$ defined by (8.5, with the weighted norm (8.4). For every n let

$$k_{jn} = O(2^{\frac{j}{2}}), \quad 2^{\frac{j_n}{2}} = n, \tag{8.8}$$

hence $K_n = O(n)$. In $\mathcal{F}_2^\sigma$, the approximation f_n of a density f is such that

$$\sum_{|k| \geq n} \alpha_{j_0 k}^2 = O(n^{-2\sigma}), \quad \sum_{j > j_n} \sum_{k \in \mathbb{Z}} \beta_{jk}^2 = O(2^{-j_n \sigma}) = O(n^{-2\sigma}).$$

For a density f of W_σ, a weighted wavelet estimator is defined as

$$\begin{aligned} \widehat{f}_{n\nu_n}(x) = &\sum_{|k| \leq n} \frac{\lambda_k}{\nu_n + \lambda_k} \widehat{\alpha}_{n, j_0 k} \phi_{j_0 k}(x) \\ &+ \sum_{j=j_0}^{j_n} \sum_{|k| \leq k_{jn}} \frac{\mu_j}{\nu_n + \mu_j} \beta_{jk} \psi_{jk}(x) \end{aligned} \tag{8.9}$$

where $\nu_n > 0$ does not depend on j and k, $\lambda_k = k^{-2\sigma}$ and $\mu_j = 2^{-j\sigma}$.

Proposition 8.3. *Under the conditions (8.8) and $\nu_n = O(n^{-\frac{2\sigma}{2\sigma+1}})$, the estimator of a density f in $\mathcal{F}_2^\sigma$ satisfies* $\mathrm{MISE}_n(f) = O(n^{-\frac{2\sigma}{2\sigma+1}})$.

Proof. The integrated squared error $\mathrm{ISE}_n(f) = \int \{\widehat{f}_{n\nu_n}(x) - f(x)\}^2\,dx$ is the sum

$$\begin{aligned} \mathrm{ISE}_n(f) = &\sum_{|k| \leq n} \Big\{ \frac{\lambda_k}{\nu_n + \lambda_k} (\widehat{\alpha}_{n, j_0 k} - \alpha_{j_0 k}) - \frac{\nu_n}{\nu_n + \lambda_k} \alpha_{j_0 k} \Big\}^2 + \sum_{|k| \geq n}^{\infty} \alpha_{j_0 k}^2 \\ &+ \sum_{j=j_0}^{j_n} \sum_{|k| \leq k_{jn}} \Big\{ \frac{\mu_j}{\nu_n + \mu_j} (\widehat{\beta}_{n, jk} - \beta_{jk}) - \frac{\nu_n}{\nu_n + \mu_j} \beta_{jk} \Big\}^2 \\ &+ \sum_{j > j_n} \sum_{k \in \mathbb{Z}} \beta_{jk}^2 + \sum_{j=j_0}^{j_n} \sum_{|k| \geq k_{jn}} \beta_{jk}^2, \end{aligned}$$

it has the expectation

$$\begin{aligned}
\mathrm{MISE}_n(f) &= \sum_{|k|\le n} \frac{\lambda_k^2\sigma_{\alpha,j_0k}^2}{n(\nu_n+\lambda_k)^2} + \sum_{|k|\le n} \frac{\nu_n^2\alpha_{j_0k}^2}{(\nu_n+\lambda_k)^2} + \sum_{|k|\ge n}^{\infty} \alpha_{j_0k}^2 \\
&+ \sum_{j=j_0}^{j_n}\sum_{|k|\le k_{jn}} \frac{\mu_j^2\sigma_{\beta,jk}^2}{n(\nu_n+\mu_j)^2} + \sum_{j=j_0}^{j_n}\sum_{|k|\le k_{jn}} \frac{\nu_n^2}{(\nu_n+\mu_j)^2}\beta_{jk}^2 \\
&+ \sum_{j>j_n}\sum_{k\in\mathbb{Z}} \beta_{jk}^2 + \sum_{j=j_0}^{j_n}\sum_{|k|\ge k_{jn}} \beta_{jk}^2,
\end{aligned}$$

it is denoted $\mathrm{MISE}_n(f) = S_{1n}+\cdots+S_{7n}$ where $S_{7n} = o(S_{6n})$. The variances of $\widehat{a}_{nk}$ and $\widehat{\beta}_{n,jk}$ have the order n^{-1} and the conditions (8.8) and (8.9) entail

$$\begin{aligned}
S_{1n} &\le \sum_{|k|\le n} \frac{n^{-1}C}{(k^{2\sigma}\nu_n+1)^2} = O(n^{-1}\nu_n^{-\frac{1}{2\sigma}}) \int_0^\infty \frac{dx}{(1+x^{2\sigma})^2}, \\
S_{2n} &= \sum_{|k|\le n} \frac{\nu_n^2\alpha_{j_0k}^2}{(\nu_n+\lambda_k)^2} < \nu_n \sum_{|k|\le n} k^{2\sigma}\alpha_{j_0k}^2 = O(\nu_n),
\end{aligned}$$

with $x^{2\sigma} = \nu_n k^{2\sigma}$ in the integral of the first bound and $S_{3n} = O(n^{-2\sigma})$. The main terms of $\mathrm{MISE}_n(f)$ are $S_{1n} = O(n^{-\frac{2\sigma}{2\sigma+1}})$ and $S_{2n} = O(n^{-\frac{2\sigma}{2\sigma+1}})$, like in the proof of Proposition 2.4. In the same way

$$\begin{aligned}
S_{4n} &= O(n^{-\frac{2\sigma}{2\sigma+1}}), \\
S_{5n} &< \nu_n \sum_{|k|\le k_{jn}} \mu_j^{-1}\beta_{jk}^2 = O(\nu_n), \\
S_{6n} &= \sum_{j>j_n}\sum_{k\in\mathbb{Z}} \beta_{jk}^2 = O(n^{-2\sigma}).
\end{aligned}$$

□

8.3 Wavelet estimators for time-dependent processes

The functional intensities of a Poisson process, the hazard rate of a jump process and the time-dependent drift and variance functions of a diffusion process are estimated by projection on wavelet bases of $L^2(\mathbb{R}_+)$, their properties rely on the convergence of the processes.

Let $(Y_t)_{t\ge0}$ be an inhomogeneous Poisson process with a functional intensity on $\mathbb{R}_+$

$$\lambda(t) = \lim_{\delta\to0} P(N(t+h) - N(t) \mid \mathcal{F}_t),$$

the infinitesimal probability of the process conditionally on the past. The cumulative intensity of the Poisson process is $\Lambda(t) = \int_0^t \lambda(s)\,ds$ and for $m \geq 0$

$$P(N(t) = m) = e^{-\Lambda(t)} \frac{\Lambda(t)^m}{m!}.$$

The expectation of $N(t)$ is $\Lambda(t)$ and the intensity $\lambda(t)$ is estimated by projection on wavelets. The restrictions ϕ^+ and ψ^+ of the functions ϕ and ψ to $\mathbb{R}_+$ define wavelets and an intensity $\lambda(t)$ of $L^2(\mathbb{R}_+)$ has an expansion

$$\lambda(t) = \sum_{k\in\mathbb{Z}} \alpha_{j_0k}\phi^+_{j_0k}(t) + \sum_{j\geq j_0}\sum_{k\in\mathbb{Z}} \beta_{jk}\psi^+_{jk}(t), \tag{8.10}$$

with the coefficients

$$\alpha_{j_0k} = \int_0^\infty \lambda(t)\phi^+_{j_0k}(t)\,dt,$$
$$\beta_{jk} = \int_0^\infty \lambda(t)\psi^+_{jk}(t)\,dt,$$

they are estimated from the observation of the process N on an increasing interval $[0, T]$. The intensity λ is approximated by

$$\lambda_T(t) = \sum_{k\leq k_T} \alpha_{j_0k}\phi^+_{j_0k}(t) + \sum_{j=j_0}^{j_T}\sum_{k\leq k_{jT}} \beta_{jk}\psi^+_{jk}(t),$$

with indices j_T, k_T and k_{jT} of order $o(T)$ and tending to infinity as T tends to infinity. The intensity λ has the estimator

$$\widehat{\lambda}_T(t) = \sum_{|k|\leq k_T} \widehat{\alpha}_{T,j_0k}\phi^+_{j_0k}(t) + \sum_{j=j_0}^{j_T}\sum_{|k|\leq k_{jT}} \widehat{\beta}_{T,jk}\psi^+_{jk}(t) \tag{8.11}$$

with the unbiased estimators of the coefficients

$$\widehat{\alpha}_{T,j_0k} = \int_0^T \phi^+_{j_0k}(t)\,dN(t),$$
$$\widehat{\beta}_{T,jk} = \int_0^T \psi^+_{jk}(t)\,dN(t).$$

The process $N - \Lambda$ is centered and its variance is Λ, it follows that the variances σ^2_{α,j_0k} and $\sigma^2_{\beta,jk}$ of the estimators $\widehat{\alpha}_{T,j_0k}$ and $\widehat{\beta}_{T,jk}$ are respectively

$$v_{\alpha,Tj_0k} = \int_0^T \phi^{+2}_{j_0k}(t)\,d\Lambda(t),$$
$$v_{\beta,Tjk} = \int_0^T \psi^{+2}_{jk}(t)\,d\Lambda(t),$$

they are $O(1)$ as T tends to infinity. The mean integrated squared error of the estimator $\widehat{\lambda}_T$ is

$$\text{MISE}_T(\lambda) = \|\lambda - \lambda_T\|_{L^2}^2 + \sum_{|k|\le k_T} E\{(\widehat{\alpha}_{T,j_0k} - \alpha_{j_0k})^2\}$$

$$+ \sum_{j=j_0}^{j_T} \sum_{|k|\le k_{jT}} E\{(\widehat{\beta}_{T,jk} - \beta_{jk})^2\},$$

the order of $\text{MISE}_T(\lambda)$ is therefore

$$\|\lambda - \lambda_T\|_{L^2}^2 + O(k_T + K_T)$$

with $K_T = \sum_{j=j_0}^{j_T} k_{jT}$ and the approximation error $\|\lambda - \lambda_T\|_{L^2}^2$ converges to zero as T tends to infinity. For an intensity of the space $\mathcal{F}_2^\sigma$, $\frac{1}{2} < \sigma < m$, the approximation error has the norm

$$\begin{aligned}\|\lambda - \lambda_T\|_{L^2}^2 &= \sum_{|k|\ge k_n} \alpha_{j_0k}^2 + \Big(\sum_{j>j_T} \sum_{k\in\mathbb{Z}} \beta_{jk}^2\Big)\{1 + o(1)\} \\ &= O(k_T^{-2\sigma}) + O(2^{-j_T\sigma}).\end{aligned}$$

Let $(N_i)_{i\le n}$ be a sample of point processes on a finite time interval $[0, \tau]$ such that $\bar{N}_n = \sum_{i\le n} N_i$ has a multiplicative intensity $\lambda_n(t) = \lambda(t)\bar{Y}_n(t)$, with λ in $L^2(\mathbb{R}_+)$. The cumulative intensity Λ_n has the estimator

$$\widehat{\Lambda}_n(t) = \int_0^t 1_{\{\bar{Y}_n(s)>0\}} \bar{Y}_n^{-1}(s)\, d\bar{N}_n(s)$$

on $[0, \tau]$ and the process $n^{\frac{1}{2}}(\widehat{\Lambda}_n - \Lambda)$ converges weakly to a centered Gaussian process with covariance function $V(s,t) = \int_0^{s\wedge t} y^{-1} 1_{\{y>0\}}\, d\Lambda$, under the condition that $\sup_{t\in[0,\tau]} |n^{-1}\bar{Y}_n(t) - y(t)|$ converges to zero in probability. The intensity λ has an expansion (8.10) on the wavelet basis $(\phi_{j_0k}^{+2})_{k\ge 1}$ and $(\psi_{jk}^{+2})_{j\ge j_0, k\ge 1}$ and it is estimated by smoothing $\widehat{\Lambda}_n$

$$\widehat{\lambda}_n(t) = \sum_{|k|\le k_n} \widehat{\alpha}_{n,j_0k}\phi_{j_0k}^+(t) + \sum_{j=j_0}^{j_n} \sum_{|k|\le k_{jn}} \widehat{\beta}_{n,jk}\psi_{jk}^+(t) \tag{8.12}$$

with the unbiased estimators of the coefficients

$$\begin{aligned}\widehat{\alpha}_{n,j_0k} &= \int_0^\tau \phi_{j_0k}^+(t)\, d\widehat{\Lambda}_n(t), \\ \widehat{\beta}_{n,jk} &= \int_0^\tau \psi_{jk}^+(t)\, d\widehat{\Lambda}_n(t).\end{aligned}$$

The variables $n^{\frac{1}{2}}(\widehat{\alpha}_{n,j_0k} - \alpha_{j_0k})$ and $n^{\frac{1}{2}}(\widehat{\beta}_{n,jk} - \beta_{jk})$ converge weakly to independent centered Gaussian variables with variance functions

$$v_{\alpha,j_0k} = \int_0^\tau \phi_{j_0k}^{+2}(t)\,dV(t),$$
$$v_{\beta,jk} = \int_0^\tau \psi_{jk}^{+2}(t)\,dV(t),$$

for all k in $\mathbb{Z}$ and $j \geq j_0$. The mean integrated squared error of the estimator $\widehat{\lambda}_n$ is

$$\text{MISE}_n(\lambda) = \|\lambda - \lambda_n\|_{L^2}^2 + \sum_{|k|\leq k_n} E\{(\widehat{\alpha}_{n,j_0k} - \alpha_{j_0k})^2\}$$
$$+ \sum_{j=j_0}^{j_n} \sum_{|k|\leq k_{jn}} E\{(\widehat{\beta}_{n,jk} - \beta_{jk})^2\},$$

the order of $\text{MISE}_n(\lambda)$ is therefore

$$\|\lambda - \lambda_n\|_{L^2}^2 + O(n^{-1}(k_n + K_n))$$

with $K_n = \sum_{j=j_0}^{j_n} k_{jn}$ and the approximation error $\|\lambda - \lambda_n\|_{L^2}^2$ converges to zero as n tends to infinity. With $k_n + K_n = o(n)$, the estimation error converges to zero as n tends to infinity. The errors have the same optimal order if

$$\|\lambda - \lambda_n\|_{L^2}^2 = O(n^{-1}(k_n + K_n)). \tag{8.13}$$

For an intensity of the space $\mathcal{F}_2^\sigma$, $\frac{1}{2} < \sigma < m$, the approximation error has the norm

$$\|\lambda - \lambda_n\|_{L^2}^2 = \sum_{|k|\geq k_n} \alpha_{j_0k}^2 + \Big(\sum_{j>j_n} \sum_{k\in\mathbb{Z}} \beta_{jk}^2\Big)\{1 + o(1)\}$$
$$= O(k_n^{-2\sigma}) + O(2^{-j_n\sigma}),$$

Proposition 8.4. *Under the condition (8.13), for the wavelet estimator (8.11) of an intensity λ in $\mathcal{F}_2^\sigma$ we have*

$$\text{MISE}_n(\lambda) = O(n^{-\frac{2\sigma}{2\sigma+1}}).$$

Proposition 8.5. *Under the condition (8.13) and for λ in $\mathcal{F}_2^\sigma$, the process $n^{\frac{\sigma}{2\sigma+1}}(\widehat{\lambda}_n - \lambda)$ converges weakly to a centered Gaussian process with finite covariance function.*

The wavelet estimator $\widehat{\lambda}_n$ applies to the estimation of the baseline hazard in proportional hazards models.

The time-dependent drift and variance functions of a nonparametric diffusion process (7.1) are also estimated by projections on the wavelet basis $(\phi^+_{j_0k})_{k\geq 1}$ and $(\psi^+_{jk})_{j\geq j_0,k\geq 1}$ of $L^2(\mathbb{R}_+)$, under the assumptions of Section 7.1. Let

$$\alpha(t) = \sum_{k\in\mathbb{Z}} a_{j_0k}\phi^+_{j_0k}(t) + \sum_{j\geq j_0}\sum_{k\in\mathbb{Z}} b_{jk}\psi^+_{jk}(t),$$

with the coefficients $a_{j_0k} = \int_0^\infty \alpha(t)\phi^+_{j_0k}(t)\,dt$ and $b_{jk} = \int_0^\infty \alpha(t)\phi^+_{jk}(t)\,dt$. They are unbiasedly estimated by

$$\widehat{\alpha}_T(t) = \sum_{|k|\leq k_T} \widehat{a}_{T,j_0k}\phi^+_{j_0k}(t) + \sum_{j=j_0}^{j_T}\sum_{|k|\leq k_{jT}} \widehat{b}_{T,jk}\psi^+_{jk}(t) \tag{8.14}$$

with indices j_T, k_T and k_{jT} of order $o(T)$ and tending to infinity as T tends to infinity, and with the estimators

$$\widehat{a}_{T,j_0k} = \int_0^T \phi^+_{j_0k}(t)\,dX(t), \quad \widehat{b}_{T,jk} = \int_0^T \psi^+_{jk}(t)\,dX(t).$$

Their variances are respectively

$$v_{a,Tj_0k} = \int_0^T \phi^{+2}_{j_0k}(t)\alpha(t)\,dt,$$

$$v_{b,Tjk} = \int_0^T \psi^{+2}_{jk}(t)\beta(t)\,dt.$$

The mean integrated squared error of the estimator $\widehat{\alpha}_T$ is

$$\text{MISE}_T(\alpha) = \|\alpha - \alpha_T\|^2_{L^2} + \sum_{|k|\leq k_T} E\{(\widehat{\alpha}_{T,j_0k} - \alpha_{j_0k})^2\}$$

$$+ \sum_{j=j_0}^{j_T}\sum_{|k|\leq k_{jT}} E\{(\widehat{\beta}_{T,jk} - \beta_{jk})^2\},$$

the order of $\text{MISE}_T(\alpha)$ is therefore

$$\|\alpha - \alpha_T\|^2_{L^2} + O(k_T + K_T)$$

and the approximation error $\|\alpha - \alpha_T\|^2_{L^2}$ converges to zero as T tends to infinity. The error $\text{MISE}_T(\beta)$ has similar properties as $E(Y^4)$ is finite. Under a boundedness condition for the sums

$$\sum_{|k|\leq k_T} v_{a,Tj_0k}, \quad \sum_{j=j_0}^{j_T}\sum_{|k|\leq k_{jT}} v_{b,Tjk},$$

the estimation errors and therefore the errors $\mathrm{MISE}_T(\alpha)$ and $\mathrm{MISE}_T(\beta)$ converge to zero as T tends to infinity.

The estimation of the expectation $\mu(t)$ and the variance function $v(t)$ of the process $Y(t)$ studied in Section 7.8 may be performed by projection on wavelets in $L^2(\mathbb{R}_+)$. Let $\mu(t) = \sum_{k\in\mathbb{Z}} \alpha_{j_0k}\phi^+_{j_0k}(t) + \sum_{j\leq j_0}\sum_{k\in\mathbb{Z}} \beta_{j_0k}\psi^+_{j_0k}(t)$ and let

$$v(t) = \sum_{k\in\mathbb{Z}^2} a_k\phi^+_{j_0k}(t) + \sum_{j\leq j_0}\sum_{k\in\mathbb{Z}^2} b_{jk}\psi^+_{jk}(t),$$

with $\alpha_{j_0k} = \int_{\mathbb{R}_+} \mu(t)\phi^+_{j_0k}(t)\,dt$, $\beta_{j_0k} = \int_{\mathbb{R}_+} \mu(t)\psi^+_{j_0k}(t)\,dt$ and

$$\begin{aligned} a_k &= \int_{\mathbb{R}_+} v(t)\phi^+_{j_0k}(t)\,dt, \\ b_{jk} &= \int_{\mathbb{R}^2_+} v(t)\psi^+_{jk}(t)\,dt. \end{aligned}$$

The functions μ and v are estimated from the continuous observation of the processes Y in an increasing interval $[0,T]$ as

$$\begin{aligned} \widehat{\mu}_T(t) &= \sum_{|k|\leq K_T} \widehat{\alpha}_{Tj_0k}\phi^+_{j_0k}(t) + \sum_{j\leq j_0}\sum_{|k|\leq K_{jT}} \widehat{\beta}_{Tjk}\psi^+_{j_0k}(t), \\ \widehat{v}_T(t) &= \sum_{|k|\leq K_T} \widehat{a}_{Tk}\phi^+_{j_0k}(t) \\ &\quad + \sum_{j\leq j_0}\sum_{|k|\leq K_{jT}} \widehat{b}_{Tjk}\psi^+_{jk}(t) \end{aligned}$$

with the estimators of the coefficients

$$\begin{aligned} \widehat{\alpha}_{Tj_0k} &= \int_0^T Y(t)\phi^+_{j_0k}(t)\,dt, \\ \widehat{\beta}_{Tjk} &= \int_0^T Y(t)\psi^+_{jk}(t)\,dt, \\ \widehat{a}_{Tk} &= \int_0^T \{Y(t) - \widehat{\mu}_T(t)\}^2\phi^+_{j_0k}(t)\,dt, \\ \widehat{b}_{Tjk} &= \int_0^T \{Y(t) - \widehat{\mu}_T(t)\}^2\phi^+_{jk}(t)\,dt, \end{aligned}$$

for K_T and $L_T = o(T)$ and tending to infinity as T tends to infinity. The variance of $\widehat{\alpha}_{Tj_0k}$ converges to a_k, the variance of $\widehat{\beta}_{Tjk}$ converges to b_{jk}, and the covariance of $\widehat{\mu}_T(t)$ and $\widehat{\mu}_T(s)$ converges to $v(s,t)$. The estimators

$\widehat{\mu}_T$ and $\widehat{v}_T$ are a.s. consistent by the ergodicity property (7.2). The weak convergence of the estimated coefficients is proved like Lemma 7.16 and it entails the weak convergence of $\widehat{\mu}_T$.

Proposition 8.6. *If $E(Y^4)$ is finite, the processes $\widehat{\mu}_T - \mu_T$ and $\widehat{v}_T - v_T$ converge weakly to centered Gaussian processes on every interval where the wavelets are finite.*

8.4 Wavelet regression estimator

For real variables X and Y, we consider the nonparametric regression model

$$Y_i = m(X_i) + \varepsilon_i,\ i = 1, \dots, n,$$

defined by a real function $m(x) = (Y_i \mid X_i = x)$ of $L^2(\mathcal{I}_X, F_X)$, and an error variable ε such that $E(\varepsilon \mid X) = 0$ and $\mathrm{Var}(Y \mid X) = \sigma^2$. The variable X has a continuous distribution F_X on the support $\mathcal{I}_X$ of X and a density probability f_X of $L^2(\mathcal{I}_X)$, with respect to the Lebesgue measure.

The space $L^2(\mathcal{I}_X, F_X)$ is the sum $L^2(\mathcal{I}_X, F_X) = V_{j_0} \oplus W_{j_0} \oplus W_{j_0+1} \oplus \cdots$ of orthogonal spaces generated by the wavelets (8.1) defined with functions ϕ and ψ of $L^2(\mathcal{I}_X, F_X)$, they are estimated from the empirical distribution of the sample $(X_i)_{i \leq n}$. Let g and h be functions of $L^2([0,1])$, the functions

$$\phi = g \circ F_X, \quad \psi = h \circ F_X \tag{8.15}$$

belong to $L^2(\mathcal{I}_X)$ and they satisfy $\int_0^1 g^2(y)\,dy = \int_{\mathcal{I}_X} \phi^2(x)\,dF_X(x)$ and $\int_0^1 h^2(y)\,dy = \int_{\mathcal{I}_X} \psi^2(x)\,dF_X(x)$. We assume that the functions g and h satisfy the conditions (8.1) on $[0,1]$. The functions ϕ and ψ are estimated from the observations by

$$\widehat{\phi}_n(x) = h \circ \widehat{F}_{nX}(x) = \sum_{i=1}^n 1_{\{X_i \leq x\}} h(n^{-1}i),$$

$$\widehat{\psi}_n(x) = g \circ \widehat{F}_{nX}(x) = \sum_{i=1}^n 1_{\{X_i \leq x\}} g(n^{-1}i),$$

they converge uniformly on $\mathcal{I}_X$ to the functions ϕ, and respectively ψ, with the rate $n^{-\frac{1}{2}}$.

A regression function m in the space

$$\mathcal{F}^{2\sigma}_{L^2(F_X)} = \{m \in L^2(\mathcal{I}_X, F_X) : N_2^{2\sigma}(m) < \infty\},$$

with $\sigma > 1$, has an expansion $m = \sum_{k\in\mathbb{Z}} a_{j_0k}\phi_{j_0k} + \sum_{j\geq j_0}\sum_{k\in\mathbb{Z}} b_{jk}\psi_{jk}$ with norm $\|m\|_{L^2(F_X)} = \sum_{k\in\mathbb{Z}} a^2_{j_0k} + \sum_{j\geq j_0}\sum_{k\in\mathbb{Z}} b^2_{jk}$. The regression function m has an approximation

$$m_n(x) = \sum_{|k|\leq k_n} a_{j_0k}\phi_{j_0k}(x) + \sum_{j=j_0}^{j_n}\sum_{|k|\leq k_{jn}} b_{jk}\psi_{jk}(x)$$

with j_n, k_n and k_{jn} tending to infinity as n tends to infinity. The coefficients $a_{j_0k} = \int_{\mathcal{I}_X} m\phi_{j_0k}\,dF_X$ and $b_{jk} = \int_{\mathcal{I}_X} m\psi_{jk}\,dF_X$ are estimated from the sample $(X_i)_{i=1,\dots,n}$ by projection on estimators of the orthonormal basis as

$$\widehat{a}_{n,j_0k} = n^{-1}\sum_{i=1}^n Y_i\widehat{\phi}_{n,j_0k}(X_i),$$

$$\widehat{b}_{n,jk} = n^{-1}\sum_{i=1}^n Y_i\widehat{\psi}_{n,jk}(X_i)$$

and the regression function m is estimated as

$$\widehat{m}_n = \sum_{|k|\leq k_n} \widehat{a}_{n,j_0k}\widehat{\phi}_{n,j_0k}(x) + \sum_{j=j_0}^{j_n}\sum_{|k|\leq k_{jn}} \widehat{b}_{n,jk}\widehat{\psi}_{n,jk}(x). \tag{8.16}$$

The convergence rate of the estimators of the coefficients is $n^{-\frac{1}{2}}$ and the mean integrated squared error in $L^2(\mathbb{R}, F_X)$ of the estimator (8.16) is

$$\mathrm{MISE}_n(m) = E\|\widehat{m}_n - m_n\|^2_{L^2(F_X)} + \|m - m_n\|^2_{L^2(F_X)}.$$

For a function m in the space $\mathcal{F}^\sigma_{L^2(F_X)}$, $\sigma > 1$, the approximation error is

$$\begin{aligned}\|m - m_n\|^2_{L^2(F_X)} &= \sum_{|k|\geq k_n} a^2_{j_0k} + \sum_{j>j_n}\sum_{k\in\mathbb{Z}} b^2_{jk} + \sum_{j=j_0}^{j_n}\sum_{|k|\geq k_{jn}} b^2_{jk}\\ &= O(k_n^{-4\sigma}) + O(2^{-2j_n\sigma}),\end{aligned}$$

and the estimation error is obtained by the same arguments as Lemma 3.3 and from the convergence rate of the estimated coefficients

$$E\|\widehat{m}_n - m_n\|^2_{L^2(F_X)} = O(n^{-1}(k_n^2 + K_n^2)).$$

As n tends to infinity, k_n and K_n are optimum as the approximation and the estimation errors have the same order, they are therefore chosen so that $K_n = O(2^{\frac{j_n}{2}})$ and

$$k_n^{-4\sigma} = O(n^{-1}k_n^2), \quad 2^{-2j_n\sigma} = O(n^{-1}2^{j_n}), \tag{8.17}$$

these conditions imply

$$k_n = O(n^{\frac{1}{2(2\sigma+1)}}), \quad 2^{j_n} = O(n^{\frac{1}{2\sigma+1}})$$

and the optimal order of the error is deduced.

Proposition 8.7. *Under the condition (8.17), the wavelet estimator (8.16) of a regression function m in $\mathcal{F}^{2\sigma}_{L^2(F_X)}$ has an error*

$$\mathrm{MISE}_n(f) = O(n^{-\frac{2\sigma}{2\sigma+1}}).$$

Proposition 8.8. *Under the condition (8.17), the process $n^{\frac{\sigma}{2\sigma+1}}(\widehat{m}_n - m)$ converges weakly to a Gaussian process with finite expectation and covariance functions.*

Cross-validation estimators of the sizes k_n, j_n and k_{jn} minimize

$$CV_n = n^{-1}\sum_{j=1}^n \widehat{m}^2_{n,j}(X_j) - 2n^{-1}\sum_{j=1}^n Y_j\widehat{m}_{n,j}(X_j),$$

where $\widehat{m}_{n,j}$ is the wavelet estimator of the regression function m defined with the sub-sample without the observation (Y_j, X_j). Equivalently, they minimize

$$\begin{aligned}
J_n = &\sum_{|k|\le k_n} \widehat{a}^2_{n,j_0k} + \sum_{j=j_0}^{j_n}\sum_{|k|\le k_{jn}} \widehat{b}^2_{n,jk}\\
&-\frac{2}{n(n-1)}\sum_{i\neq i'=1}^n\Big\{\sum_{|k|\le k_n} Y_iY_{i'}\widehat{\phi}_{ni,j_0k}(X_i)\widehat{\phi}_{ni',j_0k}(X_{i'})\\
&+\sum_{j=j_0}^{j_n}\sum_{|k|\le k_{jn}} Y_iY_j\widehat{\psi}_{ni,jk}(X_i)\widehat{\psi}_{ni',jk}(X_{i'})\Big\}
\end{aligned}$$

and they are asymptotically optimum.

The wavelet estimator (8.16) of a regression function m in $\mathcal{F}^{\sigma}_{L^2(F_X)}$ is extended to a weighted estimator like for densities, with the estimated wavelets previously defined for the estimator (8.16), let

$$\begin{aligned}
\widehat{m}_{n\nu_n}(x) = &\sum_{|k|\le n}\frac{\lambda_k}{\nu_n+\lambda_k}\widehat{\alpha}_{n,k}\widehat{\phi}_{j_0k}(x)\\
&+\sum_{j=j_0}^{j_n}\sum_{|k|\le k_{jn}}\frac{\mu_j}{\nu_n+\mu_j}\widehat{\beta}_{n,jk}\widehat{\psi}_{jk}(x) \qquad (8.18)
\end{aligned}$$

where $\nu_n = O(n^{-\frac{2\sigma}{2\sigma+1}})$, $\lambda_k = k^{-2\sigma}$ and $\mu_j = 2^{-j\sigma}$. For every n let

$$k_{jn} = O(2^{\frac{j}{2}}), \quad 2^{\frac{j_n}{2}} = n, \qquad (8.19)$$

and let $K_n = O(n)$.

Proposition 8.9. *Under the conditions (8.8), the estimator of a regression function m in $\mathcal{F}^{2\sigma}_{L^2(F_X)}$ satisfies* $\mathrm{MISE}_n(m) = O(n^{-\frac{2\sigma}{2\sigma+1}})$.

Proof. The error $\mathrm{MISE}_n(m)$ of the estimator $\widehat{m}_n$ in $L^2(F_X)$ is expanded as

$$E\int\Big[\sum_{|k|\le n}\frac{\lambda_k}{\nu_n+\lambda_k}\{\widehat{\alpha}_{nk}\widehat{\phi}_{nj_0k}(x)-\alpha_k\phi_{j_0k}(x)\}$$

$$-\frac{\nu_n}{\nu_n+\lambda_k}\alpha_k\phi_{j_0k}(x)\Big]^2\,dF_X(x)+\sum_{|k|>n}^{\infty}\alpha_k^2$$

$$+E\int\Big[\sum_{j=j_0}^{j_n}\sum_{|k|\le k_{jn}}\frac{\mu_j}{\nu_n+\mu_j}\{\widehat{\beta}_{n,jk}\widehat{\psi}_{jk}(x)-\beta_{jk}\psi_{jk}(x)\}$$

$$-\frac{\nu_n}{\nu_n+\mu_j}\beta_{jk}\psi_{jk}(x)\Big]^2\,dF_X(x)+\sum_{j>j_n}\sum_{k\in\mathbb{Z}}\beta_{jk}^2+\sum_{j=j_0}^{j_n}\sum_{|k|\ge k_{jn}}\beta_{jk}^2,$$

where the variances of $\widehat{\alpha}_{nj_0k}$, and respectively $\widehat{\beta}_{njk}$, $\widehat{\phi}_{nj_0k}$ and $\widehat{\psi}_{njk}$, have approximations $n^{-1}v_{\alpha,j_0k}+o(n^{-1})$, and respectively $n^{-1}v_{\beta,jk}+o(n^{-1})$, $n^{-1}v_{\phi,j_0k}+o(n^{-1})$ and $n^{-1}v_{\psi,jk}+o(n^{-1})$, and their covariances have similar approximations, they are $O(n^{-1})$. The orthogonality of the bases entails that the first integral of the $\mathrm{MISE}_n(m)$ has an expansion

$$\sum_{|k|\le n}\Big[\frac{\lambda_k^2}{n(\nu_n+\lambda_k)^2}\Big\{v_{\alpha,j_0k}+\alpha_k^2\int_{\mathcal{I}_X}v_{\alpha\phi,j_0k}\,dF_X$$

$$+2\alpha_k\int_{\mathcal{I}_X}v_{\alpha\phi,j_0k}\phi_{j_0k}\,dF_X\Big\}\{1+o(1)\}+\frac{\nu_n^2}{(\nu_n+\lambda_k)^2}\alpha_k^2$$

$$-2\frac{\lambda_k\nu_n}{(\nu_n+\lambda_k)^2}\alpha_kE\Big\{\alpha_k(\widehat{\alpha}_{nk}-\alpha_k)\int\widehat{\phi}_{nk}\phi_k\,dF_X\Big\}\Big]$$

$$+E\sum_{|k|\le n}\sum_{|k'|\le n,k'\ne k}\Big[\frac{\lambda_k\lambda_{k'}}{(\nu_n+\lambda_k)(\nu_n+\lambda k')}$$

$$\cdot\Big\{\widehat{\alpha}_{nk}\widehat{\alpha}_{nk'}\int_{\mathcal{I}_X}(\widehat{\phi}_{nk}-\phi_k)(\widehat{\phi}_{nk'}-\phi_{k'})\,dF_X$$

$$+2(\widehat{\alpha}_{nk}-\alpha_k)\widehat{\alpha}_{nk'}\int_{\mathcal{I}_X}(\widehat{\phi}_{nk'}-\phi_{k'})\phi_k\,dF_X\Big\}$$

$$-2\frac{\lambda_k\nu_n}{(\nu_n+\lambda_k)(\nu_n+\lambda k')}\widehat{\alpha}_{nk}\alpha_{k'}\int_{\mathcal{I}_X}\widehat{\phi}_{nk}\phi_{k'}\,dF_X\Big].$$

The bounds of Proposition (8.3) apply and we obtain

$$\sum_{|k|\leq n} \frac{\lambda_k^2}{n(\nu_n+\lambda_k)^2}\Big\{v_{\alpha,j_0k} + \alpha_k^2 \int_{\mathcal{I}_X} v_{\alpha\phi,j_0k}\,dF_X + 2\widehat{\alpha}_{nk}\int_{\mathcal{I}_X} v_{\alpha\phi,j_0k}\phi_k\,dF_X\Big\}$$
$$= O(n^{-1}\nu_n^{-\frac{1}{2\sigma}}),$$
$$\sum_{|k|\leq n} \frac{\nu_n^2}{(\nu_n+\lambda_k)^2}\alpha_k^2 = O(\nu_n),$$
$$\sum_{|k|\leq n} \frac{\lambda_k\nu_n}{(\nu_n+\lambda_k)^2} E\Big\{\alpha_k(\widehat{\alpha}_{nk}-\alpha_k)\int \widehat{\phi}_{nk}\phi_k\,dF_X\Big\}$$
$$\leq \Big\{\sum_{|k|\leq n} \frac{\lambda_k^2 v_{\alpha,k}}{n(\nu_n+\lambda_k)^2}\Big\}^{\frac{1}{2}}\Big[\sum_{|k|\leq n} \frac{\nu_n^2\alpha_k^2}{(\nu_n+\lambda_k)^2}\{1+o(1)\}\Big]^{\frac{1}{2}}$$
$$= O(n^{-\frac{1}{2}}\nu_n^{-\frac{1}{4\sigma}}\nu_n^{\frac{1}{2}}),$$

and $\sum_{|k|>n}^{\infty} \alpha_k^2 = O(n^{-4\sigma})$. By the Cauchy–Schwarz inequality, the sums on $k \neq k'$ are bounded by the products on the square roots of the sums on a single index, so the first two sums are $O(n^{-1}\nu_n^{-\frac{1}{2\sigma}})$ and the last sums is a $O(n^{-4\sigma})$, they have the same order.

The sum of the integrals depending on the coefficients β_{jk} have similar expansions and the result follows. □

The wavelet series estimators apply to the nonparametric multivariate regression models of Chapter3 and to the nonparametric generalized linear models of Chapter 4, with the estimated functions ϕ and ψ of $L^2(\mathcal{I}_X, F_X)$, the coefficients of the wavelets have empirical estimators. In models of hazard functions, nonparametric functions depending on a random covariate vector X have expansions on wavelets bases of $L^2(\mathcal{I}_X, F_X)$ and their coefficients are estimated by maximum likelihood. The cross-validation estimation of the sizes of the bases used in the estimation and weighted estimators are defined as for the nonparametric regression.

8.5 Wavelet estimators for auto-regressive diffusions

Let α and β^2 be continuous drift and variance functions on a metric space $(\mathcal{I}_X, \|\cdot\|_2)$, defining an auto-regressive diffusion process (7.10) observed on a time interval $[0,T]$. The ergodic distribution function F_X of the diffusion is consistently estimated as $\widehat{F}_{TX}$ defined by (7.24) and its weak convergence is established in Proposition (7.5).

Functions α and β^2 in $\mathcal{F}^{2\sigma}_{L^2(F_X)}$, $\sigma > 1$, have expansions on the orthonormal bases defined by the functions ϕ and ψ of (8.15)

$$\alpha(x) = \sum_{k\in\mathbb{Z}} a_{j_0k}\phi_{j_0k}(x) + \sum_{j\geq j_0}\sum_{k\in\mathbb{Z}} b_{jk}\psi_{jk}(x),$$
$$\beta^2(x) = \sum_{k\in\mathbb{Z}} \eta_{j_0k}\phi_{j_0k}(x) + \sum_{j\geq j_0}\sum_{k\in\mathbb{Z}} \zeta_{jk}\psi_{jk}(x)$$

with the coefficients $a_{j_0k} = \int_{\mathcal{I}_X} \alpha\phi_{j_0k}\, dF_X$ and $b_{jk} = \int_{\mathcal{I}_X} \alpha\psi_{jk}\, dF_X$ for the function α, and $\eta_{j_0k} = \int_{\mathcal{I}_X} \beta^2\phi_{j_0k}\, dF_X$ and $\zeta_{jk} = \int_{\mathcal{I}_X} \beta^2\psi_{jk}\, dF_X$ for the function β^2. Under the probability $P_0 = P_{\alpha_0,\beta_0}$ of the observations, the functions ϕ and ψ have the a.s. consistent estimators

$$\widehat{\phi}_T(x) = g \circ \widehat{F}_{TX}(x), \quad \widehat{\psi}_T(x) = h \circ \widehat{F}_{TX}(x), \tag{8.20}$$

then the coefficients a_{j_0k} and b_{jk} are a.s. consistently estimated from the observation of the diffusion process X on a time interval $[0, T]$

$$\widehat{a}_{T,j_0k} = T^{-1}\int_0^T \alpha(X_t)\widehat{\phi}_{T,j_0k}(X_t)\, dt,$$
$$\widehat{b}_{T,jk} = T^{-1}\int_0^T \alpha(X_t)\widehat{\psi}_{T,jk}(X_t)\, dt$$

and an a.s. consistent estimator of the function α is deduced

$$\widehat{\alpha}_T(x) = \sum_{|k|\leq k_T} \widehat{a}_{T,j_0k}\widehat{\phi}_{T,j_0k}(x) + \sum_{j=j_0}^{j_T}\sum_{|k|\leq k_T} \widehat{b}_{T,jk}\widehat{\psi}_{T,jk}(x),$$

with indices tending infinity as T tends to infinity. Under the probability P_0, let

$$Z_t = X_t - \int_0^t \widehat{\alpha}_T(X_s)\, ds = \int_0^t \beta_0(X_s)\, dB_s + o_p(1),$$

the coefficients η_{j_0k} and ζ_{jk} are a.s. consistently estimated from the sample-path of the process as

$$\widehat{\eta}_{T,j_0k} = 2T^{-1}\int_0^T Z_t\widehat{\phi}_{T,j_0k}(X_t)\, dZ_t,$$
$$\widehat{\zeta}_{T,jk} = 2T^{-1}\int_0^T Z_t\widehat{\psi}_{T,jk}(X_t)\, dZ_t,$$

they yield an a.s. consistent estimator of the function β^2

$$\widehat{\beta}^2_T(x) = \sum_{|k|\leq k_T} \widehat{\eta}_{T,j_0k}\widehat{\phi}_{T,j_0k}(x) + \sum_{j=j_0}^{j_T}\sum_{|k|\leq k_T} \widehat{\zeta}_{T,jk}\widehat{\psi}_{T,jk}(x).$$

The estimators of the coefficients have the convergence rate $T^{-\frac{1}{2}}$ and they are asymptotically Gaussian by the weak convergence of the empirical process $\nu_{T,X} = T^{\frac{1}{2}}(\widehat{F}_{TX} - F_X)$ obtained from the property (7.23) of the autoregressive diffusion process.

The mean integrated squared errors in $L^2(\mathcal{I}_X, F_X)$ of the estimators $\widehat{\alpha}_T$ and $\widehat{\beta}_T^2$ are the sums

$$\begin{aligned}\mathrm{MISE}_T(\alpha_0) &= E\|\widehat{\alpha}_T - \alpha_T\|_2^2 + \|\alpha_0 - \alpha_T\|_2^2,\\ \mathrm{MISE}_T(\beta_0^2) &= E\|\widehat{\beta}_T^2 - \beta_T^2\|_2^2 + \|\beta_0^2 - \beta_T^2\|_2^2.\end{aligned}$$

For the estimator of the function α in the space $\mathcal{F}^{2\sigma}_{L^2(F_X)}$, $\sigma > 1$, we have

$$\begin{aligned}\|\alpha_0 - \alpha_T\|^2_{L^2(F_X)} &= \sum_{|k|\geq k_T} a^2_{j_0k} + \sum_{j>j_T}\sum_{k\in\mathbb{Z}} b^2_{jk} + \sum_{j=j_0}^{j_T}\sum_{|k|\geq k_{jT}} b^2_{jk}\\ &= O(k_T^{-4\sigma}) + O(2^{-2j_T\sigma}),\\ E\|\widehat{\alpha}_T - \alpha_T\|^2_{L^2(F_X)} &= O(T^{-1}(k_T^2 + K_T^2)),\end{aligned}$$

and the order of $\mathrm{MISE}_T(\alpha_0)$ is optimum as both terms have the same order. The orders are similar for the estimator of the variance function β^2. Their optimum MISE and their weak convergence are deduced by the same arguments as for Propositions (8.7) and (8.8).

Proposition 8.10. *The wavelet estimators $\widehat{\alpha}_T$ and $\widehat{\beta}_T^2$ for functions α and β^2 in $\mathcal{F}^{2\sigma}_{L^2(F_X)}$ have mean integrated squared errors of optimum order $T^{-\frac{2\sigma}{2\sigma+1}}$.*

Proposition 8.11. *Under the conditions of optimality for wavelet estimators of the functions α and β^2 in $\mathcal{F}^{2\sigma}_{L^2(F_X)}$, the processes $T^{\frac{\sigma}{2\sigma+1}}(\widehat{\alpha}_T - \alpha)$ and $T^{\frac{\sigma}{2\sigma+1}}(\widehat{\beta}_T - \beta)$ converge weakly to Gaussian processes.*

In the diffusion model (7.29) with an unobserved stochastic volatility continuously observed on $[0, T]$, the variance of the centered process X under P_0 is $\mathrm{Var}_0 X_t = E_0 \int_0^t V_s\,ds$ and V_t is an ergodic Markov process with an invariant distribution function F_V on $\mathbb{R}_+$. The process V_t is estimated by projection of the process X_t^2 on a wavelet basis of $L^2(\mathbb{R}_+)$. The process V has the wavelet estimator

$$\widehat{V}_T(t) = \sum_{|k|\leq k_T} \widehat{a}_{T,j_0k}\phi_{j_0k}(t) + \sum_{j=j_0}^{j_T}\sum_{|k|\leq k_T} \widehat{b}_{T,jk}\psi_{jk}(t),$$

with the estimated coefficients

$$\widehat{a}_{T,j_0k} = 2\int_0^\infty \phi_{j_0k}(t)\, X_t\, dX_t,$$

$$\widehat{b}_{T,jk} = 2\int_0^\infty \psi_{jk}(t)\, X_t\, dX_t.$$

The invariant probability distribution of the ergodic process V_t is then consistently estimated by (7.31).

The drift and variance functions of the auto-regressive diffusion for the process V_t have expansions on a wavelet basis of $L^2(F_V)$, defined by functions ϕ and ψ of $L^2(F_V)$ and estimated by (8.20) with the convergence rate $T^{-\frac{1}{2}}$. The drift and variance of the diffusion process V_t have estimators similar to those of the above diffusion X, they satisfy the convergence of Propositions (8.10) and (8.11).

Chapter 9

Tests in discrete mixture models

9.1 Introduction

Let X be a random variable on a probability space (Ω, A, P_0), with values in a separable metric space $(\mathbb{X}, d, B)$. Let $\mathbb{F}$ be a family of unimodal densities with respect to a measure μ on (Ω, A) such that the density $f_0 = dP_0/d\mu$ of X belongs to $\mathbb{F}$. The family $\mathbb{F}$ is supposed $L_2(\mu)$-integrable and in $C^2(\mathbb{X})$, it is provided metric of $L_2(\mu)$ with norm $\|f\|_{L_2}$. We study the maximum likelihood test of the hypotheses H_0: the density of X belongs to $\mathbb{F}$, against the alternative of a mixture of two densities of $\mathbb{F}$ with unknown mixing proportions

$$X \sim g_{\lambda,f_1,f_2} \in \mathbb{G} = \Big\{\lambda f_1 + (1-\lambda) f_2, \lambda \in]0, \frac{1}{2}], f \in \mathbb{F}\Big\}.$$

Under the hypothesis, either λ belongs to $\{0, 1\}$ or $f_1 = f_2$ and the derivative of g with respect to λ is zero as $f_1 = f_2$ where the information matrix is degenerated. The parameter of interest for a test of H_0 reduces to $\lambda\|f_1 - f_2\|_2$, for λ in $]0, \frac{1}{2}]$, and its estimator has the constraint to be strictly positive. Several cases are considered according to the knowledge of a sub-density in the mixture. We generalize the reparametrization and the proofs of the parametric case (Pons, 2009) to functional classes of densities.

9.2 Homogeneity in contamination models

In a contamination model, the mixture density is

$$g_{\lambda,f} = \lambda f + (1-\lambda) f_0,$$

where f_0 is the known density of the hypothesis H_0 in $\mathbb{F}$, λ belongs to $[0, 1[$ and f is an unknown density of $\mathbb{F}$. Let $(X_1, \ldots, X_n)$ be a n-sample of the variable X under a probability density g. The hypothesis of the test is $H_0 : g = f_0$ and the alternative is $H_1 : g = g_{\lambda,f}$ with an unknown parameter λ distinct from 0 and 1, and an unknown sub-density f distinct from f_0. The following notations and conditions are considered

Condition 9.1.
A1. Identifiability of λ and f in $\mathbb{F}$. If there exist λ_1 and $\lambda_2 \in]0,1[$, f_1 and f_2 distinct from f_0 in $\mathbb{F}$ and such that

$$\lambda_1 f_1 + (1-\lambda_1) f_0 = \lambda_2 f_2 + (1-\lambda_2) f_0,$$

then $\lambda_1 = \lambda_2$ and $f_1 = f_2$;
A2. The $L_2(\mu)$-derivative φ_0 of f at f_0 exists in the tangent space $\dot{\mathbb{F}}$ of $\mathbb{F}$: let $u_f = \|f_0^{-1}(f - f_0)\|_{L_2(P_0)}$, then

$$\lim_{u_f \to 0} u_f^{-1} \|f_0^{-1}(f - f_0) - u_f \varphi_0\|_{L_2(P_0)} \to 0;$$

A3. The variables $\sup_{\lambda \in [0,1]} \sup_{f \in \mathbb{F}} |\log g_{\lambda,f}|$ and $\sup_{\varphi \in \dot{\mathbb{F}}} \|\varphi(X)\|^3$ are P_0- integrable;
A4. There exists a centered Gaussian process G on $\dot{\mathbb{F}}$ such that

$$\sup_{\varphi \in \dot{\mathbb{F}}} \left| n^{-\frac{1}{2}} \sum_{i=1}^{n} \varphi(X_i) - G_\varphi \right| \xrightarrow[n\to\infty]{P_0} 0.$$

The condition A4 depends on the entropy dimension of the functional class $\mathbb{F}$. Under the conditions A1 and A2 and with f_0 known, for all f in $\mathbb{F}$ and λ in $]0,1]$, the mixture density $g_{\lambda,f}$ is written as

$$g_{\lambda,f} = \lambda f + (1-\lambda) f_0 = f_0(1 + \alpha_f \varphi_f) \tag{9.1}$$

with the reparametrization

$$\begin{aligned} \alpha_f &= \lambda u_f, \\ \varphi_f &= u_f^{-1} f_0^{-1}(f - f_0) \text{ if } f \neq f_0, \\ &= \varphi_0, \text{ if } f = f_0. \end{aligned} \tag{9.2}$$

Let

$$\mathbb{U} = \{u_f; f \in \mathbb{F}\} \subset \mathbb{R}_+, \tag{9.3}$$

and let $\dot{\mathbb{F}} = \{\varphi_f; f \in \mathbb{F}\}$ be the tangent space to $\mathbb{F}$ at f_0. Under the alternative, the density g is expressed as

$$g_{\lambda,f} = f_0\{1 + \lambda f_0^{-1}(f - f_0)\} = f_0(1 + \alpha_f \varphi_f).$$

The parameters of interest for the test of a single density f_0 against the alternative of a true mixture density with f_0 and an unknown f in $\mathbb{F}$ belongs to $\{\alpha_f; f \in \mathbb{F}\}$ and the hypothesis is

$$H_0 : \sup_{f \in \mathbb{F}} \alpha_f = 0.$$

For a sample $(X_i)_{1\le i\le n}$ of X, the likelihood ratio statistic of H_0 is

$$T_n = 2 \sup_{\lambda\in]0,1]} \sup_{f\in\mathbb{F}} \sum_{1\le i\le n} \{\log g_{\lambda,f}(X_i) - \log f_0(X_i)\}$$
$$= 2 \sup_{\alpha\in\mathbb{U}} \sup_{\varphi\in\dot{\mathbb{F}}} \sum_{1\le i\le n} \log\{1+\alpha\varphi(X_i)\},$$

with the space $\mathbb{U}$ defined by (9.3). Let $l_n(\alpha,\varphi) = \sum_{1\le i\le n} \log\{1+\alpha\varphi(X_i)\}$ denote the log-likelihood ratio process parameterized with α in $\mathbb{U}$ and φ in the tangent space $\dot{\mathbb{F}}$ of $\mathbb{F}$, let Y_n be the process defined on $\dot{\mathbb{F}}$ as

$$Y_n(\varphi) = n^{-\frac{1}{2}} \sum_{i=1}^{n} \varphi(X_i)$$

and let Σ_φ be the variance matrix of $Y_n(\varphi)$, $\Sigma_{\varphi_1,\varphi_2} = \mathbb{E}_{P_0}\{\varphi_1(X)\varphi_2(X)\}$. Let $\widehat{\alpha}_n(\varphi)$ be the maximum likelihood estimator of α as φ is fixed.

Lemma 9.1. *Under the conditions (9.1),* $\sup_{\varphi\in\dot{\mathbb{F}}} \widehat{\alpha}_n(\varphi)$ *converges in* P_0-*probability to zero.*

The uniform consistency of the maximum likelihood estimator is a consequence of the concavity of l_n with respect to α and of the uniform convergence, under the conditions A1-A3, of $n^{-1}l_n$ to $l(\alpha,f) = E_0 \log\{1+\alpha\varphi_f(X)\}$ which is maximum at $\alpha_0 = 0$. The proof of the uniformity on $\mathbb{F}$ is similar to the proof for parametric densities (Lemdani and Pons, 1995).

Proposition 9.1. *Under the conditions (9.1), the statistic* T_n *converges weakly under* H_0 *to* $\sup_{\varphi\in\dot{\mathbb{F}}} Z^2(\varphi)1_{\{Z(\varphi)>0\}}$ *where* Z *is a continuous centered Gaussian process on* $\dot{\mathbb{F}}$, *with function of covariance*

$$K(\varphi_1,\varphi_2) = \Sigma_{\varphi_1}^{-\frac{1}{2}} \Sigma_{\varphi_1,\varphi_2} \Sigma_{\varphi_2}^{-\frac{1}{2}}.$$

Proof. Let φ in $\dot{\mathbb{F}}$, let $\dot{l}_n(\cdot,\varphi)$ be the derivative of $l_n(\cdot,\varphi)$ with respect to α and let $\widehat{\alpha}_n(\varphi)$ which maximizes $l_n(\cdot,\varphi)$ under the constraint $\widehat{\alpha}_n \ge 0$. There exists a penalization

$$b_\varphi = n^{-\frac{1}{2}} \dot{l}_n(\varphi, \widehat{\alpha}_n(\varphi))$$

such that $b_\varphi = 0$ if $\widehat{\alpha}_n > 0$ and $b_\varphi \le 0$ if $\widehat{\alpha}_n = 0$. For every $\varphi \in \dot{\mathbb{F}}$, let a_i or $a_i(\varphi) = \varphi(X_i)$ and let

$$R_{1n} = n^{-1} \sum_{i=1}^{n} a_i^3 (1+\widehat{\alpha}_n a_i)^{-1},$$

then

$$\begin{aligned} b_\varphi &= n^{-\frac{1}{2}} \sum_i \frac{a_i}{1+\widehat{\alpha}_n(\varphi)a_i}, \\ &= Y_n(\varphi) - n^{\frac{1}{2}}\widehat{\alpha}_n(\varphi)\{n^{-1}\sum_i a_i^2 - \widehat{\alpha}_n(\varphi)R_{1n}(\varphi)\} \qquad (9.4) \\ &= Y_n(\varphi) - n^{\frac{1}{2}}\widehat{\alpha}_n(\varphi)\Sigma_\varphi\{1+o_p(1)\} + \widehat{\alpha}_n^2(\varphi)R_{1n}(\varphi), \end{aligned}$$

assuming the properties

(a). $\sup_{\varphi\in\dot{\mathbb{F}}} |n^{-1}\sum_i[\varphi^2(X_i) - \Sigma_\varphi]|$ converges P_0-a.s. to zero.
(b). $\sup_{\varphi\in\dot{\mathbb{F}}} \widehat{\alpha}_n(\varphi)|R_{1n}(\varphi)| = o_p(1)$.

By A4, the process $Z_n = \Sigma^{-\frac{1}{2}}Y_n$ converges weakly to a centered Gaussian process with variance Σ_φ, uniformly on $\dot{\mathbb{F}}$, which proves (a). To prove (b), let $R_{1n} = R_{1n}^+ + R_{1n}^-$ with

$$R_{1n}^+ = n^{-1}\sum_{i;a_i>0} \frac{a_i^3}{1+\widehat{\alpha}_n a_i}, \quad R_{1n}^- = n^{-1}\sum_{i;a_i<0} \frac{a_i^3}{1+\widehat{\alpha}_n a_i},$$

then $0 \le R_{1n}^+ \le n^{-1}\sum_i |a_i|^3$ and it is a uniform $O_p(1)$ by A3, therefore $\widehat{\alpha}_n R_{1n}^+ = o_p(1)$ uniformly, by Lemma 9.1. For the convergence of R_{1n}^- under the constraint $\widehat{\alpha}_n(\varphi) > 0$, from (9.4) we have

$$Y_n(\varphi) + n^{\frac{1}{2}}\widehat{\alpha}_n^2(\varphi)R_{1n}^-(\varphi) = n^{\frac{1}{2}}\widehat{\alpha}_n(\varphi)\left\{\Sigma_\varphi + o_p(1)\right\}, \qquad (9.5)$$

the right-hand term is positive, R_{1n}^- is negative and $\sup_\varphi |Y_n(\varphi)| = O_p(1)$, hence $n^{\frac{1}{2}}\widehat{\alpha}_n^2(\varphi)R_{1n}^-(\varphi)$ is a uniform $O_p(1)$ and by (9.5)

$$\sup_{\varphi\in\dot{\mathbb{F}}} n^{\frac{1}{2}}\widehat{\alpha}_n(\varphi) = O_p(1).$$

By (9.4) and (A3–A4), we have

$$P_0\Big(\max_{i\le n}\sup_{\varphi\in\dot{\mathbb{F}}} a_i^2(\varphi) > C\Big) \le C^{-n}E_0^n \sup_{\varphi\in\dot{\mathbb{F}}} a_i^2(\varphi)$$

and it tends to zero as n tends to infinity, for every $C > E_0\sup_\varphi a_i^2(\varphi)$ so

$$\max_{i\le n}\sup_{\varphi\in\dot{\mathbb{F}}} a_i^2(\varphi) = O_p(1).$$

By Lemma 9.1, it follows

$$\begin{aligned} \sup_{\varphi\in\dot{\mathbb{F}}} \widehat{\alpha}_n(\varphi)|R_{1n}^-(\varphi)| &\le n^{-1}\sup_\varphi \max_{i\le n} a_i^2(\varphi)\Big|\sum_{i;a_i(\varphi)<0} \frac{\widehat{\alpha}_n(\varphi)a_i(\varphi)}{1+\widehat{\alpha}_n(\varphi)a_i(\varphi)}\Big| \\ &\le n^{-1}\max_{i\le n}\sup_\varphi a_i^2(\varphi)\,\sup_\varphi \sum_{i;a_i(\varphi)>0}\Big|1 - \frac{1}{1+\widehat{\alpha}_n(\varphi)a_i(\varphi)}\Big| \\ &= o_p\Big(\max_{i\le n}\sup_\varphi a_i^2(\varphi)\Big) = o_p(1) \end{aligned}$$

and $\sup_\varphi |R_{1n}^-(\varphi)| = O_p(1)$, which ends the proof of (b). By (a), (b) and (9.4), if $\widehat{\alpha}_n(\varphi) > 0$, then $n^{\frac{1}{2}}\widehat{\alpha}_n(\varphi) = (\Sigma_\varphi)^{-\frac{1}{2}}\{Z_n(\varphi) + \varepsilon_n(\varphi)\}$ where $\sup_{\dot{\mathbb{F}}} |\varepsilon_n(\varphi)|$ converges in probability to zero, otherwise $\widehat{\alpha}_n(\varphi) = 0$. It follows that

$$\begin{aligned} n^{\frac{1}{2}}\widehat{\alpha}_n(\varphi) &= (\Sigma_\varphi)^{-\frac{1}{2}}\{Z_n(\varphi) + \varepsilon_n(\varphi)\}1_{\{Z_n(\varphi)+\varepsilon_n(\varphi)>0\}} \\ &= (\Sigma_\varphi)^{-\frac{1}{2}} Z_n(\varphi) 1_{\{Z_n(\varphi)>0\}} + o_p(1), \end{aligned} \tag{9.6}$$

with a uniform $o_p(1)$ over $\dot{\mathbb{F}}$. By integration of R_{1n}, we get

$$l_n(\widehat{\alpha}_n(\varphi), \varphi) = n^{\frac{1}{2}}\widehat{\alpha}_n(\varphi) Y_n(\varphi) - \frac{n}{2}\widehat{\alpha}_n^2(\varphi)\, n^{-1} \sum_i \varphi^2(X_i) + R_{2n}(\varphi)$$

where

$$R_{2n} = \sum_i \int_0^{\widehat{\alpha}_n(\varphi)} \frac{a_i^3 \alpha^2}{1+\alpha a_i}\, d\alpha = \sum_i \widehat{\alpha}_n \frac{\alpha_n^{*2} a_i^3}{1+\alpha_n^* a_i}$$

with $\alpha_n^*(\varphi)$ between 0 and $\widehat{\alpha}_n(\varphi)$. As the functions $\alpha \mapsto a_i^3(1+\alpha a_i)^{-1}$ are decreasing and by the bound of R_{1n}, it follows that $\sup_{\dot{\mathbb{F}}} n^{-1}|R_{2n}(\varphi)|$ is a $o_p(1)$ uniformly over $\dot{\mathbb{F}}$ and

$$l_n(\widehat{\alpha}_n(\varphi), \varphi) = n^{\frac{1}{2}}\widehat{\alpha}_n(\varphi) Y_n(\varphi) - \frac{n}{2}\widehat{\alpha}_n^2(\varphi)\Sigma_\varphi + o_p(1)$$

uniformly on $\dot{\mathbb{F}}$. The expression (9.6) of $\widehat{\alpha}_n(\varphi)$ implies a uniform approximation of T_n as

$$T_n = 2\sup_{\varphi\in\dot{\mathbb{F}}} l_n(\widehat{\alpha}_n(\varphi), \varphi) = \sup_{\varphi\in\dot{\mathbb{F}}}[Z_n^2(\varphi)1_{\{Z_n(\varphi)>0\}}] + o_p(1).$$

□

Let A_n be local alternatives of mixture densities $g_n = f_0\{1 + \lambda_n f_0^{-1}(f_n - f_0)\}$, under probability measures P_n, with

$$\begin{aligned} \lambda_n &= n^{-\frac{1}{2}} a_n \\ f_n &= f_0\{1 + \rho_n^{-1} h_n\}, \end{aligned}$$

such that the parameter a_n converges to a non zero limit a, $\rho_n = o(n^{\frac{1}{2}})$ is the convergence rate of f_n to f_0, $h_n = \rho_n(f_n - f_0)f_0^{-1}$ converges uniformly to a function h in $\dot{\mathbb{F}}$ and $\varphi_n = \|h_n\|_2^{-1} h_n$ converges in $L^2(P_n)$ to the function φ_h of $\dot{\mathbb{F}}$, like in A2.

Proposition 9.2. *Under the assumption that the conditions (9.1) are valid under P_n, the statistic T_n converges weakly under the local alternatives A_n to $Z^2(\varphi_h)1_{\{Z(\varphi_h)>0\}}$ where $Z(\varphi_h)$ is a centered Gaussian variable with variance $K(\varphi_h, \varphi_h)$, under fixed alternatives the statistic T_n diverges.*

Proof. Under A_n, the density is

$$\begin{aligned} g_n &= f_0 + \lambda_n(f_n - f_0) = f_0\{1 + \lambda_n \rho_n^{-1} h_n\} \\ &= f_0\{1 + \alpha_n \varphi_n), \end{aligned}$$

where $\varphi_n = h_n \|h_n\|_2^{-1}$ is defined by (9.2) and $\alpha_n = \lambda_n \rho_n^{-1} \|h_n\|_2$, it converges to zero and $n^{\frac{1}{2}}\alpha_n = \rho_n^{-1}\eta_n \|h_n\|_2$ converges to zero.

Under P_n, the function φ_n converges to a non zero limit $\varphi_h = h\,\|h\|_2^{-1}$ and $\int \varphi_n(x) f_0(x)\,dx = \rho_n \|h_n\|_2^{-1} \int (f_n - f_0)(x)\,dx$ converges to zero. The process $Y_n(\varphi_n) = n^{-\frac{1}{2}} \sum_{i=1}^n \varphi_n(X_i)$ is asymptotically equivalent to $Y_n(\varphi_h)$ and it converges weakly under P_n to a centered Gaussian variable G_{φ_h}.

The log-likelihood ratio under P_n and P_0 is

$$\begin{aligned} l_n(\alpha_n(\varphi_n), \varphi_n) &= \sum_{i=1}^n \{\log g_n(X_i) - \log f_0(X_i)\} = \sum_{i=1}^n \log\{1 + \alpha_n \varphi_n(X_i)\}\} \\ &= \alpha_n \sum_{i=1}^n \varphi_n(X_i) - \frac{\alpha_n^2}{2} \sum_{i=1}^n \varphi_n^2(X_i) + o(\alpha_n^2). \end{aligned}$$

The maximum likelihood estimator is such that $\widehat{\alpha}_n - \alpha_n = O_p(n^{-\frac{1}{2}})$, and $n^{\frac{1}{2}}(\widehat{\alpha}_n - \alpha_n) = n^{\frac{1}{2}}\widehat{\alpha}_n + o(1)$, the expansion (9.4) is modified as

$$b_{\varphi_n} = Y_n(\varphi_n) - n^{\frac{1}{2}}\{\widehat{\alpha}_n(\varphi_n) - \alpha_n\}\Big[n^{-1}\sum_i a_i^2 - \{\widehat{\alpha}_n(\varphi_n) - \alpha_n\} R_{1n}(\varphi_n)\Big].$$

By (9.6), we have

$$\begin{aligned} n^{\frac{1}{2}}\widehat{\alpha}_n(\varphi_n) &= (\Sigma_{\varphi_h})^{-\frac{1}{2}} Z_n(\varphi_h) 1_{\{Z_n(\varphi_h) > 0\}} + o_p(1), \\ l_n(\widehat{\alpha}_n(\varphi), \varphi) &= n^{\frac{1}{2}}\widehat{\alpha}_n(\varphi_n) Y_n(\varphi_h) - \frac{n}{2}\widehat{\alpha}_n^2(\varphi_h)\Sigma_{\varphi_h} + o_p(1) \end{aligned}$$

Under fixed alternatives, α_f is strictly positive and $\sup_{\varphi \in \dot{\mathbb{F}}} \widehat{\alpha}_n(\varphi)$ converges in probability to a strictly positive limit, then the second term in the expansion (9.4) of the penalized log-likelihood diverges. □

As the nonparametric convergence rate differs from the parametric rate, the asymptotic behavior of the log-likelihood ratio statistic does not follow the same behavior as in parametric mixtures.

Proposition 9.1 extends straightforwardly to a test of a known densities against a mixture of p densities. For a test of a mixture of known densities as a sub-model of a mixture of densities, let

$$\mathcal{S}_p = \{(\lambda_1, \ldots, \lambda_p)^T \in [0,1]^p; \lambda_1 + \ldots + \lambda_p = 1\}$$

and let $\mathcal{S}_p^* = \mathcal{S}_p \bigcap]0,1]^p$. A mixture of $p+q$ densities of the family $\mathcal{F}$ is denoted

$$g_{p+q;\lambda,\theta} = \sum_{j=1}^{p+q} \lambda_j f_j, \quad \lambda = (\lambda_1, \ldots, \lambda_{p+q}) \in \mathcal{S}_{p+q}, \quad f_1, \ldots, f_{p+q} \in \mathbb{F}.$$

Condition 9.2.

B1. Identifiability: For every $k \leq p+q$, for all probability vectors $(\lambda_1, \ldots, \lambda_k) \in \mathcal{S}_k^\star$, $(\zeta_1, \ldots, \zeta_k) \in \mathcal{S}_k$, and for all densities $(f_1, \ldots, f_k)$ and $(h_1, \ldots, h_k)$ in $\mathbb{F}^k$ such that $j \neq j'$ implies $f_j \neq f_{j'}$, the equality

$$\sum_{j=1}^{k} \lambda_j f_j = \sum_{j=1}^{k} \zeta_j h_j$$

implies the existence of a permutation π in $\mathcal{P}_k$ such that

$$\lambda_j = \zeta_{\pi(j)},\ f_j = h_{\pi(j)},\ 1 \leq j \leq k.$$

The density of the sample $(X_i)_{i=1,\ldots,n}$ under the null hypothesis H_0 is a mixture of p known densities $f_j^{(0)}$, $1 \leq j \leq p$, of $\mathcal{F}$, with unknown mixture probabilities and the alternative is a mixture of $p+q$ densities of $\mathcal{F}$ including the p known densities. Let $f^{(0)} = (f_j^{(0)})_{1\leq j\leq p}$ in $\mathcal{F}^p$, with distinct components, the null hypothesis is

$$H_0 : \{g = g_{p;\lambda,f^{(0)}} : \lambda = (\lambda_1, \ldots, \lambda_p) \in \mathcal{S}_p^*\}.$$

A density of the alternative is

$$g_{p+q;\lambda,f} = \sum_{j=1}^{p} \lambda_j f_j^{(0)} + \sum_{k=1}^{q} \lambda_{p+k} f_{p+k}, \tag{9.7}$$

where $\lambda = (\lambda_1, \ldots, \lambda_{p+q})^T$ in $\mathcal{S}_{p+q}$ and the log-likelihood ratio test statistic for the hypothesis H_0 is

$$T_n = 2 \sup_{\lambda\in\mathcal{S}_{p+q}} \sup_{f\in\mathbb{F}^q} \sum_i \{\log g_{(p+q;\lambda,f)}(X_i) - \log g_0(X_i)\}.$$

Let $\lambda^{(0)} = (\lambda_j^{(0)})_{j=1,\ldots,p}$ be the true unknown value of the mixture probabilities in $\mathcal{S}_p^*$, the mixture density under H_0 is denoted $g_0 = \sum_{j=1}^p \lambda_j^{(0)} f_j^{(0)}$.

Example 8.1. Let $p = 1$ and $q = 2$, the density of the alternative is

$$\begin{aligned} g_{\lambda,f} &= \lambda_0 f_0 + \lambda_1 f_1 + (1 - \lambda_0 - \lambda_1) f_2 \\ &= f_0 + \lambda_1 (f_1 - f_2) + (1 - \lambda_0)(f_2 - f_0), \end{aligned}$$

the parameters of interest for the homogeneity test of a single density f_0 are $\alpha_1 = \lambda_1 \|f_0^{-1}(f_1 - f_2)\|$ and $\alpha_2 = (1-\lambda_0)\|f_0^{-1}(f_2 - f_0)\|$. The expression of $g_{\lambda,f} - f_0$ may be modified by a permutation of the indices as

$$g_{\lambda,f} - f_0 = (1 - \lambda_0 - \lambda_1)(f_2 - f_1) + (1 - \lambda_0)(f_1 - f_0),$$

the reparametrization is not unique but the expressions of the density $g_{\lambda,f}$ are equivalent.

Under H_0 and by the identifiability condition, the equality

$$\sum_{k=1}^{p} \lambda_k f_k^{(0)} + \sum_{k=1}^{q} \lambda_{p+k} f_{p+k} = \sum_{j=1}^{p} \lambda_j^{(0)} f_j^{(0)}$$

with the constraints $\sum_{j=1}^{p} \lambda_j^{(0)} = \sum_{j=1}^{p+q} \lambda_j = 1$, imply that $\lambda_k^{(0)} \geq \lambda_k$ for $k = 1, \ldots, p$, and

$$g_{\lambda,f} = g_0 + \sum_{k=1}^{p} (\lambda_k - \lambda_k^{(0)}) f_k^{(0)} + \sum_{j=1}^{q} \lambda_{p+j} f_{p+j}. \tag{9.8}$$

Let $\mathcal{C}_{q,p}$ be the set of maps from $\{1, \ldots, q\}$ into a subset of $\{1, \ldots, p\}$ such that $c(j) = k$ if and only if $f_{p+j} = f_k^{(0)}$, for k in $\{1, \ldots, p\}$ and j in $\{1, \ldots, q\}$ then

$$\begin{aligned} g_{\lambda,f} = g_0 &+ \sum_{k=1}^{p} \Big(\lambda_k + \sum_{c \in \mathcal{C}_{q,p}} \sum_{j=1}^{q} \lambda_{p+j} 1_{\{c(j)=k\}} - \lambda_k^{(0)} \Big) f_k^{(0)} \\ &+ \sum_{c \in \mathcal{C}_{q,p}} \sum_{j=1}^{q} \lambda_{p+j} (f_{p+j} - f_{c(j)}^{(0)}), \end{aligned}$$

the first sum is a mixture of p densities with unknown mixing probabilities and the second sum is zero under the hypothesis H_0. Under H_0, there exists c in $\mathcal{C}_{q,p}$ such that $\lambda_k^{(0)} = \lambda_k + \sum_{j=1}^{q} \lambda_{p+j} 1_{\{c(j)=k\}}$ and $f_{p+j} - f_{c(j)}^{(0)} = 0$. Vectors $u_f(c)$ and $\alpha_f(c)$ are defined for a map c of $\mathcal{C}_{q,p}$ by their components

$$\begin{aligned} u_j(c) &= \|g_0^{-1}(f_{p+j} - f_{c(j)}^{(0)})\|_{L^2}, \\ \alpha_j(c) &= \lambda_{p+j} u_j(c), \ j = 1, \ldots, q, \end{aligned} \tag{9.9}$$

Condition 9.3.
B2. For every $j = 1, \ldots, q$, the $L_2(\mu)$-derivative φ_0 at $f^{(0)}$ in $\dot{\mathbb{F}}^q$ is such that there exists c in $\mathcal{C}_{q,p}$ with $c(j) = k$ and

$$\lim_{u_j(c) \to 0} u_j(c)^{-1} \|g_0^{-1}(f_{c(j)}^{(0)} - f_{p+j} - u_j(c)\varphi_{0k}\|_{L_2(F(X)} = 0$$

B3. The variables $\sup_{\lambda\in\mathcal{S}_{p+q}}\sup_{f\in\mathbb{F}^q}|\log g_{\lambda,f}|$ and $\sup_{\varphi\in\dot{\mathbb{F}}^p}(\|\varphi(X)\|^3)$ are P_0- integrable.
B4. There exists a centered Gaussian process G on $\dot{\mathbb{F}}^p$ such that

$$\sup_{\varphi\in\dot{\mathbb{F}}^p}\left|n^{-\frac{1}{2}}\sum_{i=1}^{n}\varphi(X_i)-G(\varphi)\right|\xrightarrow[n\to\infty]{P_0}0.$$

Under the condition D2, there exists c in $\mathcal{C}_{q,p}$ such that the derivative φ_f is the q dimensional vector with components

$$\begin{aligned}\varphi_{fj}(c) &= u_j^{-1}(c)g_0^{-1}(f_{p+j}-f_k^{(0)})1_{\{c(j)=k\}},\ \text{if } f_{p+j})\neq f_k^{(0)}, \qquad (9.10)\\ &= \varphi_j^{(0)},\ \text{if } f_{p+j}=f_k^{(0)}1_{\{c(k)=j\}},\end{aligned}$$

for $j=1,\ldots,q$. The density of the alternative is expressed as

$$g_{p+q;\lambda,f}=g_0\{1+\alpha_f^T(c)\varphi_f(c)\},$$

this reparametrization is not unique but the expressions of the density $g_{\lambda,f}$ are equivalent under different reparametrizations, like in the examples. The null hypothesis H_0 is equivalent to

$$H_0:\sup_{c\in\mathcal{C}_{q,p}}\sup_{f\in\mathbb{F}^q}\alpha_f(c)=0.$$

The test statistic for the hypothesis H_0 has the form

$$T_n=2\sup_{c\in\mathcal{C}_{q,p}}\sup_{\alpha\in\mathcal{S}_{p+q}}\sup_{\varphi\in\dot{\mathbb{F}}^q}\sum_i\log\{1+\alpha_f^T(c)\varphi_f(c,X_i)\}.$$

A q-dimensional process Y_n is defined on $\mathbb{F}^q$ as

$$Y_n(\varphi)=n^{-\frac{1}{2}}\sum_{i=1}^{n}\varphi(X_i),$$

its the covariance matrix is denoted $\Sigma(\varphi)$. Under the conditions B, the process Y_n converges weakly to a centered Gaussian process Y with covariance Σ, uniformly on $\dot{\mathbb{F}}^q$.

Lemma 9.2. *Under the conditions (9.2), (9.3) and P_0, the maximum likelihood estimator of α is such that $\sup_{\varphi\in\dot{\mathbb{F}}^q}\|\widehat{\alpha}_{n\varphi}\|$ converges in probability to zero.*

Proposition 9.3. *Under the conditions (9.2) and (9.3), the likelihood ratio test statistic T_n converges weakly under H_0 to*

$$T_0=\sup_{\varphi\in\dot{\mathbb{F}}^q}Z^T(\varphi)Z(\varphi)1_{\{Z(\varphi)>0\}}$$

where $Z=\Sigma^{-\frac{1}{2}}Y$ and Y is a continuous centred Gaussian process defined on $\dot{\mathbb{F}}^q$, with covariance function Σ.

Proof. The proof takes up the same arguments as for Proposition 9.1, with fixed values of φ in $\dot{\mathbb{F}}^q$ and the approximation (9.6) of $n^{\frac{1}{2}}\widehat{\alpha}_n(\varphi)$

$$l_n(\widehat{\alpha}_n(\varphi), \varphi) = n^{\frac{1}{2}}\widehat{\alpha}_n^T(\varphi)Y_n(\varphi) - \frac{n}{2}\widehat{\alpha}_n^T(\varphi)\Sigma_\varphi\widehat{\alpha}_n(\varphi) + o_p(1)$$

uniformly on $\dot{\mathbb{F}}$. □

Let A_n be local alternatives of mixture densities (9.7) under probability measures P_n

$$g_{p+q;\lambda_n,f_n} = \sum_{k=1}^{p} \lambda_{nk} f_k^{(0)} + \sum_{j=1}^{q} \lambda_{np+j} f_{np+j},$$

with mixing probabilities

$$\lambda_{nk} = \lambda_{0k}\{1 + n^{-\frac{1}{2}}a_{nk}\},\ k = 1,\ldots,p,$$
$$\lambda_{np+j} = n^{-\frac{1}{2}}a_{np+j},\ j = 1,\ldots,q,$$

with a vector a_n converging to strictly positive limits a in $\mathbb{R}^{p+q}$, and for the densities f_{np+j}, $j = 1,\ldots,q$ their exists $\rho_n = o(n^{\frac{1}{2}})$ and a component of the density $f^{(0)}$ such that

$$f_{np+j} = f_{c(j)}^{(0)}\{1 + \rho_n^{-1}h_{nj}\}$$

for a map c of $\mathcal{C}_{q,p}$. The functions h_{nj} converge uniformly to functions h_j and we have

$$f_{n,p+j} - f_{c(j)}^{(0)} = f_{c(j)}^{(0)}\rho_n^{-1}h_{n,j} = O(\rho_n^{-1}).$$

The parameters $\alpha_{nj} = \lambda_{np+j}\rho_n^{-1}h_{n,j}$ are $o(n^{-\frac{1}{2}})$ for every $j = 1,\ldots,q$, the normalized functions φ_{nj} converge uniformly to non zero functions φ_{hj} depending on a normalized functions $\|h_j\|^{-1}h_j$. The proof of Proposition 9.2 extends to $\dot{\mathbb{F}}^q$ for $q \geq 1$.

Proposition 9.4. *Assuming that the condition (9.3) is valid under the local alternatives A_n and under (9.2), the statistic T_n converges on weakly $\dot{\mathbb{F}}^q$ to $Z^T(\varphi_h)Z(\varphi_h)1_{\{Z(\varphi_h)>0\}}$ where $Z(\varphi_h)$ is a centered Gaussian vector with variance $K(\varphi_h, \varphi_h)$. Under fixed alternatives, the statistic T_n diverges.*

9.3 Homogeneity test for a mixture of densities

In a mixture model of two densities, let

$$g_{\lambda,f,f'} \in \mathbb{G} = \lambda f + (1-\lambda)f',$$

be a mixture of unknown densities f and f' of $\mathbb{F}$ and let H_0 be the hypothesis of a single density of $\mathbb{F}$. By symmetry of $(1-\lambda, f')$ and (λ, f) in the distribution mixture, an identifiable model is defined under the alternative of $\lambda \in [0, \frac{1}{2}]$, without loss of generality. The hypothesis H_0 is satisfied if and only if one of the equalities holds: $\lambda = 0$ or $f = f' = f_0$. The approach of section 9.2 is modified with the parameters

$$\alpha_f = \lambda u_f,$$
$$\beta_{f'} = (1-\lambda) u_{f'},$$

and the hypothesis H_0 is equivalent to $\sup_{f\in\mathbb{F}} \alpha_f = \sup_{f'\in\mathbb{F}} \beta_{f'} = 0$. We assume the conditions

Condition 9.4.
A'1. The probability λ and the densities of $\mathbb{F}$ are identifiable: if there exist λ_1 and λ_2 in $]0, \frac{1}{2}]$, f_1 and f_2, f_1' and f_2' in $\mathbb{F}$ such that $\lambda_1 f_1 + (1-\lambda_1) f_1 = \lambda_2 f_2 + (1-\lambda_2) f_2'$, then $\lambda_1 = \lambda_2$, $f_1 = f_2$ and $f_1' = f_2'$,

with A2, A3 and A4. The mixture density is written as

$$\begin{aligned} g_{\lambda,f,f'} &= f_0\{1 + \lambda f_0^{-1}(f - f_0) + (1-\lambda) f_0^{-1}(f' - f_0)\}, \\ &= f_0\{1 + \alpha_f \varphi_f + \beta_{f'} \varphi_{f'}\} \end{aligned}$$

and the likelihood ratio test statistic is

$$S_n = 2 \sup_{f\neq f'\in\mathbb{F}} \sup_{\alpha_f,\beta_{f'}\in\mathbb{U}/2} \sum_{1\leq i\leq n} \log\{1 + \alpha_f \varphi_f(X_i) + \beta_{f'} \varphi_{f'}(X_i)\}.$$

Let $Y_n(\varphi) = (Y_n(\varphi_f), Y_n(\varphi_{f'}))$, the variance matrix Σ_φ of the vector Y_n is singular at $f = f'$ so Proposition 9.1 cannot be extended to S_n without a separation condition $|\varphi_f - \varphi_{f'}| > c$ for some constant $c > 0$. To avoid this condition, the condition A2 is replaced by a second order differentiability of the class $\mathbb{F}$, let

Condition 9.5.
A'2. The second order $L_2(\mu)$-derivative ψ of f exists in the tangent space $\ddot{\mathbb{F}}$ of $\dot{\mathbb{F}}$, with $v_{ff'} = \|\varphi_f - \varphi_0\|$ it satisfies

$$\lim_{v_{ff'}\to 0} v_{ff'}^{-1} \left\| \varphi_f - \varphi_{f'} - v_{ff'} \psi_f \right\|_2,$$

A'3. The variables $\sup_{\lambda\in[0,1]} \sup_{f\neq f'\in\mathbb{F}} |\log g_{\lambda,f,f'}|$ and $\sup_{\psi\in\ddot{\mathbb{F}}} \|\psi(X)\|^3$ are P_0-integrable, A'4. There exists a centered Gaussian variable G_ψ such that

$$\sup_{\psi\in\ddot{\mathbb{F}}} \left| n^{-\frac{1}{2}} \sum_{i=1}^{n} \psi(X_i) - G_\psi \right| \xrightarrow[n\to\infty]{P_0} 0.$$

Under the conditions, we consider v in the space

$$\mathbb{U}_2 = \{v_{ff'} = \|\varphi_f - \varphi_{f'}\|; f, f' \in \mathbb{F}\} \tag{9.11}$$

and the functions

$$\psi_{ff'} = v_{ff'}^{-1}(\varphi_f - \varphi_{f'}), \text{ if } f \neq f',, \tag{9.12}$$

in the space of the second derivatives $\ddot{\mathbb{F}}$ of $\mathbb{F}^2$, it reduces to ψ_f given by the condition A'2 if $f = f'$. By (9.12), the mixture density is now denoted

$$\begin{aligned} g_{\lambda,f,f'} &= f_0\{1 + (\alpha_f + \beta_{f'})\varphi_f + \beta_{f'} v_{ff'} \psi_{f'}\} \\ &= f_0\{1 + \gamma_{1ff'}\varphi_f + \gamma_{2ff'}\psi_{ff'}\}. \end{aligned}$$

Let $\gamma_{ff'}$ be the parameter with components $\alpha_f + \beta_{f'}$ and $\beta_{f'} u_{ff'}$, the hypothesis H_0 is now equivalent to

$$H_0 : \sup_{f,f' \in \mathbb{F}} \gamma_{ff'} = 0.$$

For a sample $(X_i)_{1 \leq i \leq n}$ of X, the likelihood ratio statistic of H_0 is

$$\begin{aligned} S_n &= 2 \sup_{\lambda \in]0,\frac{1}{2}]} \sup_{f,f' \in \mathbb{F}} \sum_{1 \leq i \leq n} \{\log g_{\lambda,f,f'}(X_i) - \log f_0(X_i)\} \\ &= 2 \sup_{\gamma_1 \in \frac{1}{2}\mathbb{U}, \gamma_2 \in \mathbb{U}_2} \sup_{\varphi \in \dot{\mathbb{F}}} \sup_{\psi \in \ddot{\mathbb{F}}} \sum_{1 \leq i \leq n} \log\{1 + (\varphi(X_i), \psi(X_i))\gamma\}. \end{aligned}$$

Let $\widehat{\gamma}_n(\varphi, \psi)$ be the maximum likelihood estimators of γ as φ and ψ are fixed. The arguments of Lemma 9.1 apply to prove the consistency of the maximum likelihood estimator $\widehat{\gamma}_n(\varphi, \psi)$ of γ, at fixed φ and ψ.

Lemma 9.3. *Under the conditions (9.4) and (9.5),* $\sup_{\varphi \in \dot{\mathbb{F}}, \psi \in \ddot{\mathbb{F}}} \|\widehat{\gamma}_n(\varphi, \psi)\|$ *converges in probability to zero under* P_0.

Replacing the function φ by the vector $(\varphi, \psi)^T$, the process Y_n is replaced by $\widetilde{Y}_n$ defined on $\dot{\mathbb{F}} \times \ddot{\mathbb{F}}$ with components Y_n and

$$Y_{2n}(\psi) = n^{-\frac{1}{2}} \sum_{i=1}^{n} \psi(X_i),$$

let $\widetilde{\Sigma}$ be the covariance function of $\widetilde{Y}_n$ and let $\widetilde{K}$ be defined from $\widetilde{\Sigma}$ like K in Proposition 9.1.

Proposition 9.5. *Under* H_0 *and the conditions (9.4) and (9.5), the statistic* S_n *converges weakly to* $\sup_{\varphi \in \dot{\mathbb{F}}, \psi \in \ddot{\mathbb{F}}} Z^T(\varphi, \psi) Z(\varphi, \psi) 1_{\{Z(\varphi,\psi) > 0\}}$ *where* Z *is a continuous centered Gaussian process on* $\dot{\mathbb{F}} \times \ddot{\mathbb{F}}$*, with covariance function* $\widetilde{K}$.

The expression of the log-likelihood has the same form as with a density (9.1) and the proof is similar to the proof of Proposition 9.1.

Local alternatives A_n to the null hypothesis are defined by densities $g_n = \lambda_n f_{1n} + (1-\lambda_n) f_{2n}$, under probability measures P_n, with

$$\lambda_n = \lambda_0\{1 + n^{-\frac{1}{2}} a_n\},$$
$$f_{kn}(x) = f_0(x)\{1 + \rho_n^{-1} h_{kn}(x)\},\ k = 1, 2,$$

where u_{fn} has the components $u_{kn} = \rho_n^{-1}\|h_{kn}\|$ and it converges to zero, $h_n = (h_{1n}, h_{2n})^T$ converges uniformly to a function h, $\rho_n = o(n^{\frac{1}{2}})$. The norms $u_{f_{kn}}$ define the derivatives $\varphi_{f_{kn}} = h_{kn}\|h_{kn}\|^{-1}$ in $\dot{\mathbb{F}}$ and $v_{f_{1n},f_{2n}} = \|\varphi_{f_{1n}} - \varphi_{f_{2n}}\|$. The parameter of the test is γ_n with components $\gamma_{1n} = \lambda_n u_{f_{1n}} + (1-\lambda_n) u_{f_{2n}} = \rho_n^{-1}\{\lambda_0\|h_{1n}\| + (1-\lambda_0)\|h_{2n}\|\} + \lambda_0 n^{-\frac{1}{2}} \rho_n^{-1} a_n(\|h_{1n}\| - \|h_{2n}\|)$ and $\gamma_{2n} = (1-\lambda_n) u_{f_{2n}} v_{f_{1n},f_{2n}}$, they converge to zero as n tends to infinity.

Proposition 9.6. *Assuming that the conditions (9.5) are valid under the local alternatives and under (9.4), the statistic S_n converges weakly to $Z^T(\varphi_h, \psi_h) Z(\varphi_h, \psi_h) 1_{\{Z(\varphi_h)>0\}}$ where $Z = \Sigma^{-\frac{1}{2}} Z_n$ is a centered Gaussian variable with variance K, under fixed alternatives S_n diverges.*

Proof. The mixture density develops in a neighbourhood of the density g_0 as

$$\begin{aligned}
g_n &= f_0 + f_0\rho_n^{-1}\{\lambda_0 h_{1n} + (1-\lambda_0) h_{2n}\} + n^{-\frac{1}{2}} a_n \lambda_0 (f_{01} - f_{02}) \\
&\quad - \lambda_0 f_0 \rho_n^{-1} n^{-\frac{1}{2}} u_n h_{2n} \\
&= f_0\{1 + \rho_n^{-1} g_{\lambda_0, h_n} + n^{-\frac{1}{2}} a_n \lambda_0 (f_{01} - f_{02})\}
\end{aligned}$$

denoted $g_n = g_0(1 + W_n)$ where $\rho_n W_n$ converges to a mixture $g_{\lambda_0,h}$ and g_{λ_0,h_n} has a second order expansion like the mixture density. The maximum likelihood estimator of the parameter γ_n is such that $\widehat{\gamma}_n - \gamma$ has an expansion like in Proposition 9.2

$$\begin{aligned}
n^{\frac{1}{2}}\{\widehat{\gamma}_n(\psi_n) - \gamma_n(\psi_n)\} &= (\Sigma_{\psi_h})^{-\frac{1}{2}} Z_n(\psi_h) 1_{\{Z_n(\psi_h)>0\}} + o_p(1), \\
l_n(\widehat{\gamma}_n(\psi), \psi) &= n^{\frac{1}{2}}\{\widehat{\gamma}_n(\psi_n) - \gamma_n(\psi_n)\}^T Y_n(\psi_n) \\
&\quad - \frac{n}{2}\{\widehat{\gamma}_n(\psi_n) - \gamma_n(\psi_n)\}^T \Sigma_{\psi_h}\{\widehat{\alpha}_n(\psi_n) - \gamma_n(\psi_n)\} \\
&\quad + o_p(1)
\end{aligned}$$

and it converges weakly. Under fixed alternatives with distinct densities f_1 and f_2, the derivatives φ_{f_1} and φ_{f_2} are distinct, and the log-likelihood ratio is parametrized like in Proposition 9.2, then the statistic diverges. □

Proposition 9.5 extends to test a mixture of p densities against an alternative of a mixture of $p+q$ densities like for contamination models, the density of the sample $(X_i)_{i=1,\ldots,n}$ under the null hypothesis H_0 is a mixture of p unknown densities of a family $\mathcal{F}$, under the alternative it is a mixture of $p+q$ unknown densities of $\mathcal{F}$ including the p densities of the hypothesis H_0, the mixing probabilities are unknown. Let $f = (f_j)_{1\le j\le p}$ in $\mathcal{F}^p$ with distinct components, the null hypothesis is

$$H_0 : \{g = g_{p;\mu,f} : \mu \in \mathcal{S}_p^*\}$$

and a density of the alternative is

$$g_{p+q;\lambda,f} = \sum_{j=1}^{p} \lambda_j + \sum_{k=1}^{q} \lambda_{p+k} f_{p+k},$$

with λ in $\mathcal{S}_{p+q}$. Like in (9.8) the density is written as

$$\begin{aligned} g_{\lambda,f} &= g_0 + \sum_{k=1}^{p} (\lambda_k - \mu_k) f_k + \sum_{j=1}^{q} \lambda_{p+j} f_{p+j} \\ &= g_0 + \sum_{k=1}^{p} \Big(\lambda_k + \sum_{c\in\mathcal{C}_{q,p}} \sum_{j=1}^{q} \lambda_{p+j} 1_{\{c(j)=k\}} - \mu_k \Big) f_k \\ &\quad + \sum_{c\in\mathcal{C}_{q,p}} \sum_{j=1}^{q} \lambda_{p+j} (f_{p+j} - f_{c(j)}). \end{aligned}$$

The density $g_{\lambda,f}$ is reparametrized with $\gamma_{ff'}$, φ_f and $\psi_{ff'}$ as $g_{\lambda,f} = g_0\{1 + \gamma_{1ff}\,\varphi_f + \gamma_{2ff}\,\psi_{ff}\}$. Propositions 9.5 and 9.6 apply to the log-likelihood ratio test statistic S_n with q dimensional vectors Y_n and Z_n.

9.4 Nonparametric high dimensional mixtures

On a probability space $(\Omega, \mathcal{A}, P)$ we consider a random variable X with value in a measurable metric space $(\mathbb{X}, \mathcal{X})$, with distribution function absolutely continuous with respect to a measure ν. On the product space, the density of $(X_j)_{j\ge 1}$ is supposed to be a mixture of a countable number of densities, corresponding to the aggregation of an infinity of sub-population in an inhomogeneous general population. For a n-sample of the variable X, the number of components $p_n = o(n)$ of the mixture is supposed to increase as n tends to infinity, at a rate sufficiently low to ensure that the sub-populations present in the sample are observed at a rate proportional to their size in the general infinite population. Since the p_n mixture probabilities sum up to one, they are indexed by n and λ_{jn} converges zero to as n tends to infinity, for every j larger than some unknown number of components j_0. The density of X of the sample is

$$g_n = \sum_{j=1}^{p_n} \lambda_{jn} f_{jn}, \tag{9.13}$$

where the densities f_{jn} belongs to the same family $\mathbb{F}$ for $j = 1, \ldots, p_n$.

The estimator and the tests have to differentiate the densities of the sub-populations present in the sample, in sub-samples of smaller dimensions. In order to avoid models with a too large number of components for the true number actually present in the sample, penalization functions are introduced in the estimation criterion. They are chosen in order to avoid too small estimators of mixture probabilities and too close estimators of densities in the L_2-distance. So, the actual sub-populations are not arbitrarily split in smaller ones. The results presented here for mixture models are different from those for a penalized likelihood with a number of parameters increasing due to the singularity of the variance matrix and they generalize results of Fan and Peng (2004).

The true density g_0 belongs to the family $\mathbb{F}_{p_n}$ of mixtures of p_n densities, and the true number k_n of distinct components with non zero coefficients is supposed to be strictly smaller than p_n in the estimation procedure. The number of components of the model is denoted $p_n = k_n + s_n$, where k_n increases with n, $k_n = o(n)$ and a number s_n of nuisance components among p_n disappears from the expression of g_0. The model is semi-parametric with a vector of parametric mixture probabilities $\lambda_{(n)} = (\lambda_{1n}, \ldots, \lambda_{p_n n})^T$ and the density functions $f_{(n)} = (f_1, \ldots, f_{p_n})^T$. For every n, the family of the mixture densities is $\mathcal{G}_n = \{g_n = g_{\lambda_{(n)}, f_{(n)}}, \lambda_{(n)} \in S^*_{p_n}, f_{(n)} \in \mathbb{F}^{p_n}\}$ defined by (9.13) with identifiable components.

9.5 Penalized likelihood

The parameter vector $\beta_n = (\lambda_{(n)}^T, f_{(n)}^T)^T$ determines the mixture density $g_{\lambda_{(n)}, f_{(n)}}$ with p_n components by (9.13), the density g_0 of a sub-model is defined by a parameter vector $\beta_n^{(0)}$ under a probability P_0. Up to a permutation of the components, the vector of probabilities in the p_n- dimensional model under P_0 is $\lambda_{(n)}^{(0)} = (\lambda_{1n}^{(0)}, \ldots, \lambda_{k_n n}^{(0)}, 0, \ldots, 0)^T$ still denoted $(\lambda_{(n)}^{0T}, 0_{s_n}^T)^T$ with $\lambda_{jn}^{(0)} \neq 0$ for $1 \leq j \leq k_n$, the densities corresponding to the s_n dimensional vector zero 0_{s_n} are not identifiable. The true parameter is $\beta_n^{(0)} = (\lambda_{(n)}^{0T}, f_{(n)}^{0T})^T$ where $f_{(n)}^{(0)}$ is the functional vector with k_n first components in F^{k_n}.

For every densities f in $\mathbb{F}$ of the sub-populations, an estimator $\widehat{f}_n$ is determined in a functional estimation space $\mathbb{F}_n$ and its expectation belongs to a spaces $\widetilde{\mathbb{F}}_n$ that may be chosen as a sub-space of $\mathbb{F}$ and such that $\limsup_n \widetilde{\mathbb{F}}_n = \bigcap_n \bigcup_{m\geq n} \widetilde{\mathbb{F}}_m = \mathbb{F}$. This sequence is an approximation of the density f, as a projection in $\widetilde{\mathbb{F}}_n$, and the convergence rate of this approximation is n^{-c} with $c \in (0, \frac{1}{2}[$. The variance of the approximation is a distance between $\widetilde{\mathbb{F}}_n$ and $\mathbb{F}_n$, it converges with the same rate n^{-2c}, for an optimal c exponent depending on the family $\mathbb{F}$. This choice of an estimating method for $\mathbb{F}$ is expressed by the following conditions. The other conditions of the model are identifiability conditions, integrability and differentiability of the family of the densities and conditions about the rate of the penalization functions with respect to n and p_n.

Condition 9.6.

D1. Identifiability: For every $k \geq 1$, for every probability vectors $(\lambda_1, \ldots, \lambda_k) \in \mathcal{S}_k^\star$, $(\zeta_1, \ldots, \zeta_k) \in \mathcal{S}_k$, and for every densities $(f_1, \ldots, f_k)$ such that $j \neq j'$ implies $f_j \neq f_{j'}$, and for every densities $(h_1, \ldots, h_k) \in \mathbb{F}^k$, the equality

$$\sum_{j=1}^k \lambda_j f_j = \sum_{j=1}^k \zeta_j h_j$$

implies the existence of a permutation $\sigma \in \mathcal{P}_k$ such that

$$\lambda_j = \zeta_{\sigma(j)},\ f_j = h_{\sigma(j)},\ 1 \leq j \leq k.$$

D2. The following variables are uniformly integrable under P_0

$$\sup_{\{f_{(n)}\in\mathbb{F}^{p_n}\}} \sup_{\{g=\lambda_{(n)}^T f_{(n)}, \lambda_{(n)}\in\mathcal{S}_{p_n}, \|\lambda_{(n)}-\lambda_{(n)}^{(0)}\|=o(1)\}} \max_{1\leq j\leq p_n} \|g^{-1} f_j\|^3(X),$$

$$\sup_{\lambda_{(n)}\in\mathcal{S}_{p_n}, f_{(n)}\in\mathbb{F}^{p_n}} \log\left(g_0^{-1} \sum_{j=1}^{p_n} \lambda_j f_j\right).$$

D3. The rate for the estimation of the functions of $\mathbb{F}$ is determined by

$$\sup_{f\in\mathbb{F}} \inf_{f_n\in\widetilde{\mathbb{F}}_n} \|f_n - f\| = O(n^{-c}),$$

$$\sup_{f_n\in\widetilde{\mathbb{F}}_n} \inf_{\widehat{f}_n\in\mathbb{F}_n} \left\|\widehat{f}_n - f_n\right\| = O_{P_0}(n^{-c}).$$

The log-likelihood at $\beta \in \mathcal{S}_{p_n} \times \mathbb{F}_n^{p_n}$ is

$$L_n(\beta) = \sum_{i=1}^n \log\left\{\sum_{j=1}^{p_n} \lambda_{nj} f_j(X_i)\right\}.$$

Penalization functions $p : [0,1] \to \mathbb{R}_+$ and $\pi : L_2(\mathcal{X}, \mathbb{R}) \to \mathbb{R}_+$ are associated to λ_j and f_j, respectively, with $p(0) = \pi(0) = 0$. They are such that p is of class C_2 and π is C_2 for the L_2-norm, i.e. the derivatives of the norm $L_2(\mathcal{X}, \mathbb{R})$ of π' and π'' are finite. For every function $f \in L_2(\mathcal{X}, \mathbb{R})$, let

$$\|\pi'(f)\| = \sup_{h \in L_2(P),\, \|h\|=1} \|\pi'(f)h\|,$$
$$\|\pi''(f)\| = \sup_{h \in L_2(P),\, \|h\|=1} \|h^T \pi''(f)h\|.$$

A penalized log-likelihood is defined by

$$Q_n(\beta) = L_n(\beta) + n\mu_n^2 \sum_{j=1}^{p_n} p(\lambda_j) + n\nu_n^2 \sum_{1 \leq j \neq k \leq p_n} \pi(f_j - f_k), \tag{9.14}$$

where the penalization constants μ_n and ν_n converge to zero as n tends to infinity. Let $\widehat{\beta}_{(n)} = (\widehat{\lambda}_{(n)}^T, \widehat{f}_{(n)}^T)^T$ be the penalized maximum likelihood estimator of β_n in $\mathcal{S}_{p_n} \times \mathbb{F}_n^{p_n}$.

Condition 9.7.
P1. The derivatives of the penalizations have norms

$$a_n = \max_{1 \leq j \leq p_n} \left| p'\left(\lambda_j^{(0)}\right) \right|, \quad A_n = \max_{1 \leq j \neq k \leq p_n} \left\| \pi'\left(f_j^{(0)} - f_k^{(0)}\right) \right\|$$

satisfying

$$\mu_n^2 a_n = O\left(p_n n^{-\frac{1}{2}}\right) \qquad \text{and} \qquad \nu_n^2 A_n = O\left(p_n n^{-c}\right).$$

P'1. For sequences of neighbourhoods converging to $V_n(\lambda^{(0)})$, for $\lambda^{(0)}$ in $\mathcal{S}_{p_n}$, and $W_n(f_{(n)}^{(0)})$, for $f_{(n)}^{(0)}$ in $\mathbb{F}^{p_n}$, the sequences

$$a_n' = \sup_{(\lambda_1, \ldots, \lambda_{p_n}) \in V_n(\lambda^{(0)})} \max_{1 \leq j \leq p_n} |p'(\lambda_j)|$$
$$A_n' = \sup_{(f_1, \ldots, f_{p_n}) \in W_n(f_{(n)}^{(0)}) \times \mathbb{F}^s} \max_{1 \leq j \neq k \leq p_n} \|\pi'(f_j - f_k)\|$$

satisfy

$$\mu_n^2 a_n' = O\left(p_n n^{-\frac{1}{2}}\right) \quad \text{and} \quad \nu_n^2 A_n' = O\left(p_n n^{-c}\right).$$

P2. The second derivatives of the penalizations

$$b_n = \max_{1 \leq j \leq p_n} \left| p''\left(\lambda_j^{(0)}\right) \right|,$$
$$B_n = \max_{1 \leq j < k \leq p_n} \left\| \pi''\left(f_j^{(0)} - f_k^{(0)}\right) \right\|$$

satisfy $\mu_n^2 b_n + \nu_n^2 B_n = o(1)$.
P'2. The sequences

$$b'_n = \sup_{(\lambda_1,\ldots,\lambda_{p_n})\in V_n(\lambda^{(0)})} \max_{1\leq j\leq p_n} |p''(\lambda_j)|,$$
$$B'_n = \sup_{(f_1,\ldots,f_{p_n})\in W_n(f^{(0)}_{(n)})\times\mathbb{F}^s} \max_{1\leq j\neq k\leq p_n} \|\pi''(f_j - f_k)\|$$

satisfy $\mu_n^2 b'_n + \nu_n^2 B'_n = o(p_n^{-\frac{1}{2}})$.
P3. the functions p'' and π'' are Lipschitz continuous and $p'_+(0) > 0$.

The conditions P are used for the convergence rates of the estimators and the conditions P' for their weak convergence. These conditions and the following results concern as well parametric families, with the rate $n^{-\frac{1}{2}}$, as nonparametric families with $c < \frac{1}{2}$, under the identifiability conditions.

9.6 Convergence of the estimators

The convergence rate of the maximum likelihood estimators of the components of λ and f is established by Pons (2009)

Proposition 9.7. *Under conditions (9.6) and P of (9.7), the maximum likelihood estimators with penalizations $\widehat{\lambda}_n$ and $\widehat{f}_{(n)}$ exist such that*

$$\|\widehat{\lambda}_{(n)} - \lambda^{(0)}_{(n)}\| = O_{P_0}(k_n^{\frac{1}{2}} n^{-\frac{1}{2}}), \quad \|\widehat{f}_{(n)} - f^{(0)}_{(n)}\| = O_{P_0}(k_n^{\frac{1}{2}} n^{-c}).$$

Corollary 9.1. *The maximum likelihood estimators without penalization have the same rates of convergence as in Proposition 9.7.*

The conditions of Proposition 9.7 are obviously satisfied without penalization, as p and π are zero. Before proving the asymptotic normality and an "oracle" property for the number of components under stronger conditions, more notations are necessary. For a vector of unknown densities $f_{(n)} = (f_1, \ldots, f_r)^T \in \mathbb{F}^{k_n}$, let $g_n = g_{n,\lambda,f} = \sum_{j=1}^{p_n} \lambda_{jn}^{(0)} f_j$, with $g_n^{(0)} = g_0$, and consider the sequences of vectors and of matrices

$$V_{(n)} = n^{-\frac{1}{2}} \sum_i g_n^{-1}(X_i)\, f_{(n)}(X_i),$$

$$W_n(f_{(n)}, h) = \left[n^{-1} \sum_i g_{n,\lambda,f}^{-1}(X_i) g_{n,\lambda,h}^{-1}(X_i) f_k(X_i) h_l(X_i) \right]_{1\leq k,l\leq k_n}.$$

Then the matrix $W_n(f^{(0)}_{(n)}, f^{(0)}_{(n)})$ is asymptotically equivalent to

$$W_0 = E_0\{g_0^{-2}(X) f_k^{(0)}(X_1) f_l^{(0)}(X)\}_{1\leq k,l\leq k_n}.$$

Finally, for $\lambda_{(n)} \in \mathcal{S}_{k_n}$, the vector of the first two derivatives of $p_{(n)} = (p(\lambda_{1n}), \ldots, p(\lambda_{k_n n}))$ are denoted $p'_{(n)}$ and $p''_{(n)}$ which is a diagonal matrix of $(p''(\lambda_{jn}))_{1\le j\le k_n}$.

Let 1_{k_n} be the vector of $\mathbb{R}^{k_n}$ when all components equal to 1 and I_{k_n} the identity matrix of dimensions $k_n \times k_n$. A k_n dimensional vector is a $o_{P_0}(1_{k_n})$ if all its components are $o_{P_0}(1)$ and a matrix $k_n \times k_n$ is $o_{P_0}(I_{k_n})$ if its columns are all $o_{P_0}(1_{k_n})$. Two vectors of $\mathbb{R}^{k_n}$ or matrices are asymptotically equivalent if their difference is $o_{P_0}(1_{k_n})$ or $o_{P_0}(I_n)$, respectively.

Proposition 9.8. *Under conditions (9.6) and (9.7), and if $p_n = o(n^c)$ the maximum likelihood estimators satisfy $\widehat{\lambda}_{k_n+1} = \ldots = \widehat{\lambda}_{p_n} = 0$, with a probability tending to one. Moreover the vector of k_n non zero components satisfies*

$$n^{\frac{1}{2}}(\widehat{\lambda}_{(n)} - \lambda_{(n)}^{(0)}) = W_0^{-1}\{V_n(f^{(0)}) - \{1_{k_n}^T W_0^{-1} V_n(f^{(0)})\}\{1_{k_n}^T W_0^{-1} 1_{k_n}\}^{-1} 1_{k_n}\} + o_{P_0}(1_{k_n}),$$

it is asymptotically equivalent to a k_n dimensional centred normal variable with variance

$$W_0^{-1} - \{1_{k_n}^T W_0^{-1} 1_{k_n}\}^{-1}(W_0^{-1} 1_{k_n})^{\otimes 2}.$$

If $p_n = o(n^{c/2})$, then

$$n^{\frac{1}{2}}(\widehat{\lambda}_{(n)} - \lambda_{(n)}^{(0)}) = W_0^{-1}\{V_n(f^{(0)}) - \{1_{k_n}^T W_0^{-1} V_n(f^{(0)})\}\{1_{k_n}^T W_0^{-1} 1_{k_n}\}^{-1} 1_{k_n}\} + o_{P_0}(1).$$

Proof. With the notations

$$\alpha_n = p_n^{1/2}(n^{-1/2} + \mu_n^2 a_n p_n^{-1}),$$
$$\alpha'_n = p_n^{1/2}(n^{-c} + \nu_n^2 A_n p_n^{-1}),$$

let $u_{(n)} = (u_1, \ldots, u_{p_n})^T$ be a vector of $\mathbb{R}^{k_n} \times \mathbb{R}_+^{s_n}$ satisfying $\sum_j u_j = 0$ and let $v_{(n)} = (v_1, \ldots, v_{k_n}, \widetilde{f}_1, \ldots, \widetilde{f}_{s_n})^T$ where $v_j \in L_2(\mathcal{X}, \mathbb{R})$ and satisfies $f_j^0 + \alpha'_n v_j \in \mathbb{F}_n$ for $1 \le j \le k_n$ and $\widetilde{f}_j \in \mathbb{F}_n$ for $1 \le j \le s_n$. The partial derivatives of $Q_n(\widetilde{\beta}_n)$ with respect to u_{k_n+j}, for $1 \le j \le s_n$, are

$$\begin{aligned} n^{-1}\frac{\partial Q_n(\widetilde{\beta}_n)}{\partial u_{k_n+j}} &= \alpha_n n^{-1}\sum_{i=1}^n \frac{\widetilde{f}_j(X_i)}{g_0(X_i)} - \alpha_n^2 n^{-1}\sum_{i=1}^n\left\{\sum_{l=1}^{k_n}\frac{u_l(\widetilde{f}_j f_l^{(0)})(X_i)}{g_0^2(X_i)}\right. \\ &\left. + \sum_{l=1}^{s_n}\frac{u_{k_n+l}(\widetilde{f}_j\widetilde{f}_{k_n})(X_i)}{g_0^2(X_i)}\right\}(1 + o_{P_0}(1)) \\ &-\alpha_n\alpha'_n n^{-1}\sum_{i=1}^n\left\{\sum_{l=1}^{k_n}\frac{\lambda_l^{(0)}(v_l\widetilde{f}_j)(X_i)}{g_0^2(X_i)}\right\}\{1 + o_{P_0}(1)) \\ &-\mu_n^2\alpha_n p'(\alpha_n u_{k_n+j})\} \end{aligned}$$

denoted $K_1+K_2+K_3+K_4$. By the weak convergence under D2, it follows

$$K_1 = O_{P_0}(\alpha_n) = O_{P_0}(n^{-\frac{1}{2}}p_n^{\frac{1}{2}}) = o_{P_0}(1),$$
$$K_2 = O_{P_0}(\alpha_n^2 p_n^{\frac{1}{2}}\|u_{(n)}\|) = O_{P_0}(n^{-1}p_n^{\frac{3}{2}}\|u_{(n)}\|) = o_{P_0}(p_n),$$
$$K_3 = O_{P_0}(\alpha_n\alpha_n' p_n^{\frac{1}{2}}\|v_{(n)}\|) = O_{P_0}(p_n^{\frac{3}{2}}n^{-\frac{1}{2}-c}k_n^{\frac{1}{2}}) = o_{P_0}(p_n^{\frac{3}{2}}n^{-c}),$$
$$K_4 = O(\mu_n^2\alpha_n a_n') = O(n^{-\frac{1}{2}}p_n\alpha_n) = 0(p_n^{\frac{3}{2}}n^{-1}) = o_{P_0}(p_n^{\frac{1}{2}}),$$

under P'1 and P4. Then $\max\{K_2,K_3\}$ is the main term in this expansion and it is strictly negative for n large enough. The partial derivatives of $Q_n(\widetilde{\beta}_n)$ with respect to u_{k_n+j} are negative with a probability tending to 1 and Q_n reaches its maximum when $\widehat{\lambda}_{k_n+j}=0$.

As the s_n last components of $\lambda^{(0)}$ are zero, the penalized log-likelihood reaches its maximum as its k_n first components have the values $\widehat{\lambda}_n$ and $\widehat{f}_n$. In the following, these estimators are reduced to their k_n first components. Under D4, P'1 and P'2, with a probability tending to 1

$$\frac{\partial Q_n}{\partial\lambda_j}(\widehat{\lambda}_n,\widehat{f}_n) = n^{\frac{1}{2}}\gamma_n, \quad 1\le j\le k_n, \tag{9.15}$$

where γ_n is the Lagrange multiplier for the constraint $\sum_1^{k_n}\lambda_j=1$.

By an expansion of (9.15) in a neighbourhood of $\lambda^{(0)}$ and from Proposition 9.7, the derivatives of the log-likelihood are

$$\gamma_n 1_{k_n} = \{V_n(\widehat{f}_n) - n^{\frac{1}{2}}\mu_n^2 p_{k_n}'(\lambda^{(0)})\}$$
$$-n^{\frac{1}{2}}\{W_n(\widehat{f}_n,\widehat{f}_n) + \mu_n^2 p_{k_n}''(\lambda^{(0)})\}(\widehat{\lambda}_n-\lambda^{(0)}) + R_{1n} + R_{2n}$$

where the last terms of the expansion for the likelihood and the penalization are respectively $R_{1n} = O_{P_0}(k_n n^{-c}1_{k_n}) = o_{P_0}(n^{-c/2}1_{k_n})$, under condition D2, and $R_{2n} = O_{P_0}(n^{\frac{1}{2}}\|\widehat{\lambda}_n-\lambda^{(0)}\|^2\mu_n^2 1_{k_n}) = o_{P_0}(n^{-\frac{1}{2}}1_{k_n})$, under P3. Then

$$\begin{aligned} n^{\frac{1}{2}}(\widehat{\lambda}_n-\lambda^{(0)}) &= \{W_n(f^{(0)},f^{(0)}) + 0_{P_0}(p_n n^{-c}I) + \mu_n^2 p_{k_n}''(\lambda^{(0)})\}^{-1} \\ &\quad\times\{V_n(f^{(0)}) - \gamma_n 1_{k_n} - n^{\frac{1}{2}}\mu_n^2 p_{k_n}'(\lambda^{(0)}) + o_{P_0}(n^{-c/2}1_{k_n})\} \\ &= W_0^{-1}\{V_n(f^{(0)}) - \gamma_n 1_{k_n}\} + o_{P_0}(n^{-c/2}1_{k_n}), \end{aligned} \tag{9.16}$$

from Proposition 9.7, D4, P'1 and P'2.

Multiplying (9.16) by $1_{k_n}^T$ and from the equality $1_{k_n}^T(\widehat{\lambda}_n-\lambda^{(0)})=0$, it follows

$$\gamma_n = \left\{1_{k_n}^T W_0^{-1}V_n(f^{(0)})\right\}\{1_{k_n}^T W_0^{-1}1_{k_n}\}^{-1} + o_{P_0}(1).$$

Therefore γ_n converges to $\gamma_0 = 1_{k_n}^T W_0^{-1}V_n(f^{(0)})\{1_{k_n}^T W_0^{-1}1_{k_n}\}^{-1}$, and the norm $\|\gamma_0\|$ is a $O(k_n^{-\frac{1}{2}})$ which tends to zero. Finally

$$\begin{aligned} n^{\frac{1}{2}}(\widehat{\lambda}_n-\lambda^{(0)}) = W_0^{-1}\{V_n(f^{(0)}) - \{1_{k_n}^T W_0^{-1}V_n(f^{(0)})\}\{1_{k_n}^T W_0^{-1}1_{k_n}\}^{-1}1_{k_n}\} \\ +o_{P_0}(n^{-c/2}1_{k_n}). \end{aligned}$$

since the norm of the vector 1_{k_n} is $k_n^{\frac{1}{2}}$, the condition for the convergence of p_n implies that $o_{P_0}(n^{-c/2}\|1_{k_n}\|) = o_{P_0}(1)$. □

9.7 Test on the number of components of the mixture

In this section, the likelihood ratio test for the hypothesis of a mixture of k_n densities from $\mathbb{F}$, against an alternative of $k_n + s_n = p_n$ components, is studied as $k_n = o(n)$ tends to infinity with n and as the mixture components are unknown under the null hypothesis. The test statistic is the likelihood ratio without penalization, and we use the notations and the approach of Section 9.3 with the estimation of the k_n first components, under the conditions (9.5).

Condition 9.8.
D4. For every f and f' in $\mathbb{F}$, tangent spaces are associated to functions f and f' by

$$h_{ff'} = \|g_0^{-1}(f - f')\|, \qquad \varphi_f = \lim_{h_{ff'} \to 0} h_{ff'}^{-1} g_0^{-1}(f - f'),$$

$$v_{ff'} = \|\varphi_f - \varphi_{f'}\|, \qquad \psi_f = \lim_{v_{ff'} \to 0} v_{ff'}^{-1}(\varphi_f - \varphi_{f'}),$$

let $\dot{\mathbb{F}} = \{\varphi_f; f \in \mathbb{F}\}$ and let $\ddot{\mathbb{F}} = \{\psi_f; f \in \mathbb{F}\}$ the tangent spaces defined by the L_2-differentiability of $\mathbb{F}$.
D5. There exist centered Gaussian variables G_φ and G_ψ with positive definite variance matrices such that

$$\sup_{\varphi \in \dot{\mathbb{F}}} \left| n^{-\frac{1}{2}} \sum_{i=1}^n \varphi(X_i) - G_\varphi \right| \xrightarrow[n\to\infty]{P_0} 0$$

$$\sup_{\psi \in \ddot{\mathbb{F}}} \left| n^{-\frac{1}{2}} \sum_{i=1}^n \psi(X_i) - G_\psi \right| \xrightarrow[n\to\infty]{P_0} 0.$$

Under the null hypothesis, the vector of the densities $f_{(n)}^{(0)}$ has distinct components and it belongs to $\mathbb{F}^{k_n}$, the mixture density is denoted g_0. By Proposition 9.8, if $p_n = o(n^c)$ and under conditions (9.6) and (9.7), the maximum likelihood estimators of $\lambda_{k_n+1}, \ldots, \lambda_{p_n}$ satisfy $\widehat{\lambda}_{k_n+1} = \ldots = \widehat{\lambda}_{p_n} = 0$, with a probability tending to one. The hypothesis and the alternative are expressed like in Section 9.3, with mixtures of k_n and respectively p_n densities

$$g_{p_n;\lambda_{(n)},f_{(n)}} = \sum_{j=1}^{p_n} \lambda_j f_j$$

with $\lambda_{(n)} = (\lambda_1, \ldots, \lambda_{p_n})^T$ in $\mathcal{S}_{p_n}$ and $f_{(n)} = (f_1, \ldots, f_{p_n})^T$ in $\mathbb{F}^{p_n}$. Let $\mathcal{C}_{s_n,k_n}$ be the set of maps from $\{1, \ldots, s_n\}$ into a subset of $\{1, \ldots, k_n\}$ such that $c(= j$ if and only if $f_{k_n+k} = f_j$, for k in $\{1, \ldots, s_n\}$ and j in $\{1, \ldots, k_n\}$. The mixture density $g_{p_n,\lambda_{(n)},f_{(n)}}$ is written as

$$g_{p_n,\lambda_{(n)},f_{(n)}} = \sum_{j=1}^{k_n} \Big\{ \lambda_j + \sum_{k=1}^{s_n} \lambda_{k_n+k} 1_{\{c(k)=j\}} \Big\} f_j + \sum_{k=1}^{s_n} \lambda_{k_n+k} (f_{k_n+k} - f_{c(k)})$$

for a map c of $\mathcal{C}_{s_n,k_n}$ and it is reparametrized with $\gamma_{ff'}$, φ_f and $\psi_{ff'}$ defined by (9.12) as

$$g_{\lambda,f} = g_0 \Big\{ 1 + \sum_{k=1}^{s_n} \gamma_{1 f_{k_n+k}, f_{c(k)}} \varphi_{f_{k_n+k}} + \gamma_{2 f_{k_n+k}} \psi_{f_{k_n+k}} \Big\} \tag{9.17}$$

Under the hypothesis H_0 of a mixture density $g_0 = g_{k_n, \lambda^0_{(n)}, f^0_{(n)}}$, we have $\lambda_{k_n+k}(f_{k_n+k} - f_{c(k)})$ for $k = 1, \ldots, s_n$, equivalently

$$H_0 : \sup_{f_{k_n+k}, f_{k_n+j} \in \mathbb{F}^2, j,k=1,\ldots,s_n} \gamma_{f_{k_n+k}, f_{k_n+j}} = 0,$$

The others parameters are nuisance parameters and the density g_0 is estimated. Let $\mathcal{G}_{(k_n)}$ be the space of mixtures of k_n densities of $\mathbb{F}$ and let $g_{(n)}$ in $\mathcal{G}_{(k_n)}$.

For a sample $(X_i)_{1 \le i \le n}$ of X, the likelihood ratio test statistic for H_0 is

$$T_n = 2 \Big\{ \sup_{\lambda_{(n)} \in \mathcal{S}_{p_n}} \sup_{f_{(n)} \in \mathbb{F}^{p_n}} l_n(p_n; \lambda_{(n)}), f_{(n)}) - \sup_{\lambda_{(n)} \in \mathcal{S}_{k_n}} \sup_{f_{(n)} \in \mathbb{F}^{k_n}} l_n(k_n; \lambda_{(n)}, f_{(n)}) \Big\}$$

where $l_n(k_n; \lambda_{(n)}, f_{(n)})$ is the log-likelihood of the mixture model with k_n dimensional parameters $\lambda_{(n)}$ and $f_{(n)}$. Under H_0, the likelihood is maximum at $\widehat{\lambda}_{(n)}$ which has the limiting distribution determined by Proposition 9.8 and $\widehat{f}_{(k_n)}$ which satisfies conditions (9.6).

Under the conditions, we consider the parameter spaces $\mathbb{U}$ defined by (9.3) and $\mathbb{U}_2$ defined by (9.11). With the reparametrization by (9.12, the test statistic is written

$$T_n = 2 \sup_{\gamma_n \in \mathbb{U}^{s_n} \times \mathbb{U}_2^{s_n}} \sup_{\varphi_n \in \dot{\mathbb{F}}^{s_n}} \sup_{\psi_n \in \ddot{\mathbb{F}}^{s_n}} \sum_{1 \le i \le n} \log\{1 + \gamma_{1n}^T \varphi_n(X_i) + \gamma_{2n}^T \psi_n(X_i)\}.$$

Lemma 9.4. *Under the conditions (9.6), (9.7) and (9.8), the maximum likelihood estimator of the parameters γ_n satisfies*

$$\limsup_{n\to\infty} \sup_{g_{(n)}\in\mathcal{G}_{(k_n)}} \|\widehat{\gamma}_{g_{(n)}}\| \xrightarrow[n\to\infty]{P_0} 0.$$

This convergence is a consequence of the P_0-uniform convergence on the image space of the new parameters and of the concavity of the expected log-likelihood under H_0 at the parameter value $\gamma_0 = 0$.

Using (9.17) and following the proofs of section 9.2, a $(2s_n)$ dimensional process $Y_n = (Y_{1n}, Y_{2n})$ is defined by its components

$$Y_{1n,k}(\varphi) = n^{-\frac{1}{2}} \sum_{i=1}^{n} \varphi_k(X_i),$$

$$Y_{2n,k}(\psi) = = n^{-\frac{1}{2}} \sum_{i=1}^{n} \psi_k(X_i),$$

for $1 \leq k \leq s_n$, its covariance matrix $\Sigma(\varphi, \psi)$ under H_0 is positive definite. Let $Z_n = \Sigma^{-\frac{1}{2}} Y_n$.

Proposition 9.9. *Under the conditions (9.6), (9.7) and (9.8), the statistic T_n converges weakly under H_0 to*

$$\sup_{\gamma_n\in\mathbb{U}^{s_n}\times\mathbb{U}_2^{s_n}} \sup_{\varphi_n\in\dot{\mathbb{F}}^{s_n}} \sup_{\psi_n\in\ddot{\mathbb{F}}^{s_n}} Z^T(\varphi,\psi)Z(\varphi,,\psi)I_{\{Z(\varphi,,\psi)>0\}},$$

where $Z = \Sigma^{-\frac{1}{2}} Y$ and Y is a centred Gaussian process with covariance Σ.

The proof proceeds by maximization of the likelihood under the constraints $\alpha_k \geq 0$, $k = 1, \ldots, s_n$, at fixed c in $\mathcal{C}_{s_n,k_n}$, like in Section 9.2.

Bibliography

Anderson, J. A. (1979). Multivariate logistic compound, *Biometrika* **66**, pp. 17–26.

Antoniadis, A. (1994). Smoothing noisy data with coiflets, *Statist. Sin.* **4**, pp. 651–678.

Antoniadis, A. (1997). Wavelets in statistics: a review, *J. Ital. Statist. Soc.* **2**, pp. 97–130.

Benhaddou, R., Pensky, M., and Picard, D. (2013). Anisotropic denoising in functional deconvolutionwith dimension free convergence rates nonparametric Laguerre estimation in the multiplicative censoring model, *Elec. J. Statist* **7**, pp. 1935–1715.

Berkes, I. and Philipp, W. (1979). Approximation theorems for independent and weakly dependent random vectors, *Ann. Probab.* , pp. 29–54.

Bickel, P. J., Klassen, C. J., Ritov, Y., and Wellner, J. A. (1993). *Efficient and adaptive estimation for semiparametric models* (Johns Hopkins Univ. Press, Baltimore and London).

Billingsley, P. (1968). *Convergence of probability measures* (Wiley, New York).

Breslow, N. and Crowley, J. (1974). A large sample study of the life table and product limit estimates under random censorship, *Ann. Statist.* **2**, pp. 437–453.

Caroll, R. J. and Hall, P. (1988). Optimal rates of convergence for deconvoluting a density, *J. Am. Statist. Assoc.* **83**, pp. 1184–1186.

Cavalier, L. (2001). On the problem of local adaptive estimation in tomography, *Bernoulli* **7**, pp. 63–78.

Cavalier, L. (2008). Nonparametric statistical inverse problem, *Inv. Problems* **24**, pp. 1–19.

Clayton, D. G. (1978). A model for association in bivariate life tables and its application in epidemiological studies of familial tendency in chronic disease incidence, *Biometrika* **65**, pp. 141–151.

Cohen, R., A. DeVore and Hochmuth, R. (2000). Restricted nonlinear approximation, *Constr. Approx.* **16**, pp. 85–113.

Comte, F., Genon-Catalot, V., and Rozenholc, Y. (2007). Penalized nonparametric mean square estimation of the coefficients of diffusion processes,

Bernoulli **13**, pp. 514–543.

Cox, D. R. (1972). Regression model and life tables (with discussion), *J. Roy. Statist. Soc., Ser. B* **34**, pp. 187–220.

Cox, D. R. and Snell, E. J. (1989). *Analysis of binary data* (2nd ed., Chapman and Hall, London).

Crain, B. R. (1974). Estimation of distributions using orthogonal expansion, *Ann. Statist.* **2**, pp. 454–463.

Craven, P. and Wahba, G. (1979). Smoothing noisy data with spline functions, *Num. Math.* **31**, pp. 377–403.

DeVore, R. A. (1989). Degree of nonlinear approximation, *Approximation Theory VI, C.K. Chui, L.L. Schumaker and J.D. Ward (eds)* **1**, pp. 175–201.

Donoho, D. L. and Johnstone, I. (1989). Projection-based approximations and a duality with kernel methods, *Ann. Statist.* **17**, pp. 58–106.

Donoho, D. L., Johnstone, I., Kerkyacharian, G., and Picard, D. (1996). Density estimation by wavelets thresholding, *Ann. Statist.* **24**, pp. 508–539.

Fan, J. (1991). On the optimal rates of convergence for nonparametric deconvolution problems, *Ann. Statist.* **19**, pp. 1257–1272.

Fan, J. and Peng, H. (2004). Nonconcave penalized likelihood with a diverging number of parameters, *Ann. Statist.* **32**, pp. 928–961.

Fan, J. and Truong, Y. K. (1993). Nonparametric regression with errors in variables, *Ann. Statist.* **21**, pp. 1900–1925.

Feller, W. (1971). *An Introduction to Probability Theory and its Applications, vol. 2* (Wiley, New York).

Gao, J., Tong, H., and Wolff, R. C. (2002). Adaptive orthogonal series estimation in additive stochastic regression models, *Statist. Sin.* **12**, pp. 409–428.

Genon-Catalot, V. and Jacod, J. (1993). On the estimation of the diffusion coefficient for multidimensional diffusion processes, *Ann. Instit. H. Poincaré, Probab.-Statist.* **29**, pp. 119–151.

Genon-Catalot, V., Jeantheau, T., and Larédo, C. (1998). Limit theorems for discretely observed stochastic volatility models, *Bernoulli* **4**, pp. 283–303.

Gill, R. (1983). Large sample behaviour of the product-limit estimator on the whole line, *Ann. Statist.* **11**, pp. 49–58.

Gill, R. D., Vardi, Y., and Wellner, J. A. (1988). Large sample theory of empirical distributions in biased sampling models, *Ann. Statist.* **16**, pp. 1069–1112.

Girard, D. (1998). Asymptotic comparison of (partial) cross-validation, gcv and randomized gcv in nonparametric regression, *Ann. Statist.* **26**, pp. 315–334.

Glidden, D. V. and Self, S. G. (1998). Semi-parametric likelihood estimation in the Clayton-Oakes failure time model, *Scand. J. Statist.* **26**, pp. 363–372.

Gottlieb, D. and Orszag, S. A. (1977). *Numerical analysis of spectral methods: Theory and applications* (Soc. Ind. Appl. Math., Philadelphia).

Haerdle, W., Hall, P., and Marron, J. (1992). Regression smoothing parameters that are not far from their optimum, *J. Amer. Statist. Assoc.* **87**, pp. 227–233.

Hall, P. (1983). Large sample optimality of least sqares cross-validation in density estimation, *Ann. Statist.* **11**, pp. 1156–1174.

Hall, P. (1987). Cross-validation and the smoothing of orthogonal series density

estimators, *J. Multiv. Anal.* **21**, pp. 189–206.
Hall, P. (1989). On projection pursuit, *Ann. Statist.* **17**, pp. 573–588.
Hall, P. and Qiu, P. (2007). Nonparametric estimation of a point spread function in multivariate problems, *Ann. Statist.* **35**, pp. 1512–1534.
Hall, P. and Titterington, D. M. (1986). On smoothing technics used in image restauration, *J. R. Statist. Soc. Ser. B* **71**, pp. 330–343.
Hastie, T. and Tibshirani, R. (1990). *Generalized additive models* (Chapman and Hall, New York).
Hastie, T. J., Tibshirani, R. J., and Friedman, J. (2001). *The elements of statistical learning* (Springer Verlag, New York).
Heywood, H. B. and Fréchet, M. (1912). *L'équation de Fredholm et ses applications à la physique mathématique* (Hermann, Paris).
Hougaard, P. (1984). Lifetime methods for heterogeneous populations: Distributions describing the heterogeneity, *Biometrika* **71**, pp. 75–83.
Hougaard, P. (1986). Survival models for heterogeneous populations derived from stable distributions, *Biometrika* **73**, pp. 387–396.
Hougaard, P. (1995). Frailty models for survival data, *Lifetime Data Anal.* **1**, pp. 255–273.
Huang, J. Z., Kooperberg, C., Stone, C. J., and Truong, Y. K. (2000). Functional anova modeling for proportional hazards regression, *Ann. Statist.* **28**, pp. 961–999.
Huber, P. J. (1985). Projection pursuit, *Ann. Statist.* **13**, pp. 435–525.
Jewell, N. P. (1982). Mixtures of exponential distributions, *Ann. Statist.* **10**, pp. 479–484.
Johnstone, I. and Silverman, B. W. (1990). Speed of estimation in positron emission tomography and related inverse problems, *Ann. Statist.* **18**, pp. 250–280.
Kai, B., Li, R., and Zou, H. (2011). New efficient estimation and variable selection methods for semi-parametric varying-coefficient partially linear models, *Ann. Statist.* **39**, pp. 305–332.
Kerkyacharian, G. and Picard, D. (1993). Density estimation by kernel and wavelets mathods: optimality of besov spaces, *Statist. Probab. Let.* **18**, pp. 327–336.
Koo, J.-Y. and Chung, H.-Y. (1998). Log-density estimation in linear inverse problems, *Ann. Statist.* **26**, pp. 335–362.
Lai, T. and Ying, Z. (1991). Estimating a distribution function with truncated and censored data, *Ann. Statist.* **19**, pp. 417–442.
Laird, N. M. and Ware, J. H. (1982). Random-effects model for longitudinal data, *Biometrics* **38**, pp. 963–974.
Larédo, C. (1990). A sufficient condition for asymptotic sufficiency of incomplete observations of a diffusion process, *Ann. Statist.* **18**, pp. 1158–1171.
LeCam, L. and Traxler, R. (1978). On the asymptotic behavior of mixtures of poisson distributions, *Z. Wahrsch. verw. Geb..* **44**, pp. 1–45.
Lemdani, M. and Pons, O. (1999). Likelihood ratio tests in contamination models, *Bernoulli* **5**, pp. 705–719.
Leroux, B. G. (1992). Consistent estimation of mixing distributions, *Ann. Statist.*

20, pp. 1350–1360.

Li, K.-C. (1987). Asymptotic optimality for C_p, C_L, cross-validation and generalized cross-validation: Discrete index set, *Ann. Statist.* **15**, pp. 958–975.

Liang, D. Y. and Zeger, S. L. (1986). Longitudinal data analysis using generalized linear models, *Biometrika* **73**, pp. 13–22.

Lin, Y. and Zhang, H. H. (2006). Component selection and smoothing in multivariate nonparametric regression, *Ann. Statist.* **34**, pp. 2272–2297.

Lindsay, B. (1983a). The geometry of mixing likelihoods, part i: A general theory, *Ann. Statist.* **11**, pp. 86–94.

Lindsay, B. (1983b). The geometry of mixing likelihoods, part ii: The exponential family, *Ann. Statist.* **11**, pp. 783–792.

Linton, O. B. and Härdle, W. (1996). Estimation of additive regression models with known links, *Biometrika* **83**, pp. 529–540.

Massart, P. (1986). Rates of convergence in the central limit theorem for empirical processes, *Ann. Instit. Henri Poincaré* **22**, pp. 381–423.

McCullagh, P. and Nelder, J. A. (1983). *Generalized linear models* (Chapman-Hall, London).

Meier, L., Van de Geer, S., and Bühlmann, P. (2009). High-dimensional additive modeling, *Ann. Statist.* **37**, pp. 3779–3821.

Meyer, Y. (1990). *Ondellettes et opérateurs I: Ondellettes* (Hermann, Paris).

Murphy, S. A. (1995). Asymptotic theory for the frailty model, *Ann. Statist.* **23**, pp. 182–198.

Nelder, J. A. and Wedderburn, R. W. M. (1972). Generalized linear models, *J.R.Stat. Soc. A* **135**, pp. 370–384.

Neuhaus, J. (2002). Bias due to ignoring the sample design in case-control studies, *Aust. N.Z.J. Stat.* **44**, pp. 285–293.

Nielsen, G. G., Gill, R. D., Andersen, P. K., and Sorensen, T. J. A. (1992). A counting process approach to maximum likelihood estimation in frailty models, *Scand. J. Statist.* **19**, pp. 25–43.

O Sullivan, F. (1986). A statistical perspective on inverse problems, *Statist. Sci.* **1**, pp. 502–527.

Oakes, D. (1986). Semi-parametric inference in a model for association in bivariate survival data, *Biometrika* **73**, pp. 353–361.

Oakes, D. (1989). Bivariate survival models induced by frailties, *J. Am. Statist. Soc.* **84**, pp. 487–493.

Pons, O. (1986). Vitesse de convergence des estimateurs à noyau pour l'intensité d'un processus ponctuel, *Statistics* **17**, pp. 577–584.

Pons, O. (2000). Nonparametric estimation in a varying-coefficient Cox model, *Math. Meth. Statist.* **9**, pp. 376–398.

Pons, O. (2009). *Estimation et tests dans les modèles de mélanges de lois et de ruptures* (Hermès Science Lavoisier, London and Paris).

Pons, O. (2011). *Funtional Estimation for Density, Regression Models and Processes* (World Scientific Publish., Singapore).

Pons, O. (2014). *Statistical Tests of Nonparametric Hypotheses: Asymptotic Theory* (World Scientific Publish., Singapore).

Pons, O. (2015). *Analysis and Differential Equations* (World Scientific Publish.,

Singapore).

Pons, O. (2017). *Inequalities in Analysis and Probability, 2nd ed.* (World Scientific Publish., Singapore).

Pons, O. and Visser, M. (2000). A non-stationary Cox model, *Scand. J. Statist.* **27**, pp. 619–639.

Rebafka, T. and Roueff, F. (2015). Nonparametric estimation of the mixing density using polynomials, *Math. Meth. Statist.* **24**, pp. 200–224.

Rebolledo, R. (1980). Central limit theorems for local martingales, *Z. Wahrsch. Verw. Gebiete* **51**, pp. 269–286.

Simar, L. (1976). Maximum likelihood estimation of a compound Poisson process, *Ann. Statist.* **4**, pp. 1200–1209.

Singh, R. S. (1987). MISE of kernel estimates of a density and its derivatives, *Statist. Probab. Let.* **5**, pp. 153–159.

Stone, C. J. (1974). Cross-validation choice and assesment of statistical prediction, *J. Roy. Statist. Soc. Ser. B* **36**, pp. 111–147.

Stone, C. J. (1982). Optimal global rate of convergence in nonparametric regression, *Ann. Statist.* **10**, pp. 1040–1053.

Stone, C. J. (1984). An asymptotically optimal window selection rule for kernel density estimates, *Ann. Statist.* **12**, pp. 1285–1297.

Stone, C. J. (1986). Additive regression and other nonparametric models, *Ann. Statist.* **13**, pp. 685–705.

Szëgo, G. (1959). *Orthogonal polynomials* (Am. Math. Soc. Coll. Pub. 32, Providence).

Tibshirani, R. (2011). Regression shrinkage and selection via the lasso: a retrospective, *J. R. Statist. Soc. B.* **73**, pp. 273–282.

Tibshirani, R. and Hastie, T. (1987). Local likelihood estimation, *J. Am. Statist. Assoc.* **82**, pp. 559–567.

Van de Geer, S. and Bühlmann, P. (2009). On the conditions used to prove oracle results for the lasso, *Elec. J. Statist.* **3**, pp. 1360–1392.

Vardi, Y. (1982). Nonparametric estimation in the presence of length bias, *Ann. Statist.* **10**, pp. 616–620.

Wahba, G. (1973). Convergence rates of certain approximate solutions to Fredholm integral equations of the first kind, *J. Approx. Theor.* **7**, pp. 167–185.

Wahba, G. (1975). Optimal convergence properties of variable knot, kernel, and orthogonal series methods for density estimation, *Ann. Statist.* **3**, pp. 15–29.

Wahba, G. (1981). Data-based optimal smoothing of orthogonal series density estimates, *Ann. Statist.* **9**, pp. 146–156.

Wahba, G. (1990). *Spline Models for observational data* (SIAM, Philadelphia).

Walter, G. G. (1977). Properties of Hermite series estimation of probability density, *Ann. Statist.* **5**, pp. 1258–1264.

Walter, G. G. and Hamedani, G. G. (1991). Bayes empirical Bayes estimation for natural exponential families with quadratic variance functions, *Ann. Statist.* **19**, pp. 1191–1224.

Woodroof, M. (1985). Estimating a distribution function with truncated data, *Ann. Statist.* **13**, pp. 163–177.

Wu, C. O. (2000). Local polynomial regression with selectionbiased data, *Statist.*

Sin. **10**, pp. 789–817.

Zeger, S. Z., Liang, K.-Y., and Albert, P. S. (1988). Models for longitudinal data: A generalized estimating equation approach, *Biometrics* **44**, pp. 1049–1060.

Zeger, S. Z., Liang, K.-Y., and Self, S. G. (1985). The analysis of binary data with time-independent covariates, *Biometrika* **72**, pp. 31–38.

Index

Printed in the United States
By Bookmasters